T0138452

The Spanish Disquiet

The Spanish Disquiet

THE BIBLICAL NATURAL PHILOSOPHY OF BENITO ARIAS MONTANO

María M. Portuondo

The University of Chicago Press   CHICAGO & LONDON

The University of Chicago Press, Chicago 60637
The University of Chicago Press, Ltd., London
© 2019 by The University of Chicago
Published 2019
Printed in the United States of America

28 27 26 25 24 23 22 21 20 19   1 2 3 4 5

ISBN-13: 978-0-226-59226-8 (cloth)
ISBN-13: 978-0-226-60909-6 (e-book)
DOI: https://doi.org/10.7208/chicago/9780226609096.001.0001

Library of Congress Cataloging-in-Publication Data

Names: Portuondo, María M., author.
Title: The Spanish disquiet : the biblical natural philosophy of Benito Arias Montano /
    María M. Portuondo.
Description: Chicago ; London : The University of Chicago Press, 2019. | Includes
    bibliographical references and index.
Identifiers: LCCN 2018051004 | ISBN 9780226592268 (cloth : alk. paper) |
    ISBN 9780226609096 (e-book)
Subjects: LCSH: Arias Montano, Benito, 1527–1598. | Biblical scholars—Spain—Biography. |
    Humanists—Spain—Biography. | Natural history—Spain—History—16th century. |
    Natural history—Religious aspects—Christianity. | Physics—Religious aspects—
    Christianity. | Science—Spain—History—16th century. | Bible and science. | Spain—
    Intellectual life—1516–1700.
Classification: LCC BX4705.A68 P67 2019 | DDC 230/.2092—dc23
LC record available at https://lccn.loc.gov/2018051004

♾ This paper meets the requirements of ANSI/NISO Z39.48–1992 (Permanence of Paper).

Conozco la grande obligacion en que me pone el sujeto presente, alientanme empero tantos varones doctos, que empleados en su alabança suplen la insuficiencia mia.

[I know the great obligation that the present subject places upon me, but I am encouraged by so many learned men who engaged in praising him make up for my shortcommings.]

FRANCISCO PACHECO,
Libro de descripción de verdaderos retratos (1599)

EL DOTOR BENITO ARIAS MONTANO.

46,

Conosco la grande obligacion en que me pone el sugeto presente, alientanme em=
pero tantos varones doctos, que empleados en su alabança suplen la insuficiencia
mia. bien veo que me obligo amucho, i que siempre fue dificil escrevir bien isto=
ria, pues cuando no uviera otra cosa, la obligacion de tratar verdad bastava.

(en

Doctor Benito Arias Montano. From Francisco Pacheco, *Libro de descripción de verdaderos retratos* (1599), fol. 90, Biblioteca Virtual de Andalucía.

CONTENTS

ILLUSTRATIONS

ABBREVIATIONS

ACDF	Archivium Sacra Congregationis pro Doctrina Fidei (Archive of the Congregation for the Doctrine of the Faith)
ACDF Indice	Subsection of the ACDF on the Sacred Congregation of the Index of Prohibited Books
ACDF Santo Ufficio	Subsection of the ACDF on the Congregation of the Holy Office of the Inquisition
AGS	Archivo General de Simancas (General Archive of Simancas)
AHN	Archivo Histórico Nacional (National Historical Archive), Madrid
Anima	Benito Arias Montano, *Liber generationis et regenerationis Adam, sive De historia generis humani: Operis magni pars prima, id est Anima* (Antwerp: Jan Moretus, 1593).
AntPoly	Antwerp Polyglot: *Biblia Sacra Hebraice, Chaldaice, Graece, et Latine: Philippi II. Reg. Cathol. pietate, et studio ad Sacrosanctae Ecclesiae usum*, edited by Benito Arias Montano, Guido Fabricius Boderianus, F. Raphelengius, Andreas Masius, Lucas of Bruges, et al., 8 vols. (Antwerp: Christophe Plantin, 1569–73).
BAV	Bibliotheca Apostolica Vaticana (Vatican Apostolic Library)
BME	Real Biblioteca del Monasterio de San Lorenzo de El Escorial (Library of the Royal Monastery of San Lorenzo of El Escorial)
BNE	Biblioteca Nacional de España (National Library of Spain, Madrid)
Corpus	Benito Arias Montano, *Naturae historia: Prima in magni operis corpore pars* (Antwerp: Jan Moretus, 1601).

DIE   Martín Fernández de Navarrete et al., eds., *Colección de documentos inéditos para la historia de España*, 113 vols. (Vaduz: Kraus Reprint, 1964–75).

HistNat   Benito Arias Montano, *Historia de la naturaleza: Primera parte del Cuerpo de la Obra magna*, ed. Fernando Navarro Antolín, Bibliotheca Montaniana (Huelva: Universidad de Huelva, 2002).

HSA   Hispanic Society of America, New York City

LibGen   Benito Arias Montano, *Libro de la generación y regeneración del hombre, o Historia del género humano: Primera parte de la Obra magna, esto es, Alma: Estudio preliminar de Luis Gómez Canseco*, trans. Fernando Navarro Antolín et al. (Huelva: Universidad de Huelva, 1999).

VFL   Vatican Film Library at Saint Louis University, Missouri

Biblical citations from various versions of the Bible are indicated as follows: Vulgate (VUL), King James Version (KJV), and the online version of the New American Bible found at http://www.usccb.org/bible/index.cfm (cited here as NAB). Citations and text from the Hebrew Bible are from the Westminster Leningrad Codex (WLC) unless otherwise indicated.[1] Arias Montano was not consistent in his transliteration of Hebrew words, nor did he follow modern conventions.[2] His transliterations are shown in small capital letters; other transliterations from Hebrew follow the Society of Biblical Literature (SBL) guidelines. Hebrew words and roots not expressly identified by Arias Montano are taken from *Strong's Concordance*. To assist readers interested in engaging directly with Arias Montano's *Magnum opus*, I have included in the footnotes references to the Plantin edition in Latin and their most recent translations into Spanish (ex. *Anima* ii; *LibGen* ii). Citations from the *De arcano sermone* are from *Libro de José o sobre el lenguaje arcano* (Huelva: Universidad de Huelva, 2006), with page numbers of the Spanish translation indicated first, followed by the corresponding page of the Latin facsimile included in that edition, and finally the page number of the 1571 edition in square brackets. In a concession to modern usage I have replaced Arias Montano's use of "man" with "humanity," except in quotations from his works or where the author specifically referred to the human male. All translations into English are my own unless otherwise indicated.

1. Digital and facsimile versions of the WLC were consulted at https://tanach.us/.

2. On Arias Montano's transliterations see the appendix by Fernández López in Arias Montano, *Comentarios a los treinta y un primeros salmos de David*, 2:†1–†10.

The historical exploration in this book is informed by some fundamental questions from the history of science. How are explanations about the natural world fashioned? How should the natural world be studied? Why should the natural world be studied at all? They inquire about the methodology, epistemology, and purpose of the enterprise labeled today as science. Historians have answered these questions in many ways, yet for me their study is most enlightening when taking into account all the contingencies that inform the very human pursuit of seeking knowledge about nature. In this book I ask these questions during a very specific moment in time and in a place undergoing profound transformations: the dawn of the early modern era in sixteenth-century Spain. My objective is twofold: to draw the outline of natural philosophical speculation in sixteenth-century Spain by asking the questions posed above, and to do so using as our guide the life and works of one of the towering figures of European humanism, Benito Arias Montano (ca. 1525–98). My aim has been to write a richly contextualized intellectual history of science, one that serves well readers unfamiliar with natural philosophical thought in late sixteenth-century Spain but that also situates Arias Montano's work within the broader discourse of early modern science. The events and ideas discussed in this story were also defined by a profound religiousness that might seem unfamiliar to us today. It was an era indelibly marked by the soul-searching that followed the Reformation and laid the groundwork for the Catholic Counter-Reformation.

The sixteenth century presents a complicated natural philosophical landscape. It was a time when an increasing distrust of Aristotelian natural philosophy—particularly in its Scholastic garb—and a growing skepticism about the validity of philosophical systems inherited from antiquity threatened to dismantle the framework through which Europeans interpreted the natural world. Yet it was also a time when knowledge gained empirically—by observing nature in newly discovered parts of the globe, carrying out trials in the alchemist's furnace, or experimenting with exotic materia medica—supplied a seemingly endless stream of new knowledge about nature. Long before Francis Bacon (1561–1626) articulated the parameters for a new science based on induction and experimentation, Spanish scientific practitioners and theorists were wrestling with the relation between natural philosophy and experience. As we will see, their concerns were a continuation of important questions raised during the Middle Ages about the role of God in nature and the interpretation of the biblical Creation account. This book draws the contours of natural philosophical thought in sixteenth-century Spain and within it identifies a community of natural philosophers and biblical scholars who shared what I call the Spanish disquiet, a preoccupation with the inability of prevailing natural philosophical systems to explain the natural world. While aspects of the broader disaffection with ancient philosophical systems are easily found throughout Europe, the "Spanish" aspect of the disquiet refers to concerns among the group of mostly Spanish thinkers included in this book, which informed the natural philosophical proposals of Benito Arias Montano. They shared two additional preoccupations: that natural philosophical systems inherited from antiquity were not wholly in concert with the ends advocated by natural theology, and that the purely empirical and theory-free understanding of nature gaining momentum in Spain during the century of discovery might be philosophically unsound and morally dangerous.

The book situates Benito Arias Montano within the Spanish disquiet and shows him as an eloquent and influential advocate for natural philosophical reform. He had to contend with a broad range of brilliant Spanish Scholastic Aristotelians—many of them Jesuits—who continued making important contributions to natural philosophy throughout the sixteenth and seventeenth centuries. Their eclecticism and their metaphysical and philosophical rigor set a high bar for others hoping to propose viable alternatives to Aristotelian natural philosophy. Arias Montano was up to the challenge. His life's course and his intellectual development had prepared him to attempt an ambitious reform of the way humanity approached the study of nature. In doing so he partook in the reformist ethos of the era that found the Scholastic Aristotelianism taught at the universities unsatisfactory. Since he was a widely read member of the European intellectual community, his motivation to reform natural philosophy also reminds us that the Spanish disquiet was just a local manifestation of concerns about natural philosophy that yielded the works of such figures as Bernardino Telesio, Francesco Patrizi, and Francis Bacon. Like them, he was well aware of what a cohesive natural philosophical system entailed, and for his project he developed epistemic approaches and methodological tools that allowed him to formulate a new metaphysics, cosmology, physics, and natural history. He presented these ideas in an unconventional and novel natural philosophical work that he wrote late in life—his self-titled *Magnum opus*.[1] It was an attempt to fashion an entirely new natural philosophy derived from first principles he identified in the Hebrew Bible, particularly in the book of Genesis. Although his approach to reform—recapturing a "lost" Mosaic philosophy—was unusual and his exegetical methodology unconventional, his motivation was in complete concert with the principal objectives of reformers of natural philosophy during the early modern era. He was the first European, I argue, to base a complete natural philosophical system on the Sacred Scriptures *and* on sensual experience and to propose it as an alternative to Aristotelian natural philosophy. The heart of my book is therefore an analysis of the *Magnum opus* and the intellectual journey that led to it.

Arias Montano's approach to studying nature framed the natural world as unfolding from a series of historical events described in the book of Genesis. He thought the Creation narrative encapsulated, albeit cryptically, metaphysical principles that undergirded a complete natural philosophical system based on the unshakable and cross-confessional authority of the Bible. He folded these ideas into theories of knowledge and of language that produced a distinctive Montanian hermeneutics of nature (theory of interpretation of nature).[2] Arias Montano claimed his approach divorced natural philosophy from systems of antiquity and set natural philosophical inquiry on firm theological ground while also encouraging the empirical investigation of nature. His biblical exegesis relied on the literal interpretation of the Hebrew Bible and on applying philological tools that when cleverly deployed squeezed from

1. The work survives in two published volumes: Benito Arias Montano, *Liber generationis et regenerationis Adam, sive De historia generis humani: Operis magni pars prima, id est Anima* (1593), and Benito Arias Montano, *Naturae historia, prima in magni operis Corpore pars* (1601); hereafter, *Anima* and *Corpus*. A third part, the *Vestis*, appears never to have been written or has been lost.

2. I rely on James Bono's conceptualization of hermeneutics of nature as "a set of interpretive practices or strategies for reading the Book of Nature" that was necessary in order for science to emerge as a distinctive discourse: Bono, *Word of God and the Languages of Man*, 72. Kenneth Howell illustrates the value of considering that the study of nature and the study of scripture sometimes share a hermeneutical lens: Howell, "Hermeneutics of Nature and Scripture."

the text the very last bit of natural philosophical meaning. First rehearsed in the treatises he appended to the monumental Antwerp Polyglot (*Biblia Sacra Hebraice, Chaldaice, Graece, et Latine*, 1569–73), his exegetical approach came to fruition in the hermeneutics of nature of the *Magnum opus*. It led to some unconventional postulates about cosmology, such as a three-part world, a fluid heaven, and a cyclical communion of terrestrial and celestial effluvia. He also posited a new theory of dualistic matter in which everything in the world was composed of various combinations of two primordial fluids. In the Montanian cosmos there was no essential distinction between the material nature of the heavens and the earth. He also found in the Bible a "natural" taxonomic scheme for classifying plants and animals. These theoretical pronouncements were often supported with experiential evidence, careful observations, and some experiments. They also rested confidently on a profusion of biblical citations.

Scholarship on Arias Montano is vast, yet it is unsatisfying for historians of science. The secondary literature has long focused on unearthing and documenting details of his biography and *entorno* (milieu), editing his correspondence, and studying specific areas of expertise—the lion's share going to his work on the Antwerp Polyglot and as a late Renaissance humanist and theologian. A significant part of this scholarship appeared in articles in journals such as the *Revista de estudios extremeños*, *La ciudad de Dios*, *Sefarad*, and *Humanistica Lovaniensia*, especially after the four hundredth anniversary of his death in 1998. There are two, now dated, book-length biographies in English, by Aubrey Bell (1922) and by Ben Rekers (1972). The most recent and comprehensive research on his life's chronology can be found in the works of Gaspar Morocho Gayo and the introductions to translations of Arias Montano's works. His natural philosophy was the subject of a monograph by Juan José Jorge López, but more recent work by Luis Gómez Canseco has revised much of this earlier scholarship and made essential contributions to our understanding of Montanian philosophy of nature. I engage further with this scholarship in the pertinent sections of this book. By necessity rather than by inclination, I have kept the number of references cited modest, especially when they relate to well-studied topics in early modern history of science and Spanish history.

Over the past fifteen years many of Arias Montano's major published works have been translated and edited in Spanish in the series Humanistas españoles and Bibliotheca Montaniana. They are impressive scholarly achievements of historical studies and of Latin and Spanish philology. Volumes from these collections in their green bindings—my precious *libros verdes*—have graced my bookshelf for years now. They are the foundation of the modern conceptualization of Montañista thought and have profoundly influenced this book. I owe a profound debt to the Montanianos who have devoted their scholarly life to the humanist from Fregenal. Although his entire personal correspondence has not yet been edited, there are comprehensive and thoughtfully annotated collections by, among others, Antonio Dávila Pérez, Baldomero Macías Rosendo, and Luis Charlo Brea. His letters tell the story of the complex social and political environment of patronage and favors, duties and desires, friendships and enemies that determined his life's course and his intellectual development. A few of Arias Montano's works remain unpublished—drafts, sermons, working notes, and some finished treatises—most of them at the Royal Library of El Escorial, the Biblioteca Nacional de España, and surely still undetected in other European archives.

Whether Arias Montano can be considered a natural philosopher has been debated among scholars of Spanish intellectual history.[3] Ramiro Flórez found signs of Arias Montano's definitive rejection of philosophy in his measured praise for Ficino, the antiphilosophical stance in the *Dictatum Christianum*, and the quiet desperation with ancient philosophies he expresses in the proem of the *Naturae historia*. But at the same time they pointed to a profound commitment to cultural humanism, a perspective that vindicated humanity's place in the world and its cultural and intellectual legacy.[4] For Flórez these two postures were dichotomous facets of Arias Montano's complex personality. Perhaps missing from this analysis is the realization that Arias Montano was rejecting not philosophy per se, but rather the answers proposed by ancient and modern philosophy. To consider him a scientist is even more questionable, if not anachronistic. My approach has been to step away from these characterizations and frame his project using his own preoccupations as guide. Arias Montano, like others who shared his disquiet, had at his disposal a vast array of approaches to understanding the workings of the natural world. These could include writing detailed descriptions of plants and animals or experimenting with pumps, activities that seem scientific to us now. But this was only one set of approaches, which I refer to in this book as "scientific practices." Another set of approaches were squarely metaphysical and philosophical, that is, they set forth the first principles and epistemic criteria for assembling knowledge of nature into a coherent corpus and thus formed a natural philosophy, un-

3. Jorge López, *El pensamiento filosófico de Benito Arias Montano*, 449–93.
4. Flórez, "Actitud ante la filosofía."

derstood as rules of thought that explained how the natural world worked. Throughout this book I use "scientific enterprise" to refer to both natural philosophy and scientific practices, all the while acknowledging that the relation between the two during the sixteenth century was far from stable.

Historical studies of science in early modern Spain now include a broad spectrum of scientific practitioners, take seriously practices considered dead ends by previous historiographies, and emphasize the diversity of approaches and alternatives to scientific questions by the early modern scientific enterprise. One of the principal developments in the field of history of science has been a shift in our current understanding of what constituted "science" during the early modern era. As a whole, the field now acknowledges the rich and variegated character of natural philosophy and its associated scientific practices without singling out for study only those aspects that contributed to the rise of modern science. Long gone is the progressivism where the benchmark of scientific merit was taken to be whether a country contributed to the developments that culminated in the new experimental science of the seventeenth century—the New Science—the cornerstone of a historical narrative known as the Scientific Revolution. What the field finds interesting about the early modern scientific enterprise is the very diversity of responses to a broadly acknowledged "crisis of doubt" or, said otherwise, to knowledge about explanations of the natural world that gradually overtook early modern European thought.[5] These ranged from revivals of ancient philosophies to the highly personal interpretations of the Aristotelian corpus that Charles Schmitt labeled eclectic Aristotelianism.[6] What this book discusses as the Spanish disquiet is just one of these responses.

And yet histories of early modern Spanish science—particularly those written in English—are often under pressure to engage with a historical thesis that maintains it was Protestant insistence on the literal interpretation of the Bible that forced Europeans to "jettison traditional conceptions of the world" and heralded a new attitude toward science.[7] While the thesis has been controversial and debated,[8] one of its derivatives insists that Catholic literal biblical exegesis was so bound to patristic authority that it simply could not accom-

5. For a general introduction to the topic, see Copenhaver and Schmitt, *Renaissance Philosophy*, 22–24, 239–60. Popkin frames it as a crisis that originated with the Reformation and quickly overtook philosophy: Popkin, *History of Scepticism*, 15–16.

6. Schmitt, *Aristotle and the Renaissance*, 89–91.

7. Harrison, *Bible, Protestantism, and the Rise of Natural Science*, 4.

8. Killeen and Forshaw, *Word and the World*, 2, 6.

modate the interpretive latitude afforded to Protestant exegetes. In response, Kevin Killeen and Peter Forshaw have explained that the "natural philosophy of the era, far from being at implacable odds with the Bible and the Church, is better characterized by its willingness and desire to marry scripturalism with the study of the natural world." As Killeen explained further, for seventeenth-century "scientists" natural theology was intertwined with the Bible in a manner and to a degree that historians of science have been slow to recognize, largely because by the eighteenth century a deliberate disassociation had taken place. Natural theology was an exercise in correlation between scriptural texts and the natural world that used a "reciprocal hermeneutics in which natural philosophy can explicate the accommodated meaning of the biblical text and conversely."[9] The same thing, I argue, held true for the direct heirs of Raymond Sebond in Spain and for Arias Montano as well.

Surveys of the history of science in Spain have long noted the contrast between the degrees of vitality of the scientific enterprise in the sixteenth and the seventeenth centuries.[10] The cosmographical and medical arts flourished during the earlier century, but during the one that followed it seemed as if the latest developments in astronomy and mechanics simply passed the peninsula by, so much so that by the beginning of the eighteenth century critics charged that Spain had been left out of the European Republic of Letters that discussed experiments, the tenets of the New Science, and the natural laws that resulted from them. By the 1740s fray Benito Jerónimo Feijoo (1676–1764) had been remarking for years about Spain's slow adoption of the Copernican thesis and Newtonian physics and the marginal status of the country's scientific enterprise.[11] During the latter half of the twentieth century, with the invention of the historiographical concept of the Scientific Revolution, the same concerns were expressed as, Why didn't Spain play a role in the Scientific Revolution?

The persistence of the "flourish, then decline" narrative was also largely due to how nicely it fit within the broader historiography of the seventeenth-century decline of Spain, a strand of which maintains that depopulation, economic decline, and the rise of religious orthodoxy hampered scientific activity. Historians have little difficulty finding many examples to illustrate both periods. The Casa de la Contratación (House of Trade) in Seville is a case in point. Whereas during the sixteenth century it was the locus of innovative solutions to cosmographic and cartographic problems, by the mid-seventeenth century

9. Killeen, *Biblical Scholarship, Science and Politics*, 23.
10. López Piñero, *Ciencia y técnica*, 372–73.
11. Carta XXIII—Sobre los sistemas filosóficos (1745), in Feijoo, "Cartas eruditas y curiosas."

these efforts had largely been abandoned and were portrayed as having gradually waned in proportion to the severe decline in silver and other remittances
coming from America. Furthermore, this narrative hints at the disproportion-
ately negative influence of recalcitrant religious authorities who, stoked by
Counter-Reformationist zeal, kept Spanish natural philosophers from par-
ticipating fully in the intellectual renewal associated with the New Science.
For José María López Piñero, the premier historian of Spanish science of the
late twentieth century, the rescuers were the *novatores*, mostly physicians and
advocates of experimental science, who began the difficult and gradual introduction of the New Science to Spain during the late seventeenth century
and who found their most vocal advocate—if not a practitioner—in Benito
Jerónimo Feijoo.

As an explanation, the narrative outlined above proved very convincing,
yet it is flawed. While it correctly identified linkages between a reduction in
certain scientific practices and broader causes of the state's decline, its main
flaw lies in framing the seventeenth century as one of scientific decline *relative*
to the New Science and the Scientific Revolution. As a result of this analyti-
cal perspective, earlier historiography tended to dismiss the alternative ways
the Spanish scientific enterprise addressed the crisis of knowledge that had
overtaken European natural philosophical thought.[12] One especially productive approach entails adjusting the historian of science's frame of analysis so
that the objective becomes explaining the thoughts, practices, and works of
Spanish practitioners in their own right rather than measuring Spain's scientific achievement against that of England or Italy, or even that of Spain one
hundred years earlier. Undergirding this approach is a fundamental change in
our understanding of the origin of modern science that rejects the positivist
view of the "New Science" as the inevitable outcome of the early modern
crisis of knowledge. In fact, for most of the seventeenth century, experimental
natural philosophy was a highly contested enterprise offering no guarantee of
its eventual success.

The "contextual turn" in historical studies has broadened the range of
questions asked beyond those associated with intellectual history. This broad
perspective makes the experience of discovering the New World central to any
analysis of the Spanish scientific enterprise. The intimate contact with a new
world and the desire to understand its nature was an unprecedented concep-

---

12. This was one of the conclusions that emerged from an international conference of historians of Spanish science titled "Beyond the Black Legend," held in Valencia in fall 2005: Eamon
and Navarro Brotóns, *Más allá de la Leyenda Negra/Beyond the Black Legend*, 27–39.

tual and methodological challenge in the history of the West, one where the extraordinary work of Spanish and Portuguese naturalists was unparalleled. This enterprise dictated new questions that required new approaches, which often meant setting aside models that had served Europeans adequately since antiquity. The enterprise of the Indies also coincided with attempts to consolidate an immense empire within the composite political structure of the Habsburg monarchy that is often referred to as "Spain," more as a shorthand than as an accurate descriptor or actors' category. The thrust of the scientific enterprise in Spain during the sixteenth century was directed *by* and *toward* understanding a new world. The bureaucratic and economic structures put in place in response to the discovery created productive spaces for specialized scientific practitioners and defined the scientific disciplines that would receive the most patronage: astronomical navigation, cosmography, natural history, and medicine. Spanish institutions demanded utilitarian results from these scientific practitioners, resulting in adjustments to the methodologies and the epistemic criteria used to ascertain natural facts in these fields.[13]

These disciplines have long attracted the interest of historians of science—the archives are rich in relevant materials, and there are decades of scholarly work to guide the novice. Recent scholarship, however, has reframed the study of these disciplines around their common epistemology rather than the utility their applied nature brought to the realm. By approaching them from this perspective, we see that the sixteenth-century developments in astronomical navigation, cosmography, natural history, and medicine share a profound commitment to empiricism. My earlier work and that of Antonio Barrera-Osorio and others has focused on practitioners of science and bureaucrats immersed in a culture of utilitarian pragmatism, who often sidestepped natural philosophical speculation if it impeded providing their patrons with useful results. Arndt Brendecke, taking an even more expansive look at the Spanish enterprise of the Indies, has argued that this reliance on empirical information was integral to the efficient running of a transoceanic empire.[14] Yet it is fair to ask, Is reframing the Spanish scientific enterprise of the sixteenth century in terms of empirical practices enough? It has been my objective in this book to explore the other half of the scientific enterprise—the flip side, if you will—the

13. By 1894, if not earlier, this observation about the Spanish scientific enterprise had become commonplace: Menéndez y Pelayo, "Esplendor y decadencia de la cultura científica española," 435.

14. Barrera–Osorio, *Experiencing Nature*; Portuondo, *Secret Science*; Sánchez Martínez, *La espada, la cruz y el padrón*; and Brendecke, *Empirical Empire*.

intellectual world of those natural philosophers for whom empiricism was an insufficient, even flawed, way of approaching the study of nature. By necessity my study is selective, and I acknowledge that my brief discussion of Spanish Jesuit metaphysicians and natural philosophers of the later sixteenth and early seventeenth centuries is modest at best. It is also possible to argue that for every Spanish natural philosopher or theologian dissatisfied with Scholastic Aristotelianism, countless others in schools and universities were perfectly content to teach the established curriculum and dance around the edges of Aristotelian paradigms. And still others, such as the cosmographers I studied in an earlier book, appear to have been wholly unconcerned about the finer points of metaphysics or philosophy.

The Spanish Disquiet follows Arias Montano's life as he makes his way in the tumultuous world of Counter-Reformation Europe.[15] From his early days in the Seville of the era of discovery and his education at the Trilingual College of the University of Alcalá, where he displayed a talent for biblical languages, poetry, and medicine, we follow him as he becomes heir to Cardinal Cisneros and Cipriano de la Huerga's school of intensely philological biblical exegesis. His induction into the European community of humanist biblical scholars took place during his sojourn in Antwerp overseeing, editing, and contributing important (and controversial) treatises to the Antwerp Polyglot. Arias Montano's contributions and stewardship embroiled him in one of the most caustic theological polemics of his time: contentions between Spanish theologians who wanted a biblical exegesis based solely on the Latin Vulgate or the Greek Septuagint Bible and those who, like Arias Montano, sought the *hebraica veritas* from the Sacred Scriptures. After defending the Antwerp Polyglot in Rome and avoiding the Inquisition in Spain, for most of his adult life he served the most powerful monarch of his day, Philip II of Spain, as emissary, informant, librarian, and confessor. He finally retired from court to devote his energies to biblical scholarship and exegesis, splitting his time between the bustle of Seville and the solitude of his beloved Peña de Alájar in the remote mountains of his native Extremadura. He spent the last years of his life on the *Magnum opus*, a work he saw as the culmination of his life's studies.

---

15. Portions of *The Spanish Disquiet* have been published as follows: chapter 2 contains some material from Portuondo, "On Early Modern Science in Spain"; chapter 6 draws on "America and the Hermeneutics of Nature in Renaissance Europe"; and chapter 11 refers to, but does not tread the same ground as, "The Study of Nature, Philosophy and the Royal Library of San Lorenzo of the Escorial."

The Spanish Disquiet tells Arias Montano's life story as the story of the books' gestation, maturation, and realization. This focus will of necessity leave unexamined many aspects of his life and thought such as his poetry, his interest in the graphic arts, and significant segments of his theological thought. It touches on some on his extensive exegetical works, but only where they discuss themes developed further in the *Magnum opus*.

Such a full life requires us to take several excursuses meant to explore more fully the genesis and development of the ideas that inform the *Magnum opus*. These include natural theological thought in Spain, the tradition of hexameral commentaries, and the development of Mosaic philosophies during the Renaissance as an expression of the *prisca sapientia* tradition. Arias Montano's encounters with the Spanish and Roman Inquisitions and their regulatory bodies, such as the Congregation of the Index, merit close scrutiny for what they tell us about the relation between science and religion—including the role of Cardinal Roberto Bellarmino in censoring his works—but also for what they reveal about Arias Montano's convictions and his commitment to his approach to biblical scholarship and natural philosophy. Throughout this book I have also endeavored to put Arias Montano's natural philosophical ideas in conversation with those of his contemporaries, paying little attention to national borders, working on the assumption that this highly regarded polymath was familiar with ongoing developments and controversies of the early modern scientific enterprise.

In undertaking this study I am conscious that earlier historical approaches to science, especially in disciplines such as astronomy or mechanics that claim modern counterparts, often tended to set scientific aspects apart from religious or theological aspects, even when these interests are intertwined in the same person. Furthermore, popular studies still subscribe to the thesis of conflict between science and religion and portray late medieval and early modern Christianity, and Catholicism in particular, as antithetical to science. I hope *The Spanish Disquiet* shows that studying the natural world through empirical investigations and natural philosophy was a central preoccupation of a significant number of theologians. And while they might have echoed, as Arias Montano certainly did, the words of Tertullian, Saint Basil, and Saint Augustine against the ancient philosophers, we recognize that these injunctions were made not against the study of nature or against natural philosophy, but rather against the blind acceptance of dogmatic positions that in their opinion only obscured the true nature of the Creation God had placed under humanity's dominion. One thing quickly becomes clear to any student of the sixteenth century: the central objective of early modern natural philosophers was finding an answer to the question, What was God's plan in nature? Everything else—the sun's position in the world, identifying the period of swinging pendulums, how mollusks make their shells, or whether seminal principles existed in matter—paled beside this question. Francis Bacon announced that the pinnacle of knowledge was finding "the summary law of nature" through which God designed the world.[16] A few years earlier, in his *Naturae historia* or *History of Nature*, Arias Montano claimed he had uncovered this very law in the pages of the Bible.

16. "*Opus quod operatur Deus à principio usque ad finem*, the summary law of nature." Francis Bacon derives the notion of the summary laws of nature from Ecclesiastes 3:11, which also suggests that such knowledge was hidden from man: Bacon, *The Advancement of Learning*, book 2.

CHAPTER 1

The Challenge Ahead

The great artificer and maker of the world, God, has declared in [the Sacred Scriptures]
the reason why he made the terrestrial orb on behalf of man; he also explained in them
all the treasures of science and wisdom that can be perceived by man concerning knowledge of nature and of the arts necessary for humankind.[1]

Sometime in 1566 or 1567 Benito Arias Montano learned about a proposal to
reedit and publish a new polyglot edition of the Bible. He probably heard the
news from Gabriel de Zayas, a royal secretary and longtime friend, who had
been discussing the project with Antwerp printer Christophe Plantin. Not
long afterward, Arias Montano received a request from Philip II of Spain;
the king wanted his opinion about the merits of undertaking such a venture.
Arias Montano's reply was prompt and enthusiastic. Such projects have always
been worthy of kings, he told Philip, and the current times more than any
other needed good kings who would guard and defend from heretics the scriptural treasures of the Catholic religion.[2] After discussing the proposal with
the Council of the Inquisition, the king asked Arias Montano to present the
project to the Faculty of Theology at the University of Alcalá, the theologian's
alma mater.

The proposal was presented to the college as a new edition of the famous
1520 Complutensian Polyglot Bible, the crowning achievement of Spanish religious humanistic studies. From the cloisters of the University of Alcalá—the
old Roman Complutum—Cardinal Francisco Jiménez de Cisneros (1436-
1517) had overseen the creation of a Bible with the Old and New Testaments
edited in their original languages—Hebrew, Greek, and Chaldean (Aramaic)—
with Latin translations based on the oldest extant manuscripts. With the in-

1. Arias Montano, "Phaleg sive De gentium sedibus primis orbisque terrae situ," in AntPoly,
8:†2r.

2. Macías Rosendo, *Biblia Políglota*, 71–73.

clusion of a final volume comprising dictionaries, grammars, and scholarly aids, this edition was designed to help humanist scholars read the Bible in its original languages. This new polyglot Bible, however, would not germinate from Spain's long tradition of scriptural studies as had the Complutensian Bible. It would be compiled, edited, and printed in Flanders, in the city of Antwerp, specifically in the publishing house of Christophe Plantin (Christoffel Plantijn; ca. 1520–89).

Not long after the doctors at Alcalá endorsed the new project, it was placed under Arias Montano's supervision. The king's request must have stunned him. The instructions—a royal order by any other name—asked him, in essence, to step into the role played by the venerable Cardinal Cisneros. Here was the opportunity to bring the definitive biblical text of the Old and the New Testaments to light again, but now in the spotlight of Post-Tridentine biblical studies and under the auspices of the Most Catholic Monarch. For the modest and relatively unknown Arias Montano, being asked to undertake such an enterprise must have seemed an exhilarating opportunity; only later would he realize what a tremendous burden it would prove to be.

Arias Montano was to preside over the printing of the Bible and to moderate this activity so that it would be in line with a very specific "order and form" that had been agreed on by the ecclesiastical bureaus involved.[3] He accepted, of course, and his life would change in unimagined ways. Antwerp would prove to be transformative personally and intellectually, but it would also cement intellectual commitments he had already developed and nurtured in his relatively quiet fifteen years as a biblical scholar in Seville and Alcalá. The stays in Antwerp and Rome were to give his voice a new inflection, a self-assured tone that would earn him devoted followers and virulent detractors. This new polyglot Bible would thrust him onto the international stage, and his voice would be heard in dozens of biblical commentaries, in books of poetry and emblems, and through the copious correspondence he maintained with Europe's leading humanists. He was to become a biblical authority whose work crossed confessional divides, earning praise for its deep devotion and extraordinary erudition. Catholic censors in Spain and Rome would take exception to some of his approaches, finding objectionable very different aspects of his work. The censors would eventually succeed in Rome, and once Arias Montano's defenders had died, censorship—initially thwarted in his native Spain—would descend on his work there as well.

3. Ibid., 78, 79.

Arias Montano's life was defined by two preoccupations. The first, perhaps the most important and best known, was biblical interpretation. In commentary after commentary, he delved into the Word trying to extract vestiges of what God had chosen to reveal to humanity. The objective of this quest was to anchor religious spirituality firmly in the Sacred Scriptures, thus confirming that a spiritual life built on a correct understanding of the Bible led to salvation. His second preoccupation was somewhat analogous to the first. He was driven by a desire to identify within the written vestiges of the divine Word an explanation of the natural world as God intended humanity to know it. Just as previous interpreters of the Bible had gotten the religious messages wrong, Arias Montano believed that previous interpreters of nature—particularly the philosophers of antiquity—had also gotten tragically wrong their explanations of the "why" and "how" of the natural world. This was much more for him than a challenging natural philosophical problem; understanding the natural world was an essential part of what it meant to be human. God had designed humanity so that its merits and potential salvation would be tested precisely while living on *this* earth. If mankind did not know how to live in concert with the natural world, how then could humanity be saved?

Toward the end of his life Arias Montano thought he had succeeded in identifying and articulating a philosophy of nature that was in complete concert with the revealed Word. It was, furthermore, also in concert with a lifetime of careful observation of the natural world. He recognized that the way the European intellectual heirs to Greco-Latin antiquity, as well as philosophers of Islam and Judaism, went about explaining the natural world was deeply flawed. The natural philosophy inherited from the past, be it Pythagorean, Platonic, Atomistic, or Aristotelian, had not served humanity well. He set out on a path of reform. As evidence for his reform program he marshaled the history of the Hebrew language and its philology, and he submitted his linguistic investigations to empirical tests that were informed equally by experience of natural phenomena, rudimentary experiments, and an internalized and perhaps unconscious reliance on the very same Aristotelian conception of the natural world he was trying to overturn. He structured his philosophy of nature in three parts: the *Book of the Generation and Regeneration of Adam*, or *Anima* (1593); the *History of Nature*, or *Corpus* (1601); and the *Vestis* (unpublished, and perhaps now lost or never written). Cognizant of the monumental task he had set for himself, he took to referring to the trilogy as his *Magnum opus*—not out of vanity, but rather out of apprehension at the magnitude of the task.

WHO WAS BENITO ARIAS MONTANO?

To historians of Spanish humanism Benito Arias Montano needs no introduction. He is universally acknowledged as one of the leading Christian Hebraists and biblical scholars of his time. He was an accomplished Latinist and mastered Greek, Hebrew, and Aramaic. Anecdotal accounts claim he knew over thirteen languages. He devoted his linguistic talents to the study of the Sacred Scriptures, but his interests extended well beyond biblical studies. He was an avid historical researcher and antiquarian and likewise a curious inquirer into the secrets of nature, as his impressive collection of *naturalia* and *artificialia* testifies. During his peripatetic years he carried his collection with him in a portable *museolum* but would eventually install it in a house near his botanical garden in Seville. To his close circle of friends, he was a reluctant participant in the turbulent commerce of the court and church, but his protestations masked his deftness at navigating the politics of those institutions. He rarely missed an opportunity to tell his correspondents how much he yearned to retire from public life, both figuratively and physically. Throughout his life he sought the solitude of the remote mountains of the Sierra Morena in southern Spain. He so cherished his homeland that he appended the surname Montano ("from the mountain") to his name. It would be there, in his retreat in the Peña de Alájar—now renamed the Peña de Arias Montano as a charming if remote tourist destination—that he would find the intimate communion with nature that would bring to light the *Anima* and the *Corpus*.

As historian Juan Gil wrote with great insight in the opening line of his book on Arias Montano, "A lot is known about Benito Arias Montano and at the same time very little" ("De Benito Arias Montano se sabe mucho y al mismo tiempo muy poco").[4] It is impossible to state definitively that all the substantive details of Arias Montano's life and lineage are known; historians continue to fill in the details of the exegete's biography and to extensively revise the historiography built on what are now acknowledged misinterpretations of key aspects. This scholarship has proved effective in demolishing two aspects of Arias Montano's biography that most English-language scholars still take as fact: his thinly vailed Judaism and his membership in and proselytizing for the *Familia charitatis* (Family of Love).[5] Disengaging his historical profile

---

4. Gil, *Arias Montano y su entorno*, 15.

5. For a synthesis of this thesis and its historiographical consequences, see Pastore, "Arias Montano, Benito." Rekers put forth the thesis of Arias Montano's membership in the Family of Love in *Benito Arias Montano*. The 1973 Spanish edition the book includes an influential

from this presumed Nicodemism makes him no less interesting, and in fact it allows us to understand in a new light the obsessions that drove his impressive output. He need not be a crypto-Jew or a secret Protestant who only outwardly played the part of a devout Catholic to be considered one of the most interesting figures of sixteenth-century European thought. These alleged aspects of his life made him interesting to a historiographical school that viewed the Spain of the Counter-Reformation in a negative light and sought to fashion possibly subversive figures into challengers of the status quo.

Moving beyond Arias Montano's possible Judaism and affiliation with the Family of Love has opened new perspectives about his spirituality and its origin. While Marcel Bataillon in his highly influential *Érasme et l'Espagne* (1937; in Spanish, 1950 and 1966) presented Arias Montano as possible heir to the Erasmian-inspired *iluminados* who advocated an interior piety—many of whom were of converso origins—recent scholarship situates him not as a disciple of Erasmian spirituality, but rather along the spiritual lines of the Dominican fray Luis de Granada (1504–88) and the Franciscan fray Francisco de Osuna (ca. 1492 to ca. 1541). They advocated, and Arias Montano shared in, a profound and orthodox piety that later found expression in what Melquiades Andrés Martín describes as the ecumenical and theological humanism of the University of Alcalá, fostered there by Cardinal Cisneros.[6]

The biographical sketch that follows collects some of the key events of his extraordinary life as we currently know them.[7] His family's roots lay in the

---

essay by Ángel Alcalá that supports the thesis: *Arias Montano* (Madrid: Taurus, 1973), 235–52. For recent criticism, see Martínez Ripoll, "Universidad de Alcalá," 13–92; Landtsheer, "Benito Arias Montano and the Friends from His Antwerp Sojourn," 39–61, and Alcalá's own reassessment of the debate, Alcalá Galve, "Arias Montano y el familismo flamenco."

6. Andrés Martín, "Espiritualidad ecuménica," 21.

7. Here I follow the latest biographical information compiled by leading Montanian scholars, as well as a flurry of publications that marked the four hundredth anniversary of his death, in particular Martínez Ripoll, "Universidad de Alcalá"; Morocho Gayo, "Trayectoria humanística de Benito Arias Montano, 1"; Morocho Gayo, "Trayectoria humanística de Benito Arias Montano, 2"; Gil, *Arias Montano y su entorno*; Gil, "Arias Montano en Sevilla"; Domínguez Domínguez, *Arias Montano y sus maestros*; and Caso Amador, "Origen judeoconverso." Although the documentary appendix remains essential, current scholarship makes significant corrections to González Carvajal's account, "Elogio histórico del Doctor Benito Arias Montano." The two book-length biographies of Bell and of Rekers, though largely outdated, continue to inform much of Montanian English-language scholarship: Bell, *Benito Arias Montano* (1922), and Rekers, *Benito Arias Montano* (1972).

mountainous region of Extremadura and in the town of Fregenal de la Sierra. His parents were Benito Arias, a familiar and apostolic notary of the Holy Office and therefore also an officer of the local royal judiciary, and Francisca Martínez (Isabel Gómez). Based largely on purity of blood proceedings made when Benito Arias Montano entered the Order of Saint James, it was believed that his family belonged to the lesser nobility, although members had to supplement the income from their modest landholdings by working for the bureaucracy of the Habsburg monarchy. The family's respectable social standing was complemented by the support of a close-knit circle of friends, family, and patrons in Fregenal and Seville. Recent scholarship, however, has debated this characterization of the family's origins. It shows that the extended Arias family from Fregenal de la Sierra worked as artisans in the textile and leather trades and in commerce. The social position of these professions would not only have excluded them from the ranks of lesser nobility (*hidalgo*) but situated them as working in occupations associated with converted Jews (*judeoconversos*). Furthermore, some members of the Arias clan had been brought before the Inquisition during the fifteenth century on suspicion of Judaizing, though none of these individuals have been traced to Arias Montano's familiar line. More evidence for Jewish origins of the clan has been supplied by the genealogical studies of the exegete's first cousin four times removed, Captain Benito Arias Montano (1588–1641).[8] The Ariases of Fregenal were apparently a family who had converted—for the most part—to Christianity during the late fourteenth and early fifteenth centuries and, like so many others, were intent on erasing any traces of this association as statutes on purity of blood intensified throughout the Iberian peninsula.

Benito Arias Montano's precise birth date is uncertain, sometime between 1525 and 1527. The eldest brother, Rodrigo Arias, apparently died young; the second brother, Juan Arias de la Mota, earned a degree in civil law and, after serving in various administrative posts in Castile, left for Peru in 1560 as part of the household of Viceroy Diego López de Zúñiga y Velasco. America was a recurrent draw for the Arias family, with a nephew and the aforementioned greatnephew of don Benito also emigrating.[9] Our Benito's early education took

8. Gil first called attention to the artisan roots of the Arias clan and *entorno*. Caso Amador expands on earlier studies with extended genealogies: "Origen judeoconverso del humanista Benito Arias Montano," 1666, 1678. He is responding to the devastating critiques of the historiography concerning Arias Montano's Jewish heritage in Martínez Ripoll, "Universidad de Alcalá," 18–20; Sánchez, "Arias Montano y la espiritualidad en el siglo XVI," 33–49; and Sáenz-Badillos, "Benito Arias Montano."

9. Martínez Ripoll, "Universidad de Alcalá," 32.

place at home under his father's guidance, where he gained an appreciation for languages and the graphic arts as well as for physics and astronomy.[10] His education continued with the local curate, Diego Vázquez Matamoros, who had been to the Holy Land and stoked the young boy's geographical imagination by drawing sketches and maps of Jerusalem and recounting stories of his travels. Historians speculate that his education must have continued in a more formal academic setting, perhaps in a school associated with a church or monastery in Seville, because by the time he enrolled in the university he was well versed in Latin and had already written a short treatise on ancient Castilian numismatics in 1541, about the time of his father's death. After his father died he seems to have alternated residence between Fregenal and Seville. In Seville, the prominent and well-off Vélez de Alcocer family welcomed him as he continued his education. He would come to regard this family of converso origins as close blood relatives, and he considered their son Gaspar his dear boyhood friend. (Gaspar left for America and became a wealthy businessman in Lima; he visited Seville infrequently until his death in 1597.)[11] Arias Montano always cherished his links to the city on the Guadalquivir, and throughout his life he sought the warmth and support of an extended family among the close circle of friends he cultivated there.

Although it is not clear precisely where he attended school before commencing his university studies, Arias Montano also came under the tutelage of the poet Juan de Quirós and the historian Pedro Mexía, both of whom he mentioned in later works.[12] In 1545, at about age eighteen, he began his liberal arts studies at the College of Santa María de Jesús in Seville. He completed the first two years of the traditional university curriculum (dialectic and logic), but after sitting through most of the third year (physics and natural philosophy), he left for the Complutensian University at Alcalá de Henares, apparently without sitting for his exam and getting the bachelor's degree in arts at Seville.[13]

The humanistic and theological studies available at Alcalá beckoned. By the end of the 1548 school year, while enrolled at the Colegio Mayor de San Ildefonso, Arias Montano completed the requirements for a bachelor's degree in arts and philosophy; he ranked eleventh in his class of seventy. He continued into the fourth year (metaphysics) and earned his licensure the following year. Meanwhile, and as an indication of where his true interests lay, he also enrolled

10. González Carvajal, "Elogio histórico," 8.

11. Martínez Ripoll, "Universidad de Alcalá," 48–49.

12. Domínguez Domínguez, *Arias Montano y sus maestros*, 20–30.

13. Ibid., 33.

in classes on Sacred Scriptures in the Faculty of Theology. But the course of study undertaken there is the best testament to what drove his departure from the more conservative University of Seville; he dived into the humanistic study of the Sacred Scriptures that the university had championed since its founding in 1528.

The creation of the Colegio de San Jerónimo, the trilingual college, at the University of Alcalá marked the effort to institutionalize the humanistic and philological approach to the study of the Sacred Scriptures advocated by Cardinal Cisneros, as a special focus of the humanistic interest in the rediscovery of the works of classical antiquity.[14] It was only the second European university (after Louvain) to teach Oriental languages. In Spain, however, it was the emphasis on Hebrew that distinguished its humanistic biblical studies. For theologians trained at Alcalá, "ad fontes" meant returning, whenever possible, to Hebrew versions of scripture and studying it using the linguistic tools and commentaries of Spain's Sephardic tradition, including Kabbalah.

The existence of the trilingual college emphasized the universitywide focus on studying the Sacred Scriptures from a philological perspective. By continuing his education in Alcalá, Arias Montano had chosen to study them as a Christian Hebraist, to favor the literal exegesis based on the *hebraica veritas* over any other, and certainly to cast aside the Scholastic approach to scriptural studies still prevalent in Seville. On arriving in Alcalá he began studying Greek and Hebrew, and he might have started studying Chaldean (Aramaic) and Arabic as well. Eventually his studies led him to the chair of Sacred Scriptures, held at the time by the Cistercian friar Cipriano de la Huerga (1510–60), a Hebraist who became a cherished mentor.[15]

Fray Cipriano de la Huerga wore the mantle of biblical studies in Alcalá in the tradition of the trilingual college, but now in the different religious climate of the Counter-Reformation.[16] The humanistic zeal that marked Spanish biblical studies of the early sixteenth century at Alcalá and Salamanca, with their focus on philological precision and extracting the literal meaning of Sacred Scriptures, had by the 1550s flourished into an exegetical approach that openly shunned the traditional fourfold sense of scriptural interpretation (literal, allegorical, moral, and analogical) and, especially in fray Cipriano's

---

14. There is a frustrating lack of primary sources that chronicle the establishment of the school or describe its curriculum, as noted by Alvar Ezquerra, "El Colegio Trilingüe de la Universidad de Alcalá de Henares."

15. Morocho Gayo, "Magnum illum vergensem," 87.

16. Fernández Marcos, "Exégesis bíblica de Cipriano de la Huerga."

case, favored exclusively the literal sense, preferably extracted from the Hebrew or Chaldean originals.[17] This hermeneutical style went hand in hand with a powerful oratorical prose that testified to the importance of rhetoric at Alcalá. The resulting analysis took philology as a starting point but aimed to deeply contextualize historically and within the history of ideas the possible meanings gleaned from the Sacred Scriptures in their Hebrew, Chaldean, and Greek originals. More often than not, patristic literature took a backseat to the works of ancient philosophers—Presocratics, Neoplatonists, Hermetists—whom fray Cipriano often relied on in his exegesis.[18] (This was an aspect of his teacher's work that Arias Montano did not overtly emulate.) His vast curiosity and erudition led him also to embrace the notion of a *prisca sapientia* (ancient knowledge or sacred wisdom) and the authority of the *prisci theologi* (ancient theologians). We will see how his students, a veritable constellation of Spain's biblical scholars of the later sixteenth century, including Arias Montano, fray Luis de León, Juan de Mariana, and Pedro de Fuentidueña, followed their master. In this pedagogical environment of Alcalá, the Sephardic medieval tradition, with its emphasis on literal exegesis and philology as exemplified by Rabbi David Kimhi (Radak), found resonance. The translations of his works by Alfonso de Zamora (who held the chair of Hebrew) reinforced a pedagogical approach to the study of the Bible and the preferred method of exegesis taught in the school: the literal method. This was the environment where Arias Montano spent his formative years and learned to study the Sacred Scriptures. It also helps explain why he undertook such an ambitious role in the Antwerp Polyglot. As Sergio Fernández López points out, Arias Montano, drawing from a long Spanish tradition of literal interpretation, wanted others to understand the simple (*llano*) or literal meaning of the Bible.[19]

Although the years in Alcalá were a period of tremendous intellectual growth—Arias Montano was awarded the prize of *poeta laureatus*—sometime during 1551 or 1552 he became very ill with a gastrointestinal condition that manifested itself in a *bilis negra* (black bile or bloody stool) and seems to have also left him emotionally drained.[20] The illness became chronic, and he suffered periodic attacks throughout his life, often accompanied by emotional malaise. After having retired to the Peña de Alájar for a while during 1552, in

17. Fernández Marcos and Fernández Tejero, *Biblia y humanismo*, 15–25.

18. Paradinas Fuentes, "Cipriano de la Huerga y la filosofía del Renacimiento," 36, and Domínguez Domínguez, *Arias Montano y sus maestros*, 63–71.

19. Fernández López, *Alfonso de Zamora y Benito Arias Montano*, 29–32.

20. Martínez Ripoll, "Universidad de Alcalá," 21–22.

1553–54 he traveled to the University of Salamanca, possibly to teach Greek and Hebrew.[21] Arias Montano had left Alcalá without completing the examinations that would have granted him a doctoral degree in theology, but by 1558 he was using or was referred to by the title of *maestro*, or *doctor*.[22]

It is also not clear when and where he took holy orders. Some records, now lost, refer to a don Benito Arias Montano serving about 1554 in the parish of Castaño de Robledo, a town not far from his precious Peña. He was in the area during that time; years later he fondly recalled the many friends and family who came calling, including Gaspar Vélez de Alcocer, who spent four months there while Arias Montano taught him astrology and Hebrew. This close friendship morphed into a teacher-student relationship that led Arias Montano to style his *Rhetoricorum libri IIII* as a discourse to Gaspar. It was also during this time that he began writing the geographical material that would become part of the appendix or *Apparatus* of the Antwerp Polyglot.[23]

In 1559 Arias Montano accepted an invitation from his friend and surgeon Francisco de Arceo (ca. 1494 to ca. 1575) to visit Llerena to preach during the Easter festivities. It was far from a casual invitation, for Arceo secured letters of invitation from Llerena's city council, the provincial governor, and the local inquisitors he worked for.[24] Arceo apparently never earned a medical degree; he came from the surgical tradition and had practiced in the renowned hospital of the Real Monasterio de Guadalupe in Cáceres. In a preface he wrote to Arceo's book on the art of surgery, *De recta curandorum vulnerum ratione libri II* (1574), Arias Montano recounted how during the four months he spent with Arceo the old surgeon had given him lessons on a variety of surgical techniques.[25] Arias Montano had studied some medicine while at Alcalá, and as he noted in the preface, had found little of value in the Scholastic approach to curing, preferring the practical and experience-based techniques Arceo brought to the art. Once in Antwerp and having taken on the task of shepherding Arceo's book to press, Arias Montano asked a Spanish doctor, Álvaro Núñez, to gloss each chapter, taking care to point out how Arceo's methods compared with Hippocratic and Galenic teachings, in effect giving the practical manual the learned academic polish or *ornamentum* Arceo was unable to provide. The

21. Morocho Gayo, "Magnum illum vergensem," 98–99.

22. Domínguez Domínguez, *Arias Montano y sus maestros*, 143–45.

23. Morocho Gayo, "Trayectoria humanística, 1," 180–81.

24. González Carvajal, "Elogio histórico," 29.

25. Arceo de Fregenal, *Método verdadero de curar las heridas*, 109–10, and Domínguez Domínguez, *Arias Montano y sus maestros*, 145–52.

book became one of the first popular surgery textbooks, precisely because it integrated the essential elements of empirical practice within the theoretical medical framework of early modern Europe.[26]

After Arias Montano returned to the Peña, an enemy from Fregenal, whom the documentary records identify only as "Morales," made serious allegations against him; he was arrested on 9 June 1559, although the exact nature of the charges remains something of a mystery.[27] Arias Montano was taken to Seville and imprisoned in the archbishop's jail. Diego Díaz Becerril, a lifelong friend and an in-law of the Vélez de Alcocer family, swore to a substantial bond of five thousand ducats so Arias Montano would not be shackled and could move freely within the jail. The charges were swiftly dismissed by Juan de Ovando, then *provisor general* of the cathedral of Seville, an influential friend of the accused who would later serve as president of the Council of the Indies. Arias Montano countercharged "Morales," presumably for libel. It seems this latter case came to naught. Arias Montano's imprisonment occurred a few months before a roundup by the Inquisition of some high-profile preachers in Seville who were accused of being Lutherans. This coincidence has given rise to the belief that the inquisitorial accusation rested on matters of faith, although there are no records indicating this.[28]

Soon after these events Arias Montano petitioned to join the religious order of Santiago (Saint James), thereby becoming a *caballero de la Orden de Santiago*. The review process or *prueba* of his petition was a comprehensive affair. Held in Fregenal, it involved compiling a substantive *información* based on questioning eight sworn witnesses.[29] The witnesses responded to a series of questions aimed at ascertaining Arias Montano's and his family's history, their reputation, and whether "the said Master Arias Montano and his parents and grandparents have been and are known and taken for gentlemen, old Chris-

26. Pascual Barea, "El epitafio latino inédito de Arias Montano."

27. The inquest does not seem to have taken place, and the one surviving record does not clearly specify the nature of the offense. In the place where a description of the offense would appear, the document has a scribble, which some have interpreted as the *pecado nefando*, or sodomy, while others suggest it reads "RF," or "*relapsus in fide*." Gil, "Arias Montano en Sevilla," 275–80; Morocho Gayo, "Trayectoria humanística, 1," 185n179; and Macías Rosendo, *Correspondencia de Arias Montano con Juan de Ovan do*, 37n41.

28. Martínez Ripoll, "Universidad de Alcalá," 64–65.

29. The often-cited documents are copies supposedly based on the original at the Convent of San Marco de León but made almost one hundred years later: "Información para entrar a la Orden de Santiago," 13 February 1560, in González Carvajal, "Elogio histórico," 123–31.

tians without having the race nor a mix of Jew, Moor, or *converso* &c."[30] The responses consistently asserted the *hidalgo* and old Christian reputation of the family and the moral virtue of maestro Arias Montano. Arias Montano's case moved extraordinarily quickly, at the insistence of the prior of the convent of San Marcos de León, for in May 1560 he was professed and joined the community at that convent. He was now under the protection of a powerful order and could continue his studies.

HINTS OF DISQUIET

The relaxed rule the order followed gave him plenty of time for study, and he seemed to settle contently into his novitiate. In his earliest surviving letter from this period of his life, a missive between Arias Montano and the Augustinian friar Luis de León (1527-91), he reveals the type of questions that engaged him during the months just before he entered the order.[31] When we look beyond the affective language of humanistic friendship, it is still evident that this is a letter between close friends, who knew each other's intellectual concerns, encouraged each other's work, and could be trusted with delicate questions. He wanted fray Luis as a matter of routine and in a synthetic way to collect notes and references on a series of topics *de quibus in nostro tempore disputatur* (that are discussed in our times) and that they, as interpreters of the Sacred Scriptures, should bear in mind and must be prepared to discuss.[32] The topics fell into two categories: the religious and the natural philosophical. The religious topics were a veritable survey of the pressing theological issues of the time, and in researching these Arias Montano asked fray Luis to collaborate but also to keep the inquiry to himself. (Voicing such concerns could be a problem in the fraught post-Tridentine environment where Arias Montano's frank conversations with fray Luis took place.)[33]

30. "El dicho Maestro Arias Montano y sus padres y abuelos han sido, y son habidos y tenidos por hijos–dalgo, christianos viejos sin tener raza ni mezcla de judío, moro ni converso &c," in González Carvajal, "Elogio histórico," 123.

31. BNE MS 8588, fols. 129-31v, B. Arias Montano to Luis de León, n.d. Published first in J. López Toro, "Fray Luis de León y Benito Arias Montano," and recently edited and translated by Domínguez Domínguez, "Carta de Arias Montano a fray Luis de León."

32. Domínguez Domínguez, "Arias Montano a fray Luis de León," 308.

33. When in 1572 fray Luis was brought before the Inquisition on suspicion of holding heretical views, he admitted in response to the testimony of fray Diego de Zuñiga to having denounced in 1560 a certain manuscript book, possibly in Italian, that Arias Montano had read to him. The denunciation took place in 1562, but it seems not much came of it. (It has not been

The list began with the issues of predestination, election, and reprobation, the justification of fallen mankind, faith, and obedience *sine operibus*. It continued with the infusion of the Holy Spirit, the human capacity for sin or virtue, and whether this virtue is God-given or can be acquired through prolonged practice and repetition. The list next turned to the topic of grace: Who are the sanctified and confirmed by grace? And then to the virtue and efficacy of the sacraments: Where do they come from? Where do their virtues reside? In certain matters initiated by prayer, by Christ's anointing and power as given by the Father, or in those sacraments, such as circumcision, whose external representation discloses an internal one?[34] Throughout his life, Arias Montano repeatedly returned to these topics and delved into them in his exegetical works. In a later chapter we will see that, especially on the issue of grace and despite how carefully he maneuvered around censors, his works were ultimately unable to avoid expurgation.

In contrast to the guarded way he presented the religious topics occupying him, when Arias Montano turned to the natural philosophical points he wanted fray Luis to keep tabs on, he openly discussed his views. (He wrote on these issues in the first-person plural, which might suggest that fray Luis shared his views, but it could just as easily be that Arias Montano was using the *pluralis modestiae*.) He asked fray Luis to collect opinions from both ancient and modern authors, either poets or prose writers, on two cosmological subjects. Fray Luis should also note whether these authors agreed or disagreed with the opinions Arias Montano and fray Luis held, since these contradicted commonly held views on the subjects. The first issue was the elemental nature of fire, the second the number of celestial orbs. He stated his position unambiguously:

> For neither do we maintain fire to exist as an element in nature, nor do we reckon the number of celestial orbs extends to eleven or twelve, for we are happy with three and dispense with the others. And in fact, we call the first orb from here to the planets "air," next we place all of the stars including the wandering and fixed, and finally the seat of God and the blessed spirits.[35]

---

found in inquisitorial records.) However, when fray Luis was forced to bring up the subject during his trial years later, it placed Arias Montano under additional inquisitorial scrutiny. See fray Luis de León's response to fray Diego de Zuñiga, in DIE, 10:373–81 and 11:192. For suggestions on what this book might have been, see Prosperi, *L'eresia del Libro grande,*" 382–85.

34. Domínguez Domínguez, "Arias Montano a fray Luis de León," 308, 311.

35. Ibid., 309.

In these two sentences Arias Montano was dismissing two essential tenets of Aristotelian-Ptolemaic cosmology, and in doing so he also undermined a host of other cosmological and physical precepts. He was happy with a radically new cosmology: heaven is not composed of a series of nested orbs clearly delimiting the world. There were only three: a region above the earth (the "here" above) that extended to the planets, another that housed all the planets (wandering stars) and stars (fixed), and finally an outer region where God and the angels resided. This arrangement displaced the traditional location assigned to elemental fire. Fire, as the lightest of the four Aristotelian elements (earth, water, air, and fire), occupied the uppermost region of the sublunary orb, a space above elementary air and nearest to the first planet (understood then as the moon). He now placed "air" in the space traditionally occupied by elemental fire.

Whereas Arias Montano had asked fray Luis to keep the investigation into the religious questions secret, "for you are the person whom I have chosen as guide, confidant, and partner in all my studies," he encouraged him to share the inquiry into these natural philosophical questions with others inclined toward this type of study. The distinction between subjects that could be discussed openly and those best contemplated prudently is very telling. Clearly the religious topics were sensitive, and an eager investigation of these questions could be misunderstood by suspicious eyes. But the same eyes would find little or nothing to be concerned about with the cosmological questions. The letter indicates for the first time a relative openness to alternative natural philosophical theories and signals the radical cosmology that Arias Montano would place at the center of the *Magnum opus*.

Sometime before 1562 he earned his doctoral degree, presumably from the Complutensian University, and was asked to join the delegation of Bishop Martín Pérez de Ayala of Segovia, a member of the Order of Santiago, to the third and final session of the Council of Trent. Arias Montano formed part of a large cadre of theologians who, though not permitted to vote, were called on to argue and deliberate on certain issues and to issue opinions. He rose to speak during two sessions. In session 21/5 of 20 June 1562 he spoke on the subject of granting lay members of the church Communion under the two species (host and wine). Arias Montano argued that the Bible seemed to mention no obligation to take Communion in the two species, and thus the church could decide who should be permitted to do so. He also clarified that children were not under obligation to take Communion, since it implied an action of thanksgiving that they were incapable of acknowledging. He supported all his observations with textual evidence drawn from the Bible. (Years later Arias Montano would recall that his comments had been received well, even with applause.) He also spoke again in session 24/8 almost a year later, on the sacrament of matrimony. He would later describe his intervention as difficult, and he acknowledged being inspired by the Holy Spirit to finally find something heartfelt to say. Sadly, we have no record of his statements.[36]

Pleased with Arias Montano's service at Trent, Bishop Ayala asked that he be given a pension of two hundred ducats, a complicated transaction that he managed to receive only in 1567.[37] The events of his life in the years immediately after Trent are not fully known. He might have traveled throughout Europe; in a later letter he mentions having been in Venice after his stay in Trent. Royal tutor Honorato Juan proposed him as a royal chaplain, and he was awarded the post in 1566. His star was clearly rising despite his not yet having published any religious works. Over the next few years he wrote a series of works that would come to light from Plantin's press only during the 1570s. In March 1568, probably while fulfilling his duties as royal chaplain at the court in Madrid, he read attentively the king's instructions about the publication of a new version of the Complutesian Polyglot Bible. Did Arias Montano realize he was about to step onto the public stage? Did he meet this opportunity with excitement or apprehension? We do not know, but his life would never be the same.

THE CHALLENGE AHEAD

While exploring every aspect of his multifaceted life is beyond the scope of this book, the rich corpus he left allows us to pose a series of questions with a reasonable expectation of answers: What drove Arias Montano to embrace the unusual cosmology he discussed with fray Luis? On what basis was he willing to overturn fifteen hundred years of received wisdom about the constitution of the universe? Perhaps it would be more accurate to ask, Why did Arias Montano devote so much of his life to composing a new history of mankind and of nature? What compelled him to conceive, propose, and compose such a wildly ambitious project? To approximate an answer to these questions we have to explore our author's intent—his sense of purpose—and try to elucidate the reasons that compelled him to take on this project. What drove Arias Montano in this enterprise?

We have seen in his letter to fray Luis de León that he was already comfort-

36. Fernández Nieva, "Extremeño en Trento," 963–66.
37. Alcocer, *Felipe II y la Biblia de Amberes*, 14.

able discussing cosmological notions that subverted the Aristotelian-Ptolemaic paradigms. Although twenty-odd years had passed since the publication of Copernicus's *De revolutionibus orbium coelestium* (1543), astronomers and natural philosophers held firmly to the notion of a celestial realm distinct from the earthly one and arranged as a series of concentric crystalline spheres. Only a handful embraced the Polish astronomer's heliocentric proposal—a proposal that, it must be noted, only timidly questioned the existence of the celestial spheres.[38] Yet despite this apparent conservatism, the panorama of sixteenth-century natural philosophy was marked by episodes of intellectual effervescence that would result in an array of alternative proposals.

During the late sixteenth century the cornerstone of natural philosophical studies at universities in Spain and throughout Europe continued to be Aristotle's books on nature. This corpus was taught following Scholastic methods and discursive practices, giving rise to what historiography refers to as Scholastic Aristotelianism. In some universities, principally in Italy, Renaissance Platonism jockeyed with Aristotelianism as a natural philosophy. But even the legacies of Marcilio Ficino and Pico della Mirandola held little sway in most universities' curricula. And yet an influential segment of the intellectual elite of Europe had begun to voice dissatisfaction with the prevailing ways of explaining the natural world. The scholarship on this period has largely interpreted this dissatisfaction as the result of humanists' reengagement with a variety of natural philosophical systems from antiquity. In the hands of Ficino and his followers, Neoplatonism and material from the *Corpus hermeticum* became Renaissance Platonism and stood as the first viable challenge to Scholastic Aristotelianism. They posed alternatives to the tired Aristotelian natural philosophy as taught at universities. Epicurean atomism and Pythagorean philosophies were also discussed, and whereas astronomers warmly embraced an emphasis on mathematics, they rejected other aspects as too tainted by their pagan origins. A good dose of skepticism entered the arena about 1562 with the rediscovery and Latin translations of Pyrrhonism. Renaissance thinkers also took a serious second look at Academic skepticism by engaging with the ancient sources themselves and shedding the philosophy's Augustinian garb. Its reacquaintance with skepticism led Richard H. Popkin to describe the whole era as subsumed in a crisis of doubt.[39]

38. Westman, *Copernican Question*, 215.

39. As a general introduction to the topic, see Copenhaver and Schmitt, *Renaissance Philosophy*, 22–24, 239–60. Popkin frames it as a crisis that originated with the Reformation and quickly overtook philosophy: Popkin, *History of Scepticism*, 15–16.

For some, doubt is disquieting if not profoundly dangerous. Those who could not embrace the notion that it was impossible to speak with certainty about the natural world (or about ethics, for that matter), or who found little comfort in the profusion of ancient natural philosophical alternatives, found in the intellectual climate of the sixteenth century an opportunity to set forth their own proposals. And propose they did! Spanish thinkers of the sixteenth century, as we will see in the next chapter, did not lack for new approaches or alternatives to the systems being proposed throughout Europe. It was in this climate of new theorizing that Arias Montano set out to write his *Magnum opus*. He shared in the Spanish disquiet: a deep-seated discomfort with doubt, with the threat it poses to human salvation, and with incomplete or ill-construed natural philosophical proposals. This disquiet also defined the parameters of what it was possible to expect from natural philosophical systems conjured by human beings. In Arias Montano the disquiet served as the motivation to write the *Magnum opus*. He grew convinced that he had found within the pages of the Bible a complete and unambiguous natural philosophy, one that was not tainted with the paganism of the ancient systems or corrupted by the passage of time and that—because it came straight from the Word of God—was infallible.

CHAPTER 2

The Spanish Disquiet

Toward the end of his life Arias Montano wistfully recalled spending his youth deeply engrossed in the study of natural philosophy. In a verse proem to the second book of the *Magnum opus*, the *Corpus*, or *History of Nature*, he explained the life course that drove him to undertake the project. The poem is in essence a lamentation—a votive elegy—for the years he spent studying the theories of the ancient natural philosophers. Astronomy in particular had absorbed hours with computations and hard work. Now these theories seemed to him only "empty words that produce nothing but words."[1] He had come to realize that the natural philosophy he had pursued so diligently in his younger years was corrupted and confusing. It had earned him praise then, but now it all seemed a waste of time:

> And those things I remember learning long ago, as a young man with many others, and read in various books. I remember repeating them; I remember them, yes, and now regret the great effort and long fatigue as in time I have ventured down dark paths. I was not yet fifteen years old, and others were already saying of me that I was not ignorant of nature. I learned the movement of the stars and of the heaven, the phases of the moon, and the unequal times of the solstices: how the double day made the hours of light and darkness equal for all lands. To say a lot about the beginnings, the end and limits, as well about the forms, was for me a diversion. Yes, I now realize it was certainly

1. "Haec sibi dum fingunt, et inania condere verba/Soliciti intendunt, nil nisi verba ferunt.": "Elegia Votiva," *Corpus*.

a diversion, and I now mourn and lament the damage caused by so much lost time.[2]

Similar lamentations about the futility of studying the philosophy of the ancients were becoming commonplace among the more radical natural philosophers of the era; we hear them first from the fringes in the words of Miguel Servet, Agrippa, and Paracelsus. By the seventeenth century they had become a trope in the works of Francis Bacon, Jan Baptist van Helmont, and René Descartes. In Arias Montano the lamentation is not filled with the contempt or anger found in those philosophers but acts as a preamble to a moment of deep reflection that was interrupted when an otherworldly voice revealed to him the futility of pursuing the study of nature simply for its own sake or while subservient to existing natural philosophical systems. To fully understand this message and what it instructed Arias Montano to do requires a glance at the panorama of sixteenth-century Spanish approaches to the study of nature, especially since the dissatisfaction he expressed in the votive elegy was not his alone.

Other Spanish natural philosophers and scientific practitioners shared his concern about the state of natural philosophy, its purpose, and the aims of its associated pursuits. The most pressing matter seems to have been a tension between the study of nature and the purpose of the endeavor. The perception prevailed that natural philosophical systems and associated practices no longer functioned as suitable tools for producing new knowledge. They not only failed to yield satisfactory explanations of the natural world, but also—and of utmost importance to some—derailed the ultimate destiny of the practitioner. This chapter and the next will explore the contours of this attitude toward prevailing natural philosophical systems and delineate the Spanish disquiet. The disquiet was the result of a three-way tension between prevailing natural philosophies (seen by some as an increasingly flawed way of learning about and organizing knowledge of nature), a natural theological approach to the study of nature popular in sixteenth-century Spain, and a purely empirical methodology and didactic organization of knowledge that seemed to grow increasingly disconnected from philosophy and theology. I first turn to the long history in Spain of writing about nature in the natural theological tradition and identify within this tradition a particular attitude toward the study of the natural world. Attention then shifts to the growing importance of the empiricist approach during the century of discovery and to the study of nature and the range of alternative systems that resulted from this approach. Finally,

2. "Elegia Votiva," *Corpus*, and *HistNat*, 97.

I return to the otherworldly voice Arias Montano heard cautioning him about pitfalls in the study of the natural world.

NATURAL THEOLOGY IN SPAIN

Historical writing about the relation between natural theology and science has long been dominated by Protestant perspectives, understandably so given the importance of the field for eighteenth- and nineteenth-century history of science. But exploring the world from the perspective afforded by natural theology has much earlier roots. For Plato, clues about the nature of the transcendent Demiurge could be found by attentively studying his creation, and within the Judeo-Christian tradition the Bible presented the study of nature in a similar light. In the book of Wisdom and the Psalms nature is portrayed as a herald of God's glory: "The heavens declare the glory of God; the firmament proclaims the works of His hands" (Ps. 19:2 NAB). It is in books 6–8 of Saint Augustine's *City of God* that the possibility of a *theologia naturalis* was first fully articulated within a Christian context.[3] During the Middle Ages, the Franciscan tradition framed the study of nature as a contemplative and active spiritual exercise that enlightened the faithful about the nature of the Divinity. In his *Breviloquium* (1257), Saint Bonaventure (1221–74) describes the creation as like a book ("quasi quidam liber," 2.12.1) that made manifest the nature of God. But this was an insight postlapsarian humanity has been deprived of; humans can only approximate knowledge of God through the study of nature, apprehending through its wonders a faulty contemplation of God. God left in nature a divine footprint (2.11.2). For Bonaventure and the legions of Franciscans who followed him the study and contemplation of the visible world brought insights into the nature of God.[4]

It is through Augustine and the vast commentary tradition his works spawned that natural theology was appropriated in the Catholic Hispanic tradition. Particularly during the sixteenth century, natural theology in Spain developed into a viable, orthodox program of study that helped humanity understand its relationship to God, the nature of the divine, and the creation. The object of study was the natural world, but "as the product of God's active, creative nature, the natural world pointed beyond itself to the principle which had created it, so that knowledge of it could reveal aspects of the nature of God."[5]

3. Collins, "Natural Theology and Biblical Tradition," 2–3.

4. French and Cunningham, *Before Science*, 209–12.

5. Methuen, "Interpreting the Books of Nature and Scripture, 189."

This teleological argument for the existence of God also proved particularly useful in legitimizing the study of nature. Natural theology endowed observing and contemplating nature with a virtuous purpose; even the pure empiricist could describe inquiries into nature as worthwhile with the argument that the purpose of the endeavor was ultimately to elucidate the nature of human beings' relationship with the Divinity or to learn about the divine nature. The metaphorical book of nature could be read—*had* to be read—to better understand the natural laws that God had set down for its orderly governance and through which the nature of God could be discerned.[6]

Spanish natural philosophers and theologians embraced this heuristic approach to theology, with the Majorcan theologian and missionary Ramón Llull (1232–1316) setting forth an innovative method for deciphering the book of nature inspired by Saint Bonaventure and Franciscan mysticism. Llull proposed a unified theological and philosophical approach meant to explain the nature of the Divinity to non-Christians. Central to his art was the acknowledgment that the predicates describing what is good in the natural world are an imperfect reflection of the attributes that reside—perfected—in God. His thought followed the mystical tradition influenced by Neoplatonism and the affirmative approach to knowledge of the divine set forth by Pseudo-Dionysius in *The Divine Names*.[7]

It was Llull's follower, the Catalan Raymond Sebond (also known as Raimundo Sibiuda or Ramón Sabunde, ca. 1385–1439), who inaugurated the genre of the early modern natural theological tract with his *Theologia naturalis, sive Liber creaturarum* (1485), the first printed book to refer to a "natural theology."[8] Sebond's book later reached a large popular audience through Montaigne's French translation (1569) and a number of smaller compendiums such as the *Viola animae* (1499) by Peter Dorlant and the *Oculus fidei* (1661) by Comenius. A 1549 Castilian edition, *Violeta del anima*, published by Francisco Fernández de Córdova, popularized the text to a lay audience in Spain.[9] The book was reprinted dozens of times despite censorship of its prologue in 1559.[10]

6. On the earliest patristic references to reading the book of nature, see Groh, "Emergence of Creation Theology," 30–33.

7. Mayer, "Llull and the Divine Attributes in 13th Century Context," 141–43, 152–53.

8. Woolford, "Natural Theology and Natural Philosophy," 150–53.

9. Sánchez Nogales, *Camino del hombre a Dios*, 45–55. For a discussion of Sebond via Montaigne as a precursor to Bacon's natural theology, see Peterfreund, *Turning Points in Natural Theology*, 1–16.

10. Martínez de Bujanda, "L'influence de Sebond," 80; Puig, *Sources de la pensée philosophique de Raimond Sebond*, 65. Sebond's book appeared in the Index of Paul IV (1559),

Sebond's objective was to propose a new theological science whose purpose was to uncover the existence, attributes, and agency of God. He built on the notion of God's two books to put forth the idea that the book of nature written by God through the act of Creation was open for all people "to read" according to their abilities. The ultimate purpose of this activity was unequivocal: to learn about the nature of the Divinity. But in doing so the follower of natural theology would also learn about the nature of human beings— principally his or her own nature—and discover how to live in concert with natural and divine law. We read in the preface:

> In the book of nature a man can study by himself, and without a teacher, the doctrine that he needs. God created this whole visible world for himself and gave it us as his own natural and infallible book, written with the finger of God, wherein the creatures are as it were letters devised, not by the will of man but by the wisdom of God, to convey to man wisdom and teaching necessary for his salivation. No man indeed can see or read by himself in this book [always opened] though it be unless he be enlightened by God and purified from original sin; and so none of the ancient heathen philosophers could read this knowledge, because they were blinded so to the knowledge of themselves, although in that book they did read some knowledge, and all which they had they derived thence.[11]

Sebond presented natural theology as a self-directed course of study that stood proudly apart from the Scholastic curriculum and offered a pleasant— but also orthodox—path to the reconciliation between scripture, faith, and reason. Its promise was a journey buttressed by grace and faith that led to greater knowledge of God. Only those enlightened and free from original sin could undertake this journey, however, which was the reason ancient philosophers and pagans had not been able to learn this science. Yet for the faithful this science could be studied without need of the liberal arts or knowledge of physics or metaphysics. It supplied arguments based on "true experience"

but by the second printing a few months later only the prologue was banned. The reason for censoring the prologue remains unclear, with authors positing several possibilities: the emphasis placed on the study of nature alone as enough to reveal God to humanity; the resulting downplaying of the role of the Bible and its revelation; the emphasis on a rational approach to tenets of faith; or the downplaying of the need for church fathers and tradition to explain the faith or references to the end of time that might have seemed millenarian: Sánchez Nogales, *Camino del hombre a Dios*, 187–99.

11. As translated in Webb, *Studies in the History of Natural Theology*, 297.

(*vera experientia*) of the creatures themselves and their natures. Thus the evidence for the resulting theology emanated not from scripture but from "seeing" nature and from sensory experience.

The *Natural Theology* gave plenty of examples of how the experiences of followers of this science reflected upon their observations of nature. The observation of nature was simply a starting point leading to a theological precept about the nature of God and culminating in a meditation on an aspect of the Divinity. For example, Sebond explained, just as the superior bodies rule inferior bodies, so it is with humanity. Just as the sun gives life and light to all its celestial companions and sends its influence above and below it, so in humans the heart furnishes life and heat to all the parts and likewise, because it is in the middle of the body, spreads its influence throughout the body. This type of parallel could then be noticed in things of a higher nature. Just as seeds planted in fertile ground result in trees with branches and fruit, so a small amount of seed from our parents produces our bones, eyes, teeth, and hearts. Should not contemplating this tell us that we are a work of great God?

Sebond's natural theology did not prescribe a methodology for how nature was to be observed or "read," let alone studied. Nature was meant to be observed as it was and interpreted as a tangible manifestation of the act of creation. Observing and experiencing nature were conceptualized as a purely passive activities, more contemplative than active, from which the follower of this science could discern vestiges of the Divinity in nature. What was perfectly clear, however, was the reason one "read" the book of nature: to get to know God and earn salvation. For it was through observing nature, understanding humanity's place in its hierarchy, and undertaking the arduous process of deciphering the theological message in God's writing therein that humanity could understand its place in creation.

Sebond's natural philosophical hermeneutical lens was, not surprisingly, clearly Aristotelian; yet it was inflected with the Platonic belief that a perfect analog of nature existed elsewhere. In this Christian context perfection dwelt in the divine. Yet natural theology was not wedded to a particular natural philosophical system for its interpretive framework; it sufficed to look carefully at the world as it was and to reflect about the Creator who had placed it at humanity's disposal. This philosophical and methodological ambiguity of natural theology became an opening for others to propose alternative ways to "read" the book of nature. On the one hand Sebond's natural theology could be used to give theory-free empiricism a justifiable and theologically sound purpose, and on the other it could ascribe a similar purpose to natural philosophical systems with dubious or ambiguous reputations. We must bear in mind that during the sixteenth century natural theological treatises were not meant to be repositories of knowledge about the natural world as were natural histories along Plinian lines or cosmographies like that of Johannes de Sacrobosco; they were a distinct genre of writing about nature, more devotional manual that natural philosophical exploration.

An audience inclined toward reflection or mysticism could turn to another style of natural theological treatise. In these the objective was to uncover wondrous aspects of nature that gave testament to an omnipotent God. Faith followed from awe. Here again, finding the awe-inspiring in nature, and recognizing what was beautifully complex yet perfectly ordered within it, was achieved only through careful observation and by attentively experiencing the world. In these types of natural theologies the role of natural philosophy as hermeneutical lens was little more than marginal. The author's objective was to draw the faithful through the linear sequence of largely devotional steps: observe nature, feel awe, believe in God, and worship him.

In Spain, among the best known of these natural theologies was the *Introducción del símbolo de la fe* (Salamanca, 1583–85) of fray Luis de Granada, OP (1504–88). Here the author selected for commentary only those works of the six days of Creation that best reflected God's power and providence.[12] Written in an affective language and peppered with personal experiences and anecdotes, it searched for the attributes of the Creator in both the macrocosm and the microcosm, with added emphasis on the wonder and beauty of remarkably small animals. Granada introduced each topic with a flurry of rhetorical questions: "So, what can I say of the ants' abilities and of the subtlety of the spider's web and of the republic of bees with their king—so well organized—and the craftsmanship of silkworms, which are all for the ornament of the world?"[13] Thus we learn that just as ants foresee the coming winter and take care to store food, humans should also build up a storehouse of good works while on earth in preparation for the next life.[14] We are also asked to wonder, Who taught a spider to weave a web or to cover its burrow with fine webbing? Granada explained, "The smaller and viler they are, so much more they declare the omnipotence and wisdom of the Lord which in such small bodies placed such strange abilities."[15]

The great expanses in the natural world also gave Luis de Granada a chance to revel in the beauty of nature and to make natural theological arguments

---

12. Granada, Introducción del *Símbolo de la fe*," 50.

13. Ibid., 112.

14. Ibid., 339.

15. Ibid., 351.

based on aesthetics. "Who will explain the beauty of the sky? How delightful it is to see in the middle of a still summer night the full moon, so clear that its brilliance hides all the stars!"[16] God, the sovereign painter (*soberano pintor*), placed them there for our admiration, wonder, and awe. Even a child who by some misfortune happened to grow up in a jail without ever seeing the sky would at his first glance of a starry summer sky exclaim, "Who could paint such a beautiful domain with so many different flowers, unless it was the most beautiful and powerful Maker?"[17] Luis de Granada encouraged curiosity about the natural world, frequently citing the expressions of wonder and curiosity of ancient philosophers as a way to encourage his readers. That they did not recognize the Judeo-Christian God seemed not to worry him; what mattered for his natural theological argument was that they had recognized intuitively through nature the existence of a single maker. Thales, Plato, Aristotle, and his favorite, Seneca, had all—in his opinion—exhorted their fellows to study nature in order to understand the nature of the divine. From Seneca's letter 65 to Lucilius, Luis de Granada borrowed a quotation that has often been interpreted as a defense of studying nature but is really a call to study lofty things: "Do you forbid me to have a share in heaven? In other words, do you bid me live with my head bowed down? I was born to a greater destiny than to be a mere chattel of my body."[18] As Luis de Granada correctly interpreted it, Seneca was defending his right to tear his mind from things perceived only by the senses and turn toward the study of mysteries that lay beyond them.

For someone like Alejo Venegas de Busto (ca. 1498–1562), a popular lay doctor of theology at the Real Universidad de Toledo and later in Madrid,[19] the sense of purpose communicated by a natural theology could be used to imbue Aristotelian natural philosophy with a virtuous purpose. His *Primera parte de las diferencias de libros que ay en el universo* (First part of the different books that are in the universe, Toledo, 1540) is a case in point. It was a very popular book—with five editions during the sixteenth century alone—intended for a literate audience and written in very accessible Spanish. The book's premise is that all there was to know about humanity's place in the universe had been written by God in four "books": Original or Divine (on the archetypal world known only to God); Natural (on the visible world); Reason (on the purpose and use of reason); and Revealed (on what is learned in the Sacred Scriptures).

16. Ibid., 185.

17. Ibid., 186.

18. Ibid., 132. English translation from Seneca, *Ad Lucilium epistulae morales*, ed. Richard M. Gummere, 1:457.

19. Zuili, "Algunas observaciones," 17–29.

The burden of reading the latter three books—the first being unintelligible to humanity—and through them learning the nature of the divine Venegas placed squarely on the shoulders of every human being regardless of capabilities. As he explained, echoing Sebond, "Of the books we must read, because they were written for us, the first is the book of nature; it is the university of all creatures. It is so open that a human being who cannot read it simply has no capacity for reasoning, because it is so clear and legible that—if read attentively—it can be read without an external teacher to instruct him."[20] Saint Anthony had done just this; when asked how he could subsist in the wilderness without a library, he exclaimed that the book of nature was the entire library he needed because it was written by the hand of God. The principal lesson to glean from reading the book of nature was that there is a single Creator who stands as the final cause of everything and whom humanity is not excused from getting to know about, thank, and worship. In fact Venegas titled the second book "Natural Book on Philosophy: Where It Is Declared How to Read the Omnipotence, Wisdom, and Kindness of God in Created Things."[21] His objective in the second book was to point out "the marvels of the world" (*las maravillas del mundo*) so that through them we can learn about its creator (*tener noticia de su criador*), and not to dispute (*disputar*) what the schools of the philosophers have to say about the elemental and celestial regions.[22] Throughout the second book his purpose was to present a rational framework through which the works of creation could be admired and understood and their creator revered.

For Venegas, however, the *fabrica mundi* was best explained with the help of Aristotle; the entire second book is a course on Aristotelian physics with a good dose of Ptolemaic astronomy and Plinian natural history. Venegas was skeptical about some of this received wisdom, such as Pliny's claim that salamanders extinguish fire or that the appearance of Saint Elmo's fire is a portent (he prefers a purely naturalistic explanation).[23] He was equally critical of a certain "famous author" who goes by the name Dizque ("So-they-say"), who seems to be the source cited for all manner of folktales, common knowledge, and lies.[24] He gladly questioned what "los philosophos" have to say, leading readers from theoretical precepts such as the nature of the four elements to their material manifestations on earth as lightning, wind, dew, rain, and dirt.

20. Venegas, *Primera parte de las diferencias de libros* (1540), fol. 35r. This preface was never censored.

21. Ibid., fol. 34v.

22. Ibid., fol. 47v.

23. Ibid., fols. 98v, 95r.

24. Ibid., fol. 86v.

LIBRO II.

¶ *Eſtando aqui el Sol a onze de Iunio es el dia mayor.*

FIGURE 2.1. Celestial spheres. Alejo de Venegas, *Primera parte de las differencias de libros que ay en el uniuerso* (1569), fol. 102v. Biblioteca Histórica de la Universidad Complutense de Madrid (BH FLL 8712).

Among the disciplines, he pointed to cosmography as being particularly well developed in Spain and singled out for praise the works of royal cosmographer Alonso de Santa Cruz.[25] We find the celestial spheres described as being nested like the skin of an onion, each so close to the next that a mustard seed would not fit between them. The explanation about the thickness of the spheres of planets, with their eccentric orbits and epicycles, calls for an analogy to the thicker skin of an orange[26] (fig. 2.1).

25. Ibid., fols. 56r–v, 80r.
26. Ibid., fols. 42r, 53r.

Often lengthy explanations about natural phenomena, particularly meteorological or optical—Venegas's favorite topics—go by without a religious
lesson attached. But in the final chapters the author returned to the second
book's stated objective. Among other things, he reminded readers that learning about the movement of the planetary spheres carries an important lesson.
The circuit of the five planets should remind us of the trajectory humanity
follows from knowledge of created things to knowing their maker. Students
of nature should imagine themselves as orbiting planets. When at the most
distant point of the cycle we should be reminded of our lowly nature; as we
cycle west we should think about our sins; while turning on our epicycle we
should recall how little we have harvested of what we have been given; and
finally, turning upward we present ourselves to God, having assumed complete
responsibility for our actions. The very motions of the sun and moon also
delivered different lessons. The moon has no light of its own but receives light
from the sun: its waxing and waning should be a lesson about our willingness
to receive God's grace. The sun also has no epicycle: this should remind us
that perfect humans do not waver in their proximity to God.

Whereas Sebond and Luis de Granada had embraced a contemplative approach to nature at the expense of a philosophical one, Venegas found a place
for the explanatory power of natural philosophy within the natural theological scheme. By framing his book as "reading the book of nature" he gave
his vernacular version of Aristotelian natural philosophy a suitable purpose
and placed it on par with the other two books humanity could hope to read:
the book of reason and the book of revelation. In Venegas we also notice a
departure from the contemplative hermeneutics of nature of Sebond and Luis
de Granada. Venegas's hermeneutics of nature pushed the boundary of inquiry beyond what is purely apparent to the senses and sought within natural
philosophical explanations additional sites for exploring the nature of the divine and humanity's relation to it. But what about a hermeneutics of nature
that pushed the boundaries of inquiry into nature as Venegas's did, but solely
for the sake of pushing them, without the ultimate purpose imposed by the
science of natural theology?

The Empirical Perspective

For some scientific practitioners, particularly those who stayed away from
natural philosophical pronouncements or left no trace of them, their participation in the Spanish disquiet becomes apparent only after careful examination
of their proposals and the way they addressed the shortcomings they found in

prevailing systems. For other practitioners, the inability of Aristotelian natural philosophy to give satisfactory explanations about the natural world became the very explicit motivation for revisionist work. Their solutions ranged from modest adjustments to Aristotelian philosophy to novel and idiosyncratic systems that sought to build a new natural philosophy on empiricist epistemic criteria. Many of these approaches took the form of conversations with ancient systems of thought—Neoplatonism, Atomism, Skepticism, Stoicism—that in Spain, as elsewhere in Europe, were seen as possible alternatives to Scholastic Aristotelianism. Early modern natural philosophers rarely adopted these alternatives blindly; they wove them together with other ideas, often yielding highly idiosyncratic and eclectic perspectives. Rather than simply slapping the label "syncretist" on them, it is far more productive to inquire about the nature of the problems these philosophers and scientific practitioners were addressing and to think of eclecticism or syncretism as the response to a perceived inadequacy in a particular philosophical program. This approach opens up a number of interesting avenues of inquiry: What did the author find inadequate and in need of reforming?

For an astronomer of the stature of the Valencian Jerónimo Muñoz (ca. 1520–91), who held chairs of mathematics at both the University of Valencia and Salamanca, cosmological uncertainties required particular attention. Muñoz challenged the notion of the incorruptibility of the celestial spheres by relying largely on physical arguments and astronomical observations. After observing the "new star" of 1572, he wrote a short but important treatise, *El libro del nuevo cometa* (Valencia, 1573), arguing that the celestial phenomenon was a comet engendered in the supralunary region.[27] In his unpublished commentary on Pliny's *Natural History* he went even further and argued against the solidity and existence of the celestial spheres based on visual judgment (*oculorum judicium*) and natural reasons (*naturalis ratio*).[28] After systematically considering the opinions of his predecessors, he favored the notion of the heavens as a continuous fluid body; its fluidity was the result of its "watery" nature, as had been suggested by Thales and by Moses and thus, Muñoz added, must be accepted by reason of faith. Muñoz was not content to leave the discussion on this point. He explained that his objective was to inquire about the natural reasons behind what was observed in the heavens, and these reasons supported the notion that the celestial realm was made of air, forming one

27. Navarro Brotóns, "La ciencias en la España del siglo XVII."

28. Navarro Brotóns and Rodríguez Galdeano, *Matemáticas, cosmología y humanismo*, 361, 364.

continuous region from the surface of the earth to where the stars resided.[29] This led him to reject eccentric orbits and epicycles as constructs of mathematicians and philosophers, putting forth instead a geocentric planetary theory where planets move "without orbs or poles by their natures cutting the air."[30]

Muñoz's commitment to observation and "natural reasons" illustrates a broader allegiance to empiricism (in all its gradations) that fed the Spanish disquiet. Speaking generally, for an empiricist, epistemic authority resided not in the natural philosophical system that interpreted the observation or experience, but in the credibility of the observation and its witness. Although an empiricist could admire the historical Aristotle as a keen observer of nature, the natural philosophical system he devised was not considered infallible. The concern extended not just to the Scholastic manifestation of the Stagerite's natural philosophy but to some of his metaphysical and physical pronouncements as well. The great skeptic Francisco Sanches wrote, "I believe Aristotle to be the most acute among the observers of Nature and the most extraordinary mind produced from our infirm human nature. But I also affirm that he did not know many things and doubted many others."[31] Furthermore, Aristotelian natural philosophy had accrued a Scholastic apparatus that had come to define it to early modern students. Scholastic Aristotelianism dictated that experience acquired by the senses and the resulting perception of reality had to be interpreted through a very specific interpretive lens so as to fit within a view of reality determined by a set of a priori principles. After the Aristotelian-Scholastic machine had ruminated on an observation, what was ultimately held up as knowledge was a version of it that harmonized with first principles taken as certain verities. For someone who placed great value on experience and considered it a perfectly suitable foundation for new knowledge, frustration with the system was hard to dissimulate. As Oliva Sabuco explained, "I do not want them to believe me, but instead [to believe] the experience and the truth of the thing."[32]

Empiricists also wanted to attain some measure of predictability and, perhaps, control over nature. This presented a problem to the more thoughtful empiricists; understanding particulars might account for a given instance, but it might not prove convincing when this experience was extended to other

29. Navarro Brotóns and Rodríguez Galdeano, *Matemáticas, cosmología y humanismo*, 364–67, 394–95.
30. Ibid., 398, 568.
31. Sánchez, *Quod nihil scitur*, "Ad lectorem."
32. Sabuco de Nantes, *Nueva filosofía*, 205r.

cases. At a minimum, the empiricist sought to *dar regla* (create rules) about how something worked; better yet if the predictive power of mathematics and geometry could be brought to bear. The royal cosmographer, Andrés García de Céspedes (ca. 1545–1611), explained that he had shown all the mathematical proofs necessary in his art of navigation because "he who makes the rules has the obligation of making them accurate and just."[33] On other subjects, such as the deviation of the compass needle owing to magnetic declination, he chose not to speculate: "Although some have sought to make rules about the deviation of the needle from north, because—as we will explain later—this is an uncertain thing, here I am simply giving some reports of what I know about it."[34] After considering various opinions about why magnetized needles point north, he exclaimed, "About physical things, he who wishes to strive will always find a slippery spot through which to flee; which is why we will embrace mathematical arguments."[35] After giving a mathematical explanation that showed there could not be a fixed magnetic point on earth or in the heavens, he concluded, "So let us be happy with experience, applying it to the service of humanity."[36] He chided those who peddled great secrets about the magnet and made even greater promises about determining longitude using the compass; they should be given credit only after they demonstrate what they say through deeds (*mostrando por obra lo que dizen*).

The empiricist approach to collecting, organizing, and making sense of natural knowledge was often accompanied by a pragmatic and utilitarian attitude. For a pragmatist, the criterion of truth lay in the merit of the practical outcome of a conceived notion rather than the logical integrity of its theoretical explanation. For example, if understanding the trajectory of a cannonball as comprising mixed motions—a violation of Aristotelian dicta—helped to reliably predict the landing site of the shot—a good practical outcome—the pragmatist would find this the better notion. The Spanish artillery manuals of Collado (1592), Prado y Tobar (1603), and Ufano (1612) offer clear examples of this attitude. We find that most scientific practitioners who were pragmatists operated in ways that, if not entirely divorced from Aristotelian natural philosophy, were certainly far removed from its Scholastic version, or quite simply in the theory-free zone of radical empiricism. Their pragmatic attitude could be attributed to utilitarianism and a preference for empirical practices, but it

33. García de Céspedes, *Regimiento de navegación*, 73v.
34. Ibid.
35. Ibid., 77r.
36. Ibid., 79r.

can also be ascribed to dissatisfaction with the systems of knowledge available to address—let alone answer—fundamental questions about nature. If so, this places Spanish scientific practitioners alongside their European counterparts wrestling with the crisis of knowledge that had overtaken early modern Europe. In fact, Jeremy Robbins has argued that this pragmatism prepared the way for the open reception of the experimental sciences in Spain during the latter half of the seventeenth century.[37]

Natural philosophers and theologians friendly to the empiricist approach nevertheless pointed to several problems. Some complained that it was untenable to have everyone come up with a different explanation. Equally problematic were empiricists who simply collected facts with no other purpose than finding some utility in them: danger lurked in knowledge for its own sake or solely for material gain. These attitudes could make the soul teeter toward materialism and away from God. Among the critics of a purely empiricist approach was Juan Luis Vives (1493–1540), who, although uncertain about the human capacity to unravel the secrets of nature, called for a moderate empiricism in his *De tradendis disciplinis* (1531):

> The first precept in the contemplation and discussion of nature, is that since we cannot gain any certain knowledge from it, we must not indulge ourselves too much in examining and inquiring into those things which we can never attain, but that all our studies should be applied to the necessities of life, to some bodily or mental gain, to the cultivation and increase of reverence. . . . Nor does such excessive application leave time for any thought of God, and even if a man attempts it, his own investigation, in which he is totally wrapped up, presents itself instead before his eyes. Thus the contemplation of nature is unnecessary and even harmful unless it serves the useful arts of life, or raises us from knowledge of His works to knowledge, admiration and love of the Author of these works.[38]

Conversely, Vives (and later Montaigne) also reminded Scholastics that "the world of empirical reality could not be ignored without disastrous results to knowledge and science."[39] Thus Vives urged teachers to guide students through the thicket of ancient natural philosophy but to use experience as the

---

37. Robbins's exploration of the cultural manifestations of this pragmatic attitude is evident in the use of *ser/parecer* and *engaño/desengaño*, in "Arts of Perception," 5, 18, 225–53.

38. Vives, *On Education*, 166–67.

39. Haydn, *Counter-Renaissance*, 238–39.

basis of sound judgment, "so the studious will become accustomed to give assent to reason, rather than to human authority."[40] A natural philosophical preparation was necessary, insisted Vives, to learn how to balance knowledge of nature and knowledge of the divine: "We must not examine nature by the poor and bad light of heathen knowledge but by the brilliant torch, which Christ brought into the darkness of the world."[41]

Although Vives spent his most productive years outside Spain, many others, especially those associated with Spanish universities, sought to correct ambiguities or shortcomings in established natural philosophical systems, especially when integrating empirical evidence into a particular canon. Yet simply adding a layer of empirical data to what was known about the natural world was methodologically questionable given that the philosophical approach of the times dictated that natural "effects" had to be explained using deductive reasoning derived from first principles or incontrovertible axioms. Rather than discard the existing epistemic and methodological apparatus of Scholastic Aristotelianism, they chose to build on it. Why? As we will see in what follows, by doing so they retained the search for causality as the ultimate objective of the natural philosophical enterprise. In its Scholastic embodiment, the search for causality in natural phenomena—inherent to Aristotelian natural philosophy—culminated in evidence of the actions of the Creator as a final cause. The one had become inextricably tied to the other. This knot turned Aristotelian natural philosophy into a legitimate and virtuous pursuit by countering the threat posed by a systemless quest into the secrets of nature or, worse yet, by one guided by an unsound natural philosophy with unclear objectives. These approaches harbored perils that separated humanity from God and could potentially undermine the religious truths Christianity was based on. The stakes were very high.

Revisions to Aristotle and New Alternatives

In an era scarred by the Reformation, wary of doctrinal dissension, and grappling with an evangelical effort unprecedented since the days of Constantine the Great, Catholic thinkers saw the empirical turn and the revival of some ancient natural philosophies as threats to humanity's salvation. By the mid-sixteenth century these concerns motivated some formidable intellects to revisit Scholastic Aristotelianism and try to address some of its shortcomings.

40. Vives, *On Education*, 213.
41. Ibid., 172.

Alejo de Venegas was only one of a host of thinkers with similar projects: Domingo de Soto (1494–1560), Diego de Zuñiga (1526–98), Benito Pereira (1535–1610), and Francisco Suárez (1548–1617)—the *doctor eximius*—whose work is finally receiving the attention it deserves from historians of philosophy.[42] The Spanish effort to rehabilitate Scholastic Aristotelianism continued throughout the seventeenth century and into the eighteenth, if with somewhat less momentum.

For the Jesuit Francisco Suárez, the goal was to reestablish Aristotelian natural philosophy on a metaphysical foundation that was in concert with reality as experienced through the senses and that pointed toward a theology where a better understanding of God was achieved through his role as creator.[43] This perspective offered the proper point of departure for interpreting the natural world; in putting it forward, Suárez was also responding to what he saw as a dangerous misinterpretation of the Aristotelian corpus, or radical departures from that corpus that seemed to be overtaking natural philosophy. As Benjamin Hill points out, "Indeed much of Suárez's philosophical work is best viewed as a reaction against the philosophical naturalism prominent in the Averroist interpretations of Aristotle that were bouncing around the universities in Italy during the sixteenth century."[44] Understanding metaphysics to be the starting point of any philosophical program, Suárez composed his most influential work, the *Metaphysical Disputations*, or *DM* (1597), as a program of study that broke with the commentary tradition, taking the style of a modern philosophical treatise.[45] At first glance the profuse citations of authorities in the *DM* and its intractable Scholastic language might suggest another Thomistic recitation, but under its Scholastic trappings are some of the most radical reformulations of Aristotelian natural philosophy to take place during the early modern era. Suárez proposed restricting metaphysics to the study of real beings (both finite and infinite that have a real essence), privileging the efficient and material causes over the formal (while making God the creator always the final cause) and denying a real distinction between essence and existence.[46] Furthermore, he restricted the operation of the substantial form to bodies and

---

42. "Suárez seems to have slipped between the cracks in our historiographical taxonomy." Hill and Lagerlund, eds., *The Philosophy of Francisco Suárez*, 2, and Schwartz, *Interpreting Suárez*.

43. The purpose of his metaphysics is to lay the foundation determining the relation between being (*ente*) and creation: Bauer, "Francisco Suárez," 244–45.

44. Hill and Lagerlund, *Philosophy of Francisco Suárez*, 3.

45. Doyle, *Francisco Suárez*, 7.

46. Ibid., 8–11.

thus reduced the formal cause to an operation: joining the substantial form with the body.[47] Suárez used common experiences as supporting evidence for most of these changes in an effort to bring Scholastic Aristotelianism into line with the empiricism that characterized his era. The *DM* was read widely by Catholics and Protestants alike and had immense influence on the natural philosophers who addressed the metaphysical aspects of the New Science, such as Descartes and Leibniz.[48]

The Galenic medicine taught at the universities was another target for reformers. One trend among Galen's critics was to challenge the Aristotelian theory of four elements, as Pere d'Oleza (ca. 1495–1531) did with his version of Averronian corpuscularianism in his *Summa totius philosophiae et medicinae* (1536).[49] Gómez Pereira, meanwhile, relied on a rejection of Aristotelian prime matter to support the controversial ideas on mechanistic animal behavior of his *Antoniana Margarita* (1555).[50] Later in the century, Oliva Sabuco's *Nueva Filosofía* (1587) proposed a complete reformulation of theories of the body and disease that ran counter not only to Galenic medicine as taught at the Schools but also to some fundamental aspects of Aristotelian natural philosophy.[51] The author found that only two elements of the Aristotelian four (earth and water) sufficed, and this because they functioned analogously to the sun and moon in maintaining life.[52] Although Galen bore the brunt of these reformers' critiques, many of them also questioned the ability of Aristotelian natural philosophy to explain the fundamental building blocks of the human body.

As elsewhere in Europe, some Spanish natural philosophers also sought alternatives in natural philosophical systems rediscovered during the Renaissance. Renaissance Platonism is often used as a case in point, with the principal mode of transmittal of Neoplatonic ideas being Marsilio Ficino's *Platonic Theology* (1482).[53] In the nineteenth century, historian Marcelino Menéndez Pelayo argued that several factors seemed to mitigate the Italian's influence in Spain: among them were Vives's poor opinion of Ficino—whom he called a

---

47. Hattab, "Suárez's Last Stand for the Substantial Form," 118.

48. Abellán, *Historia crítica*, 3:529, and Doyle, *Francisco Suárez*, 18–20.

49. Barona, *Sobre medicina y filosofía natural*, 77–103.

50. Pereira, *Antoniana Margarita*, 111.

51. The question of Oliva's authorship continues to be debated, with some scholars attributing the work to her father Miguel Sabuco, an apothecary. Gianna Pomata has recently discovered that Sabuco took her (or his) thematic cues from Francisco Valles. See Sabuco de Nantes, *True Medicine*, 35–37.

52. Ibid., 328–30.

53. Copenhaver and Schmitt, *Renaissance Philosophy*, 161–62.

filosofastro—the fact that Neoplatonic ideas were already integrated into the religious mysticism prevalent in many Spanish religious orders, and familiarity with an alternative in the operative Platonism of Llullian philosophy. Recently Susan Bryne has reassessed this historiography, finding that Ficino's works were consulted and debated in sixteenth-century Spain; his *De triplici vita, Corpus Hermeticum, or Pimander*, and *Platonic Theology* were particularly influential. Ficino was cited in astrological and medical texts, but his influence was most strongly felt in the incorporation of Hermes Trismegistus into doxographies as an ancient sage.[54] Three figures cited as exemplars of Renaissance Platonism in Spain flourished outside Spain: the exiled Sephardim León Hebreo (Yehuda Abrabanel, ca. 1460–1521), author of the influential *Dialoghi d'amore* (1535); the heretic—to both Catholics and Protestants—Miguel Servet (1509 or 1511 to 1553); and Sebastián Fox Morcillo (ca. 1526 to ca. 1560).

Fox Morcillo, Sevillian by birth and of *converso* origins, is considered among Spain's most committed students of Platonism during the sixteenth century. After early studies in Latin and Greek in Seville and possibly Alcalá, Fox Morcillo continued his humanistic studies at the University of Louvain, including mathematics with Gemma Frisius and philosophy with Cornelius Valerius.[55] After returning to Seville in 1556, presumably to take a position as teacher of pages in the royal court, during a Carranza inquest by the Inquisition he was named as someone involved with heterodox circles in Louvain. This and his brother's troubles likely motivated Fox Morcillo's departure from Seville and led to his subsequent death in a shipwreck. His few works give us some indication of how "rediscovered" philosophical systems were recruited to address what the philosopher saw as shortcomings of the prevailing Aristotelian natural philosophy. His project, as put on display in his *De naturae philosophia, seu De Platonis et Aristotelis consensione* (1554), was to reconcile Plato and Aristotle by evaluating and extirpating what he considered to be emendations and misinterpretations accrued through the ages in either one's philosophy. In the *proemium* dedicated to Philip II, Fox Morcillo complained about the inadequate, complicated, and boring way the natural philosophical works of Plato and Aristotle were taught, though he recognized how necessary this knowledge was for the republic. He placed the blame for his having to

---

54. Byrne, *Ficino in Spain*, 50–63.

55. Fox Morcillo's biography follows Pike, "Converso Origin of Sebastián Fox Morcillo," as well as the important corrections concerning his death and that of his brother, Francisco, who died in a 1555 *auto da fe* in Seville, accused of Lutheranism: Cantarero de Salazar, "Reexamen crítico de la biografía del humanista Sebastián Fox Morcillo."

undertake his project squarely on the shoulders of instructors who were servile and dim-witted (*serviles y tozudos de ingenio*) and who made the subject unnecessarily complicated.[56] The stated purpose of his compendium was to put forth clearly the philosophers' original ideas and explain how these had been corrected by church doctrine. Furthermore, he claimed that in this exercise he had found a universal philosophy that could reconcile many of the warring schools of thought.[57]

His poorly dissimulated objective was a trifecta: to build on the strength of the original ideas of ancient philosophers a unified philosophical system that was also in concert with empirical observations and that did not challenge church doctrine. For example, in an attempt to reconcile their metaphysics, he mapped Plato's agents, the Demiurge, Idea, and Receptacle, onto Aristotle's Privation, Form, and Matter, explaining differences in their capabilities and functions but ultimately settling for compromises.[58] When the opinions of the two philosophers could not be reconciled, Fox Morcillo turned to the Bible and to church doctrine. In another example, we see him trying to reconcile Plato's claim that the heavens are made of fire with Aristotle's notion of ethereal spheres. As a way of solving the apparent contradiction, he investigated the meaning of the Hebrew word for "firmament" (רָקִיעַ, *rāqîaʿ*) used in Genesis 1:6. He found the word could also mean "diffusion," "extension," or "solidity," so he settled on "substantia naturae tenuis," an observable quality shared by both fire and the ethereal quintessence, as the best descriptor of celestial matter.[59] (It helped his argument that this interpretation had first been put forth by Saint Basil.)

Fox Morcillo was attempting to propose a solution to what he perceived as the poorly understood and intellectually fraught mélange of ideas about the workings of the natural world typical of his era. He gave empiricism an important epistemic role in his philosophy, but his untimely death robs us of knowing how he would have engaged with the utilitarianism so characteristic of Philip II's court. Would his syncretic system have gained traction at court or at the universities? What is clear, however, is that the work passed muster with the censors and went through several editions. Arias Montano knew of this promising intellect and purchased his library—Latin, Greek, and miscellaneous books—in Seville sometime after 1558.[60]

56. González de la Calle, *Sebastián Fox Morcillo*, 95.

57. Fox Morcillo, *De naturae philosophia*, preface, fol. a5r.

58. Ibid., fol. 9.

59. Ibid., fol. 47v, and González de la Calle, *Sebastián Fox Morcillo*, 115–16.

60. Gil, *Arias Montano y su entorno*, 50.

Also entering the natural philosophical fray were contemporary reinterpretations of skepticism and Stoicism. The work of Portuguese *converso* Francisco Sanches with his *Quod nihil scitur* (1581) is often taken as the point of origin of early modern Spanish skepticism. It is a strident manifesto—a rant, really—that grew out of the author's frustration with the dissonance between experience and what natural philosophical systems taught. I suspect—with the caveat that more work is needed in this area—that rather than being guided by a revival of skepticism with the reception of Sextus Empiricus, the many readers of *Quod nihil scitur* embraced Sanches's rant because it echoed their frustration with Scholastic Aristotelianism and the confusion caused by the many alternative natural philosophical systems circulating at the time.

By the latter part of the sixteenth century, Stoicism emerged as another alternative to Scholastic Aristotelianism.[61] Its advocates called for rejecting dialectic, rhetoric, and logic as the basis of argumentation and for a rational understanding of both morality and nature. As Francisco Sánchez de las Brozas ("El Brocense," 1523–1601) makes clear in a passage he interjected in his translation and commentary on the *Enchiridion*, he claims that

> [Epictetus] reprimands the philosophers of his era (What would he do if he saw ours!), who spend all their time trying to understand Aristotle, and it is all a matter of giving *in scriptis*, accumulating opinions and never trying to make better students, but only sophistries, and with this they go about with puffed chests and try to appear learned. They owe a lot to Aristotle for having written so obscurely, because had he been clearer, they would not have a subject on which to measure their worth.[62]

Also competing with these scholarly natural philosophies were knowledge systems that addressed one of the vexing questions behind the crisis of knowledge I have been discussing: the nature of bodies and their energies. The alchemist (whether preparing medical remedies or seeking the philosopher's

61. Robbins, "Arts of Perception," 7–8. For, whereas the influence of Neostoic moral philosophy in Spain is undeniable, Stoicism's influence on natural philosophy is less clear. The principal authority was Seneca, and to a lesser extent Cicero and Diogenes Laërtius. Yet only certain sections of the *Questiones naturales* where Seneca describes Stoic natural philosophy have survived, and these give only a cursory view of Stoic cosmology. Despite this limitation, Seneca's message concerning the relationship between humanity and nature comes across clearly: humanity must live according to the rules of nature and face it with fortitude. For Stoics, as a consequence, the rules set by nature had to be investigated and understood.

62. Sánchez de las Brozas, *Doctrina del estóico filósofo Epicteto*, 221.

stone), the astrologer (practicing either licit or judiciary astrology, or both), and the magus or professor of secrets (whether Hermetist or charlatan) relied on systems of knowledge that, once mastered, helped to make sense of the world and perhaps control it. Recent historical work has made it quite clear that Spain had a significant community of alchemical practitioners, often working under royal patronage and building on an autochthonous tradition that held as preceptors the Pseudo-Ramón Llull, Arnau de Vilanova, and John of Rupescissa. These practitioners operated loosely within a number of philosophical systems of thought and sought in their emanantist forces and occult properties the ideas that gave their alchemical trials coherence.[63] In pharmacy, the role of American plants and the gardens of simples at Aranjuez and El Escorial brought together alchemical distillation practices, the iatrochemical practices of empirics, and the practices of the most learned physicians.[64]

Astrology and natural magic had a long tradition in Spain and were invigorated by the translations of the *Picatrix* and other works during the reign of Alphonse X the Wise. By the sixteenth century they had emerged as viable sources for alternative explanations about nature.[65] In particular, the tradition of Arabic and Hebrew alchemy and magic had a strong presence, and for some practitioners they had become an important alternative to Scholastic Aristotelianism. For magic, the dialectic between advocates and denouncers came to a head during the seventeenth century. While historians have shown that manuscript copies of the *Clavicula salomonis* circulated widely in Spain,[66] the popular books of Martín Del Rio (1551-1608) and Hernando Castrillo (1585-1667) tried to bring magic and witchcraft out of the shadows—and to the attention of the inquisitors.[67] As Del Rio remarked in his prologue to the *Disquisitionum magicarum libri sex* (1595), "demons might find a home in heretics."[68] He was driven by "how widespread magic has become in recent times. Evil spirits are on the loose seeking to take possession of foolish and deluded souls. . . . I have read that, as a result of the Moorish occupation of Spain, the magical arts were virtually the only subjects being taught in Toledo, Seville, and Salamanca."[69]

63. López Pérez, "Ciencia y pensamiento hermético en la Edad Moderna Temprana," 65-69.

64. Rey Bueno, *Señores del fuego*, 59-90.

65. Fernández Fernández, "'Arte mágica,'" 77.

66. Caro Baroja, *Vidas mágicas e inquisición*, 1:138-47.

67. Ibid., 1:168-70.

68. Del Rio, *Disquisitionum magicarum*, prologue.

69. Del Rio, *Martín Del Rio: Investigations into Magic*, 27-28.

By the beginning of the seventeenth century all these systems of knowledge were in frank competition with new systems entering the arena as the century progressed. The system-building efforts in Spain during the sixteenth century were matched only by Italy's, where the naturalism of Bernardino Telesio (1508–88) and its empiricist variants, the Platonism of Francesco Patrizi (1529–97), and several versions of reformed Scholastic Aristotelianism were widely discussed. Later in the century the systems hatched in Italy and Spain would be joined by the Christianized atomism of Gassendi, Boyle's corpuscularianism, Cartesian mechanism, and the New Science of experimental natural philosophy. The range of proposals was astounding: from the conservative—mostly Aristotelian—natural philosophy of the Jesuits' *Ratio studiorum* and its course of study as laid out in the *Cursus conimbricensis* (1591–92) to the Christianized magical naturalism of Paracelsus as interpreted by Robert Fludd (1574–1637). These must have been either very exciting or very troubling times to be a natural philosopher. Yet only when we divest ourselves of the notion that the triumph of the New Science was inevitable can we appreciate the variety and richness of these system builders' projects and appreciate the uncertainty of the whole enterprise. What is certain is that the doubts and responses that arose in Spain placed its thinkers in the midst of the pan-European debate about the need to develop new ways of explaining the workings of the natural world.

Is Knowledge of Nature Even Possible?

As we know, the whole of the early modern project of natural philosophical reform, at least during the sixteenth and seventeenth centuries, was never free from theological considerations of tremendous importance. We find most natural philosophers—from Agrippa to Newton, from Vives to Feijoo—grappling with the same question: What was the purpose of natural philosophy? In Spain, as in the rest of Europe, the most authoritative arguments about the purpose of the inquiry into nature had been provided by Saint Augustine. He defined natural history and geography as a mode of writing that aimed to collect all that was known about the nature of things. As he explained in *On Christian Doctrine* (2.29.45),

> There is also a type of narrative resembling description which points out to the ignorant facts about the present rather than about the past. To this class belong things that have been written about the location of places, or the nature of animals, trees, plants, stones, and other objects. We have spoken of these

writings above where we taught that they were valuable for the solution of enigmas in the Scriptures.[70]

This kind of descriptive material was useful for explaining scripture and thus worthy of pursuit. The relationship thus outlined became known as the "science as handmaiden" justification for pursuing natural philosophy.[71] Yet Saint Augustine proposed that the study of natural things should proceed with a good dose of skeptical caution, a caution informed by human limitations: "Let him recall that he is human and is investigating the works of God to the extent we are permitted."[72] So, whereas the pursuit of science could be justified by its role in explaining the Bible, another source of doubt struck at the fundamental premise of such an enterprise. Were human beings capable of knowing enough about the natural world to fashion a suitable natural philosophical system?

The response to this question hinged on the status of the human condition after Adam's fall. Was human nature so corrupted by sin that it could no longer recapture the complete knowledge of nature God had bequeathed to Adam? Even if the Fall was not determinative, were human beings ever meant to fully understand the works of the Creator? By the sixteenth century we can discern two schools of thought on this subject. The question existed as a dialectic between Augustinian and Thomistic perspectives of postlapsarian human beings' ability to read the book of nature.[73] Saint Augustine's skeptical attitude about humans' ability to grasp the true knowledge of nature often conflicted with the Thomist-Aristotelian view, which was confident it was possible. The Augustinian view was informed by an episteme that defined true knowledge in a Platonic sense; it meant knowledge of the ideal forms that by definition lay beyond humanity's reach. Furthermore, it held that the Fall had irreparably altered humans' capability to attain this knowledge. The Thomist view rested on the conviction that postlapsarian human beings' ability to understand the natural world was not compromised by the Fall and that therefore knowledge of nature was possible. However, complete knowledge of some regions of the world that lay beyond our sense perception, such as the heavens or the regions beyond the ninth sphere, might not be attainable. Following the Thomist line, Jesuits in particular embraced the empiricism of Aristotle's *Physics*, *On Gener-*

70. Augustine, *On Christian Doctrine*, 65.

71. Lindberg, "Medieval Church Encounters the Classical Tradition," 18–19.

72. Augustine, "On the Literal Interpretation of Genesis, an Unfinished Book," 167.

73. Harrison, *Fall of Man and the Foundation of Science*, 44–54.

ation and Corruption, and *History of Animals*—albeit framed with the virtues of a natural theological inquiry—to pursue a vigorous inquiry into natural knowledge.

*

As we will see, Arias Montano took a via media between these positions concerning humans' ability to gain true knowledge of nature. He repeatedly acknowledged the Augustinian view that human beings had lost the complete knowledge of nature that God had given Adam, but he held firm to the notion that, given the proper tools, they could recapture most of it. Yet he also understood the grave implications of undertaking this task without the proper foundation. At the very least it could waste time and cause frustration and confusion, but most important, "such dangers not only fool us and damage our minds, but also hasten our end."[74] In the votive elegy that introduces his *History of Nature*, Arias Montano admitted to having been misled by ancient philosophies and sciences. He had become enamored of their elegant systems, but now he realized that these simply disguised the emptiness of their propositions. Filled with remorse and overwhelmed by years of futile study, he turned to prayer, and while contemplating a painting of *facies verique bonique* (the face of truth and the good), he saw the image come to life and heard it say, "Seek here, child, understand all principles."[75] The "here" in the phrase was the Sacred Scriptures, while "all principles" were the first principles on which to build all knowledge. This voice told him he would be guided, led slowly as one would lead a child, through multiple doors of knowledge: knowledge he should share with others when he deemed them ready. In the Bible, the voice directed, he would find a new way of examining the properties of the recently created world, its force, and what drives it. But a greater prize awaited those who followed this path to knowledge: "God himself will soon give you what will satiate you": a constant yearning for God.[76]

The poetic format of votive elegy provided a lenient genre for staging the revelation as a "gift" or a mystical event without having to justify whether it really happened in this manner. Furthermore, Arias Montano's exposition also carried the unmistakable Hermetic overtones of initiation, gradual enlightenment, and ascent. To an unsuspecting reader this might have looked like just

---

74. Votive elegy, *Corpus*, lines 53–54, *HistNat*, 98.
75. Ibid., line 82, *HistNat*, 98.
76. Ibid., line 102, *HistNat*, 99.

another popular trope; among his close circle of friends, however, the biblist was reinforcing something he shared with a select few: he believed he had received a divine revelation that guided him to the true source of knowledge of nature. It resolved the tension between the Augustinian and Thomist positions concerning human beings' ability to gain knowledge of nature. In Arias Montano's thought, whereas humanity might have "lost" knowledge of nature after the Fall, it could recapture it by extracting a new natural philosophy from existing revelations—those already present in the Bible.

As we will see, Arias Montano sought guidance for this new natural philosophy in the philological study of the Sacred Scriptures, the careful observation of nature, and ultimately in divine revelation. The Bible held the clues about this natural philosophy, but they were frequently obscured by the arcane language of the Mosaic texts. Therefore he marshaled Hebrew, Greek, and Latin philology and an impressive command of the biblical text to find a new set of first principles, mostly in the book of Genesis, and clues to a complete natural philosophical system in the rest of the Bible. He was driven by the desire to present a whole system where the only arbiters were experience and the Word of God. Yet, while also deeply embedded in the natural theological tradition of Spain, he also found inspiration in the natural beauty surrounding his mountain retreat, an inspiration that would direct his biblical exegesis toward explaining the human condition and the natural world. Arias Montano's objective in the *Magnum opus* was to put forth a proposal that would assuage the Spanish disquiet.

CHAPTER 3

Faith and Nature

Inter omnes verò Hebraica lingua primatum quendam obtinuit, et in quadam veluti sacra arce collocata est.[1]

Any disquiet Benito Arias Montano felt about the means of attaining knowledge of the natural world paled compared with his concern for humanity's salvation. For him and for other theologians of his time, the very potential of salvation was inextricably tied to how well human beings understood not only their own nature, but also that of the world around them. Yet human nature was immensely complex and involved constant negotiation between the lowly, material appetites of the corporeal self, the mental capacity for rational thought, and a spiritual self that held the potential for eternal life. Understanding their own nature was one of the greatest challenges human beings faced, second only to getting to know the nature of God. This was the same God who had endowed humans with the ability to learn what they needed to know to attain salvation and live in concert with nature. Convinced that a kind God would not have deprived humanity of the ability to live beyond the state of confusion and deception that seemed to periodically overwhelm it, Arias Montano put forth a positive theory of knowledge and found its justification in the Sacred Scriptures, in the interstices of the Hebrew language in which they were originally written, and in the events of Creation.

From Arias Montano's perspective, an essential part of achieving the type of self-awareness that led to God was defined by the fact that humans were

---

1. "In truth the Hebrew language has attained among all a certain primacy and has been placed in a fortress as something sacred.": *Benedicti Ariae Montani Hispalensis in Latinam ex ebraica veritate Veteris Testamenti interpretationem. Ad christianae doctrinae studiosos praefatio*: AntPoly, 7:†1r, Arias Montano, *Prefacios*, 84.

created to live in the world; their corporeal, material nature had been created from the dust of the earth itself. Therefore, in order to acknowledge this part of their nature—and in gratitude to God for being able to rise above it—humans also had to learn the nature of all things. In Catholic theology, humanity's time on earth is seen as a trial. God created humans to live out the mortal portion of life in this world, enveloped by nature, where they would have to prove their worth before attaining salvation. So along with a deep sense of self-awareness, the promise of salvation was tied to the understanding of nature and humanity's place in it. Drawing from the Delphic and Socratic traditions of seeking to "know thyself," Arias Montano, in the opening chapters of the *Anima*, lets Persius's acrid third Satire (66–68) speak for him: "Learn, you miserable ones, and discover the causes of things: what we are, what we are born for, and where we stand in the order of things."[2]

Yet Arias Montano's concept of this quest for knowledge of nature was not a contemplative exercise like the one proposed earlier by the natural theology tradition and popularized by his contemporary fray Luis de Granada. It was instead an active enterprise through which human beings would come to understand nature and their place in it and thus also learn to make good use of the natural world placed under their stewardship. Arias Montano made this notion central to the *Anima*, *Corpus*, and *Vestis* trilogy. He considered that this enterprise could be undertaken by two *viae*, or paths, the first by way of revelation and the other through investigation of nature. Both *viae* could achieve the same end: knowledge of God. Investigation of nature, for him, consisted of an "agitation of the mind," which was "intent to rediscover from the observation and examination of what is clearly seen in the sky and the world some other aspect of the supreme nature of God."[3] By proposing the *via* of investigating nature by observing and examining the world, he was drawing from a long tradition in Christian theology that saw the natural world as a book mirroring the handiwork of the Creator, which Raymond Sebond had synthesized in his *Natural Theology*. The other *via* of revelation, generally speaking, had placed two tools at humanity's disposal: the Sacred Scriptures and the unmediated personal revelation of God. Whereas the latter was strictly of matter of faith and grace, wholly in God's hands, God had made the former accessible to all.

For Arias Montano and some of the Mosaic natural philosophers discussed in this chapter, the *via* of revelation by way of the Sacred Scriptures presented a secure path for attaining the knowledge of nature and thereby could po-

2. *Anima*, Praefatio, fol. 3r; *LibGen*, 85.
3. *Anima*, 1; *LibGen*, 101.

tentially fulfill one of the requirements for assuring salvation. Underpinning this approach was the unshaken notion that because the Bible was a revealed text—although written in a manner that accommodated its past readers and therefore confusing to current ones—it was intrinsically truthful, instructive, and ultimately decodable. Where the Bible spoke about the natural world, particularly in the Pentateuch, Arias Montano believed it contained all that was necessary for humans to know about the natural world in order to understand their place in nature and their role as its stewards. By setting the *via* of investigation of nature on the secure foundation afforded by the Bible, he proposed to bring the interpretation of scripture together with the study of nature—not as the contemplative exercise advocated by natural theology, but as an active project that encouraged the faithful to engage in interpreting scripture and the natural world. As we will see later in studying the *Magnum opus*, this approach allowed him to put aside the disquiet concerning the validity of natural philosophical systems that originated in antiquity and placed the active, empirical investigation of nature on sound theological footing.

This chapter explores the consequences to the study of nature that followed from the notion of the Pentateuch's Mosaic authorship and its availability in Hebrew—the language of Adam. While these notions ensured the Bible's reliability as a source of natural knowledge, they also allowed the exegete significant interpretive freedom and a venue for expounding on alternative visions of nature. As we will see, the notion that the Bible described the real world served as the epistemological foundation of two important genres: the very old genre of hexameral commentary, and a product of the sixteenth century, the Mosaic philosophy. Arias Montano drew on these genres when composing the *Magnum opus*, and while his work would differ substantially from them in style and even in objective, they are worth considering as important models.

IN THE LANGUAGE OF ADAM

One conviction shared by authors of natural theologies, early modern hexameral commentaries, and Mosaic philosophies of the sixteenth and seventeenth centuries was an unflinching faith in the veracity of the Genesis account. For theologians, its Mosaic authorship made the book of Genesis fundamentally unquestionable, although it was certainly open to scrutiny and clarification. (Doubts about Moses's authorship would not be raised in a meaningful way until Spinoza's *Tractatus theologico-politicus* [1670]). For theologians with a humanistic love for antiquities, that the Hebrew biblical text had been preserved unchanged since the days of Moses was further testament to its authority. Added to this was the widespread belief that Hebrew had been the language of the earliest Israelites, and even of Adam himself.[4] (Most exegetes understood Chaldean to be a later development.) In fact, the primacy of the Hebrew language would not be questioned until the mid-seventeenth century.[5] Humanistic philology could then be brought to bear on the analysis of the Genesis narrative, and many generations of Christian Hebraists made this their mission. In the heady days before the Council of Trent, two generations of humanists—who acknowledged publicly that Saint Jerome's translation in the Vulgate had its faults—attempted to recapture the *hebraica veritas* they believed could be found in the Hebrew Old Testament. This approach also promised to unravel the linguistic and cultural accommodations Moses had resorted to when writing the Bible for the ancient Hebrews. Although the belief that Moses had certainly not intended to write the Pentateuch as a natural philosophical treatise had been widely held since the days of Saint Augustine, the general understanding among theologians was that the prophet had communicated truthfully the events that took place at Creation in a manner and language that would have been intelligible to the early Jewish people.[6]

In Arias Montano, the language of the Hebrew Bible is always taken as an *arcanus sermone*, a veiled, mysterious, obscure narration. He would explore this notion at length in the appendix or *Apparatus sacer* of the Antwerp Polyglot and in the unfinished trilogy of the *Magnum opus*. In fact, unraveling the meaning of this language became the cornerstone of his exegetical approach.[7] Yet here also Arias Montano was drawing on a long tradition, reinvigorated during the sixteenth century at the universities of Alcalá and Salamanca, of studying Hebrew etymologies, onomastics, and toponymy as exegetical tools.[8]

4. See, for example, Steuco, *Cosmopoeia*, 58; Bouwsma, *Concordia Mundi*, 104–6.

5. On the notion of Hebrew as the original language of humanity, see Demonet, *Voix du signe*, 19–23, 26, and Formigari, *History of Language Philosophies*, 90.

6. On the hermeneutics of accommodation and the Scientific Revolution, see Snobelen, "In the Language of Men."

7. The work of Fernández Marcos is essential in order to understand Arias Montano's conception of Hebrew as the language of Creation; see his "Lenguaje arcano y lenguaje del cuerpo." Perea has studied the many instances where Arias Montano repeated and developed this idea; see Perea Siller, "Capacidad referencial e historia de la lengua hebrea en Benito Arias Montano."

8. Formigari, *History of Language Philosophies*, 45–46. In Spain, for example, this notion also informed the work of Francisco Sánchez de las Brozas. See Ashworth, "Traditional Logic," 156.

The sustaining justification for this approach was the belief that God created through the Word. As James J. Bono has beautifully explained,

> The "Word of God" is, of course, the Bible. But it was *also* the *Logos*, God's creative Word as manifest in His creation of the universe. This link between the Bible, biblical language, the creative Word of God, and the created universe fostered the idea of a divine language that, in turn, reflected the divine ideas "contained" in the divine mind (*mens*). These ideas were the essences or principles (*logoi*) that constituted the divine archetypes for created things. God's "creative Word" instantiated these divine ideas, bringing them into existence in material form.[9]

Thus, by the very fact that it derived from this divine language, the human language Moses used to describe this divine process of creation had to reveal something essential about its referents.

The language of the Bible, and specifically the language of Adam, was thus understood by many to be a natural language that reflected the true nature of things. This relationship gave Hebrew a special status among languages. Some understood its letters and glyphs as imbued with special significance; their manipulation and interpretation became an exegetical tool for Jewish kabbalists and Christians adept in the art. Interest in recapturing this Adamic language was behind some Christian Hebraists' fascination with Kabbalah, as in the case of Giovanni Pico della Mirandola and Johannes Reuchlin. For others the succinct three-letter roots of Hebrew words were their *dictiones essentiales* and a route toward deciphering the original relation between sign and signifier.[10] Yet the characterization of the Adamic language as "natural" was not without its detractors, among them Luther and, as we will see below, Francisco Valles de Covarrubias. Scholars and theologians debated how well this language reflected the true essence of things: whether knowledge of this language persisted after the Fall (or the Flood, or Babel) and whether it could be recaptured at all. It also led others to attempt somewhat of a reverse strategy; if the Adamic language had been lost, could reading the book of nature help one find the original language used in the design of nature? But reading the book of nature in turn presented exegetical challenges. Could the book of nature be interpreted literally, or was the mysterious nature of the Divinity hidden behind symbols and correspondences that had to be read allegorically?

9. Bono, *Word of God*, 54–64.

10. Demonet, *Voix du signe*, 74–75, 146.

And if it is possible to read the book of nature literally, what could be expected from such a reading? Here the expectation was not to recapture a lost Adamic language but to find, as in the tradition of natural theologies, evidence of a divine order and purpose in nature.[11]

Although well informed about these debates, for his justification for the existence of such a language, Arias Montano needed to go no further than the famous passage describing Adam naming the beasts:

> So the LORD God formed out of the ground all the wild animals and all the birds of the air, and he brought them to the man to see what he would call them; whatever the man called each living creature was then its name. The man gave names to all the tame animals, all the birds of the air, and all the wild animals; but none proved to be a helper suited to the man. (Gen. 2:19–20 NAB)

His unshaken conviction of an Adamic language was based on an exegetical tradition that interpreted this passage as evidence that the origin of a natural language began with an Adamic language where the sign and signifier were one and the same. Adam carried out the art of naming to perfection for the simple reason that before the Fall he had enjoyed complete knowledge of everything seen beneath the heavens and on the sphere of the earth, and also of their "times, opportunities, and benefits."[12] Whatever Adam called a beast stood for its name and therefore for the exact nature of each animal. Similarly, it had been traditional in antiquity, Arias Montano explained, to choose names that precisely reflected the virtue and efficient power of things. In the *De arcano sermone*, Arias Montano would emphasize that the Adamic act of naming had allowed subsequent generations to also know the nature of things, although perhaps not as clearly as when Adam had spoken.[13] He found a convincing biblical justification for the direct correspondence between subject and referent in the very etymology of the word SEM (שֵׁם, šēm; *nomen* or "name"):

> Among other meanings of the word SEM, that is, "name," a unique and note-worthy one is observed: the destiny, authority, faculty, force, and efficacy of a person, thing, and place and, as doctors say, is indicative of its virtue and

---

11. For a synthesis of these debates and early modern theories of natural language, beginning with Ficino's interpretation of Plato's *Cratylus*, see Bono, *Word of God*, 29–30, 42–47, 53–84, and Demonet, *Voix du signe*, 44–45.

12. "Temporum, opportunitatum, & commoditatum": *Corpus*, 374–75; *HistNat*, 493–94.

13. Arias Montano, *Libro de José*, 100, 409, [1].

property. Because these [names], having been inserted by God as the author of nature or founder of its gifts, virtues, and duties, which are transmitted by men endowed with authority, when the thing corresponds to the name, stand for its definition and demonstration.[14]

This perspective also considered the historical reality that, as a result of the Fall, humanity gradually forgot the original relation between word and thing. Yet Arias Montano argued—as Agostino Steuco had done earlier—that traces of this knowledge of nature was still extant in Adam's closest descendants, particularly Noah. When God instructed Noah to select a male and a female of every animal and provide for their sustenance in the ark, it signaled to the exegete that vestiges of this Adamic knowledge still survived—diminished, yet in sufficient measure to allow humanity the knowledge it needed to utilize the fruits of the earth for its survival. From the episode of Noah, Arias Montano gleaned a dual message: that God intended humanity to use all things on earth for its sustenance, but also that careful observation and study of nature could provide lessons for how humanity should live.

Furthermore, Arias Montano's theory posited that language was a gift God gave exclusively to humanity. He expressed this clearly in the short treatise *Adam, sive Humani sensus*, where he explained that humanity, in addition to having been given the curiosity to inquire about the nature of things both hidden and visible, was also given the ability to communicate these ideas through language.[15] At the Creation the first human was given the ability to reason and to use language, along with knowledge of the nature of things. But the complexity of language developed in response to humans' need to communicate and describe the world to each other. Arias Montano found it reasonable to assume that the first language—Hebrew—retained in its lexicon and grammatical structure vestiges of this early need to communicate using names to express the essence of things. Thus the very structure of language itself—its grammar, etymologies, and lexicon—reflected the way human beings had communicated about everything in the world.

Perhaps nowhere in the Sacred Scriptures is the power of language to designate the nature of things more evident than in the book of Genesis. As in the work of centuries of exegetes before him, in Arias Montano's work the six days

14. *Corpus*, 184, and *HistNat*, 290.
15. Arias Montano, *Tratado sobre la fe*, 92. The treatise had remained unpublished as BNE MS 149, fols. 1–11, in a fair copy by Pedro de Valencia. It might have been intended as part of the *Apparatus sacer*.

of Creation would feature prominently. But whereas for him the work of the six days of Creation, or hexamera, held the seeds of a divine metaphysics and natural philosophy, the long tradition of biblical commentaries that preceded him had largely seen their challenge as circumscribed to interpreting the creation of the world according to Moses. To understand the novelty and distinctive approach Arias Montano brought to his interpretation of the hexamera, in particular in his *Magnum opus* trilogy, it is worth delineating the outlines of the hexameral commentary as a genre and investigating them through some examples of the explanations put forth about the natural world.

HEXAMERAL COMMENTARIES

When the authors of natural theological tracts referred to the events of the six days of Creation in their exaltation of the natural world and found in the Creation itself irrefutable testimony of God's omnipotence, they were borrowing examples from the long tradition of hexameral commentaries, a subset of biblical commentaries that focused on the book of Genesis. Within the Judeo-Christian tradition the first three chapters of Genesis were predominantly thought to present a historical account of the orderly creation of the world. As different exegetical modes and their associated hermeneutics came and went over the centuries, the ways of interpreting the events of Creation and the arguments about the nature of the world it described varied. Usually structured as long glosses on each line of the first two books of Genesis, the general objective of the genre was to make intelligible the events of Creation and to extract what God, through the mediation of Moses, had wanted human beings to understand from them.

For most of the history that concerns us, it was thought the Moses had written the book of Genesis based on the divine revelation that took place on Mount Sinai (Exodus 19–20). It was thus inevitable that the events of the six days would be interpreted in light of prevailing natural philosophies.[16] Within this tradition it is possible to identify a subset of authors who used this genre as a way of discussing alternative theories about the natural world and who relished comparing the conclusions posited by different natural phil-

---

16. For a synthetic survey of early interpretations of the six days of Creation, see Lewis, "Days of Creation," 433–55. For examples of the renewed interest in the relation between the Bible and natural philosophy in the sixteenth and seventeenth centuries that take into consideration both Catholic and Protestant approaches, see Killeen and Forshaw, *Word and the World*, and van der Meer and Mandelbrote, *Nature and Scripture in the Abrahamic Religions*.

osophical schools. During the sixteenth century most explanations relied on Aristotelian natural philosophy, but some injected a good dose of Platonic or Neoplatonic ideas into their explanations of the events of Creation. Especially in the hands of a novel thinker such as Arias Montano, a hexameral commentary could be the platform for proposing radically new ideas about how the world worked.

The earliest example of hexameral literature dates to the first-century works of Philo of Alexandria. In his figurative exegesis the events of the first day involve the creation of perfect mental objects or archetypes, while the other five days are devoted to creating objects that can be perceived by the senses.[17] As a Jewish exegete, Philo's blend of various strands of Greek natural philosophy, particularly Platonism, made his writing appealing to early Christian writers. The genre entered the patristic tradition with Saint Basil's *Hexaemeron*,[18] followed by the writings of Saint Ambrose. One interest their treatises touched on was the true nature of heaven and the existence of celestial waters, and they explored questions such as What exactly was heaven made of? Was it composed of some sort of celestial waters? And if so, were these in the region of the firmament or beyond? Were the heavens of an aqueous nature but in the form of solid crystalline spheres as described by Ptolemy? Did the separation of the waters suggest the Aristotelian separation of the celestial and terrestrial realms? Saint Basil, for example, rejected the notion of a solid firmament. Instead, the firmament stood as some sort of sieve that separated different airy natures; more humid and tenuous ones passed on to the heavens, and coarser ones fell below—a model that resonated with Arias Montano.[19]

With Saint Augustine's *De genesi ad litteram libri duodecim* (just one of his five attempts to explain the meaning of Genesis), the writings of the patristic fathers exerted a determinative influence on the genre. In his earliest unfinished commentary on the six days of Creation, Saint Augustine called for studying "obscure" natural things, albeit within the parameters of Christian faith and for the ultimate purpose of discovering God as creator.[20] For him the first instance of creation caused everything to be created ex nihilo but in its perfect form; only their *appearances* would unfold temporally through the expression of the Neoplatonic seminal reasons. He explained, "Although God, who has the power when he wills, makes without a length of time, natures still produce in

17. Philo, *On the Creation of the Cosmos according to Moses*, 52–55.

18. Robbins, *Hexaemeral Literature*, 27–31, 42–44.

19. Randles, *Unmaking of the Medieval Christian Cosmos*, 4.

20. Augustine, "On the Literal Interpretation of Genesis," 145.

time their temporal motions."[21] Thus the days of Creation described only the eventual order of nature rather than an actual sequence of creation.

The aporetic style of both of Saint Augustine's literal expositions of the six days helped define how the genre was adopted by followers.[22] He explored the questions that arose from the deeds of Creation as an often contradictory litany of questions, yet he left most unresolved. As he explained—in an opinion fully informed by his skepticism about human beings' ability to fully understand the natural world—"We should affirm none of these opinions rashly, but carefully and moderately discuss them all."[23] The hexameral commentary thus entered the Western canon as a genre where questions about nature could be posed, be discussed, and be equally either resolved or left unresolved. After reading his Augustine, the exegete of the Middle Ages was invited to continue expounding on possible natural and supernatural explanations of the deeds of Creation, but without the burden of having to speak categorically on a given interpretation. Biblical scholars took on the challenge, as in the commentary on Genesis by the Venerable Bede (d. 753) and the *De divisione naturae* of Johannes Scotus Erigena (ca. 815 to ca. 877).[24] Among the topics discussed in medieval hexameral commentaries was the Augustinian notion of simultaneous creation versus the notion of a gradual, sequential creation. Saint Augustine's exposition on the subject of celestial waters was the starting point for most subsequent debates on this topic. In what follows I will return repeatedly to the existence of celestial waters to illustrate the evolution of the way natural philosophical questions were addressed in hexameral commentaries.[25]

By the thirteenth century the Aristotelian-Thomist synthesis reframed these questions into a Scholastic format and restructured the responses posed by Saint Augustine (and many others) into logical trees and quodlibetal expositions. Saint Thomas Aquinas's approach became the model for Scholastic explorations of the subject. His most extensive commentary on the six days can be found in questions 65–74 of the *Summa theologiae*. The question of celestial waters is raised by the events of the second day, and Aquinas discusses them in Q. 68 (1a.68.1). Aquinas poses the following questions:

21. Ibid., 164.

22. Ibid., 33.

23. Ibid., 166.

24. Robbins, *Hexaemeral Literature*, 74–77.

25. For a survey of the opinion of some early modern writers on the possible fluid nature of the heavens, see Donahue, "Solid Planetary Spheres in Post-Copernican Natural Philosophy," 256–59; Grant, *Planets, Stars, and Orbs*, 96–105; and Randles, *Unmaking of the Medieval Christian Cosmos*, 1–57.

1. Was the firmament made on the second day?
2. Are there waters above the firmament?
3. Does the firmament separate some waters from others?
4. Is there only one heaven, or are there many?

To answer these questions he considers the opinions of the patristic writers, but also those of natural philosophers of antiquity. The second and third articles presented particular problems—first rehearsed in the third homily of Saint Basil's *Hexaemeron*—that stemmed from the vagueness of the word "firmament" and the phrase "waters above the firmament." So whereas Saint Basil had maintained the existence of celestial waters solely on the authority of scripture ("Let us understand that by 'water,' water is meant"),[26] Aquinas dismissed the possibility of celestial waters purely on the grounds of Aristotelian natural philosophy. Water, he explained, has weight, and thus its movement is downward, not upward. Furthermore, water is fluid, "but a fluid cannot remain stable on a spherical body." And finally, water is an element, and as such its tendency is to form a compound, and these can happen and exist only *supra terram* (on earth) but below the heavens. And yet the Sacred Scriptures read, "And God made the firmament, and divided the waters which were under the firmament from the waters which were above the firmament: and it was so" (Gen. 1:7 KJV). To explain the passage, Aquinas reframed the questions so they hinged on three issues:

1. What is the "firmament"?
2. What type of waters are these?
3. Was Moses using the word "water" to refer to both the type of air and the type of water found on earth?

As he explained (1a.68.3):

1. If the firmament is taken as the sidereal heaven, the waters above it are not of the same species as those below. If the firmament is taken to mean the cloudy region, then both waters are of the same kind. In the second case two places are assigned to water for different reasons: the upper place is where they are formed, whereas the lower is where they come to rest.
2. If the waters are understood as being different in species, the firmament said

26. Basil, *St. Basil: Letters and Select Works: Hexameron*, 71.

to separate some waters from others does not do this as a cause bringing about the separation, but rather as a boundary common to them.

3. Since air and similar bodies are invisible, Moses includes all bodies of this type under the name "water." And thus it is clear that no matter how the term be taken, waters are found on either side of the firmament.[27]

Aquinas thus left the matter fully investigated in the Scholastic sense, but essentially unresolved. His explanation depended on whether by "firmament" one understood a particular sidereal place, which implied that it did not contain terrestrial elements and allowed for the possibility of another kind of water, or whether "firmament" was understood as an extension of the terrestrial region, in which case the passage referred to the place where clouds are formed and rain originates. Finally—and borrowing from Philo[28]—he also considered the possibility that "firmament" meant a solid boundary separating two kinds of water.

Later exegetes also engaged with the possible existence of celestial waters, which often led to discussions of the composition, rigidity, disposition, and even reality of celestial orbs. By the late Middle Ages the notion of a fiery, fluid heaven prevailed, yet with the acceptance of the Aristotelian-Ptolemaic cosmology during the thirteenth century, the notion of solid orbs gained more currency—solid, however, in the sense of *plenus* not as a hard or impenetrable body or bodies.[29] Though nicely argued and impeccably reasoned, the exegesis especially of Scholastic followers of Aquinas appeared to humanists as if the disciples of the Angelic Doctor were working with one hand tied behind their backs. They relied on terminology from the Vulgate version of Genesis, did little or no philological analysis, and did not bring to bear any arguments based on firsthand observation.

As Edward Grant explained, the medieval theologian-natural philosopher resisted the temptation to create a "Christian science" based on biblical texts, despite the intellectual freedom the commentary or treatise genres afforded. Expositors were not given to constructing arguments that could be resolved only by appealing to divine authority; instead they used the genres to discuss and compare alternative explanations.[30] By the Renaissance, the hexameral

27. Aquinas, *Summa theologiae*, 10:87.
28. Philo, *On the Creation of the Cosmos according to Moses*, 54.
29. Grant, *Planets, Stars, and Orbs*, 324–48.
30. Grant, *Foundations of Modern Science in the Middle Ages*, 175.

genre had developed into large compendiums of *opiniones* where the literal exegesis of the six days of Creation was compared with and informed by the vast commentary and natural philosophical literature that preceded it.[31] For humanist interpreters, however, the historical nature of the Creation account became particularly relevant and thus provided a crucial interpretive perspective that favored the literal interpretation of the Sacred Scriptures, so allegorical interpretations, if done at all, were considered secondary. Increasingly the genre was informed by a tendency to rationalize and provide natural philosophical explanations without resorting—at least not hastily, as Aquinas advised—to explanations that relied on the miraculous and on the omnipotence of God. For most of the sixteenth century the genre also served as an important platform for discussing changing natural philosophical explanations concerning the nature of the heavens. Largely absent in these considerations, however, are rhapsodic digressions like the ones found in natural theologies.

The editions by Erasmus of the hexameral works of Saint Ambrose (1527) and Saint Basil (1532) brought renewed interest to the genre during the first half of the sixteenth century. But even before then, the Creation account attracted humanists who took to the biblical narrative with their philological tool kits, and in particular with their knowledge of Hebrew. Giovanni Pico della Mirandola in his *Heptaplus de septiformi sex dierum Geneseos enarratione* (1489) proposed an interpretation of the six days without the doxographic apparatus of *opiniones* typical of contemporary commentaries, instead remaining true to the notion that an explanation for the events could be found in "the nature of things as they are observed, or to the truth ascertained by the better philosophers."[32] While Pico relied on Plato and Aristotle to supply explanations for natural phenomena, he also turned for answers to authorities from an alternative doxographic tradition, one that added Jewish philosophers to Hermetists. He situated Moses alongside the Egyptian sages and as the founder of a doxography of natural knowledge that includes Hermes and Pythagoras.[33] Leaving no room for doubt about the Mosaic authorship of Genesis, Pico pointed directly to the hexamera as the most ancient repository of knowledge:

> It will now be easy to believe that if he [Moses] treated anywhere of the nature and making of the whole world, that is, if in any part of his work he buried the

31. Few early modern sources can surpass Giovanni Battista Riccioli's extensive historical review of the question as it stood in 1651; see Riccioli, *Almagestum novum*, 218–24, and Williams, *Common Expositor*, 22–23.

32. Pico della Mirandola, "Heptaplus," 74.

33. Ibid., 68–69.

treasures of all true philosophy as in a field, he must have done so most of all in the part where avowedly and most loftily he philosophizes on the emanation of all things from God, and on the grade, number, and order of the parts of the world.[34]

In Pico's opinion Moses wrote the account in a style that seemed to be "screened and covered lest it hurt the bleary-eyed."[35] He was not referring just to the simple language of an accommodated text, but to a language informed by the "hidden alliances and affinities of nature" that tied together all regions and entities and described the terrestrial as well as the celestial acts of creation.[36] He found that the key to the Genesis account was in realizing that Moses's account of the six days referred to the different realms of the sensible and nonsensible world as understood within a Neoplatonic notion of reality. Pico was not interested in simply supplying another literal explanation of the Creation account that described the physical world; he was fascinated by the prospect of using the philological tools at his disposal to unravel the cryptic language of the Genesis account. For example, in the appendix he expounded on the significance of the phrase "In the beginning" (בְּרֵאשִׁית, bĕrē'šît), in which he used kabbalistic letter permutations of the Hebrew to elucidate its meaning.[37] In Kabbalah he found, as he had found in Hermeticism, a font of mystical teachings and tools that gave him greater insights into the Creation account. They were his keys into the nonsensible world that could provide access through magic to spiritual beings and through them to the Godhead.

Another Neoplatonic notion informed the basis of Pico's theory of interpretation or hermeneutics, the notion of the earth as a microcosmic reflection of a perfect macrocosm that existed elsewhere. He therefore interpreted the Creation account as describing a series of parallel creations that he labeled the intellectual, celestial, elemental, and human worlds.[38] But even for this committed Platonist, the Aristotelian framework was impossible to abandon. In the first exposition of the *Heptaplus* he interpreted the "heaven and earth" of Genesis 1:1 as visible manifestations of Aristotelian efficient and material causes.[39]

34. Ibid., 71.

35. Ibid., 73.

36. This is Pico's concept of "mutual containment" and a cornerstone of the notion of macrocosm/microcosm relationships and parallels between the celestial and terrestrial regions: ibid., 79, 94–95.

37. Ibid., 170–74.

38. Ibid., 127.

39. Forshaw, "Genesis of Christian Kabbalah," 133.

In Pico's interpretation, for example, when Moses described the division of the waters on the second day the prophet was compressing into one terse narrative events pertaining to the unfolding of four parallel worlds. Consider the first two. In the "intellectual world" (the world perceived by human intellect), the division of the waters indicated that the sublunary world was divided into three regions. First and uppermost, a fiery/ethereal region of pure, unmixed elements, followed by the firmament, where birds fly and "in which the celestial phenomena appear: rain, snow, lightning, thunder, comets, and the like," and last, below this, a region "where there are no pure elements (not even a pure sensible element), but all things are mixtures composed of the dregs and grosser parts of the body of the world."[40] Meanwhile, the same division of the waters in the context of Pico's "celestial world" described a different arrangement of the celestial realm. This realm was composed of an uppermost realm, or first heaven, from which emanates "the treasuries of light." Below this realm was the region of the firmament where the fixed stars reside, and last came the lowest level where the seven planets are.[41] Novel as Pico's concept of the heavens might seem, he was building on a medieval tradition that gave tremendous latitude to interpretations of the events of the second day, making the composition of the heavens, as we have seen, a topic of vigorous debate.

Agostino Steuco (Eugubinus, 1496/97–1548), absentee bishop of Kisamos (Crete) and librarian of the Vatican, followed Pico's Neoplatonic bent but differed greatly in his approach. His *De perenni philosophia libri X* (Lyon, 1540) cemented his reputation as a Christian Hebraist and hexameral writer whose greater project was to recreate a concordant perennial philosophy that brought together what he thought were significant commonalities between ancient philosophies and religions, especially those of Zoroaster and Hermes, and the earliest roots of the Judeo-Christian faith.[42] Steuco's work was informed by a unitary vision between reality and knowledge—the created world was meant to be known by humans—with the source of this knowledge residing in the earliest relationship between God and Adam.[43] This knowledge formed a *prisca theologia* and a *vera philosophia*, a component of which was an intimate and unmediated knowledge of nature. For Steuco, this knowledge had not been lost after the Fall and survived the Flood to be transmitted thereafter by

40. Pico della Mirandola, "Heptaplus," 90.

41. Ibid., 99.

42. Schmitt, "Perennial Philosophy," 516–17; Granada, "Agostino Steuco y la perennis philosophia, 23–38, 26–27; and Vasoli, "Note su tre teologie platoniche," 90–94.

43. Muccillo, "'Prisca theologia' nel 'De perenni philosophia,'" 56–57, and Schmidt-Biggemann, *Philosophia Perennis*, 36.

Noah's sons to the Chaldeans, the earliest inhabitants of Mesopotamia. They in turn communicated it to the Egyptians and Greeks, with the concomitant and unavoidable degradation. Only now and in the light of the Christian Gospels, Steuco explained, could the essential truths of this ancient philosophy be rediscovered. He was convinced that Moses had revealed—if obscurely—the true causes of nature in what he refers to as a "Mosaic philosophy," or *philosophia Mosaica*.[44] His favorite hermeneutical tools were philology and comparative history. Were there perhaps forgotten meanings of Hebrew words that could be used to elucidate the Bible? Had ancient figures (such as Zoroaster, Asclepius, and Hermes) repeated versions of the Mosaic Genesis?

In his first book on the subject, *Cosmopoeia vel de mundano opificio* (Lyon, 1535), Steuco delved into the question of the existence of celestial waters during the second day.[45] His opinion on the matter would resonate with later natural philosophers—Bernardino Telesio, Francesco Patrizi, Tommaso Campanella, and Giordano Bruno among others—who posited unconventional cosmologies.[46] Steuco, however, sought in the Hebrew Bible clues for the true nature of these waters, examining the meanings of the words for *caelum* throughout the Genesis account (Gen. 1:8). His investigation—aspects of which he claimed mirrored and elaborated on the works of Hermes Trismegistus—led him to posit a three-part heaven and to argue for the existence of primordial celestial waters that the power of the sun "separated" into air and terrestrial water.[47] He also found, for example, a concordance between Aristotle and the Mosaic account if the word "heaven" in Genesis was understood to mean "the place of waters." In this case the heaven of Genesis corresponded nicely with Aristotle's "place of air and water," or what we now consider the terrestrial atmosphere. Steuco found that the Mosaic names also declared this. If רָקִיעַ (*rāqiaʿ*) and שָׁמַיִם (*šāmāyim*) were taken to mean "air," the former as an "air that pours" (*aer effusus*), the latter as an "air where water is generated," it would explain the cycles of evaporation and precipitation of the atmosphere.[48] He also saw the Creation not as an all-encompassing sudden event, but as unfolding gradually from matter that had been made ex nihilo. Yet God in these subsequent creations acted only insofar as he provided the force that brought matter together through the agency of secondary causes such as heat, light, and wind.

44. Steuco, *Cosmopoeia*, 72.

45. Ibid., 67.

46. Granada, "New Visions of the Cosmos."

47. Granada, "Agostino Steuco y la perennis philosophia," 30n29.

48. Steuco, *Cosmopoeia*, 63, 66.

Steuco's interpretations were not without critics; the *Cosmopoeia* came to the attention of the Spanish censors and was placed in the 1583 Quiroga Index as "banned until expunged." The expurgation appeared the following year.[49] The offending passage that was deleted debated the nature of the empyrean heaven, when and whether it had been created. Steuco presented the arguments of those who thought that since the empyrean heaven is the dwelling place of God, it could *not* have been created *in principio*. Others maintained that if this empyrean heaven was not made of water, then it might have not been created at all, but have always existed. The censored passage concluded, "And if heaven is made of water, it was created after the waters by necessity; the empyrean heaven seems to be uncreated heaven, neither does it seem to be following Moses. I say, it seems to me, but I may be mistaken and I do not defend the error."[50] His disclaimer makes it clear that he was aware that this was approximating an Averronian interpretation, but the censors were unconvinced and did not allow it.

Steuco's work found some resonance among the Spanish Christian Hebraists of the mid-sixteenth century, including Antonio de Honcala (1484–1565) and fray Luis de León. Both drank avidly from Steuco's fountain, as would Arias Montano. Honcala published an extensive hexameral commentary during the first half of the sixteenth century as part of his *Commentaria in Genesim* (1555).[51] Educated in the University of Salamanca's trilingual tradition and later working as magister of theology in Ávila, Honcala presented an exegesis based on the literal sense, relying where necessary on the Hebrew text, especially where he found the Latin translation in the Vulgate was not faithful to the Hebrew Old Testament. In his work we see evidence of the philological methodology that Arias Montano would employ to fashion his natural philosophy. For Honcala, the division of the waters in Genesis 1:7 was an opportunity to discuss the nature of the heavens. He first presented an etymological analysis of the word *caelum* taken from Varro (*De lingua Latina*, book 5, chap. 3:4): *caelum* derives from *cavum*, which means a concave or hollow thing. In contrast, explained Honcala, the Hebrew etymology of the

---

49. The 1584 expurgation read: "*Pag. 29 in primum caput Genes. in illa verba.* In principio creavit Deus Caelum & terram, *expungatur ab illis verbis.* Quare, & ab Hebreis, & Latinis, *usque ad illa eiusdem pag.* Errorem non defendo.": Quiroga, *Index librorum expurgatorum*, fols. 7v–8r. The preface was banned in the 1596 Clementine Index ("nisi fuerit ex emendatis"). See Schmitt, "Perennial Philosophy," 525.

50. Steuco, *Cosmopoeia*, 29.

51. Honcala, *Commentaria in Genesim*. For a study of this work, see the introduction by Fuente Adánez, *Exégesis para el siglo XVI*, 69–78.

word for "heaven" (שָׁמַיִם, *šāmāyim*) speaks not so much to the shape of the heaven as to its nature. It is a word that because of its ending can be either singular or plural, depending on usage, but that had been taken as singular in the Latin translation. He considered other possible origins for the word but repeated Steuco's interpretation following "the Hebrews" that stated it was a compound word derived from the words for fire (אֵשׁ,*ʾēš*) and water (מַיִם, *mayim*). Thus, Honcala concluded, the nature of heaven can be understood as one that shines but also heats and cools inferior regions because they share in the qualities of these two elements.[52] Its watery nature explains the crystalline nature of the celestial spheres—since they could be made of compacted water—that in turn separate the celestial waters from the waters below.[53]

Whereas Pico had attempted to accommodate the hexamera to Neoplatonism and kabbalah, the ruptures with philosophical systems of antiquity were more noticeable in Steuco and his followers. Honcala, like Steuco, let the Hebrew meaning of key words of the Genesis narrative drive his interpretation of the kind of reality the events of the six days described, whether or not it agreed with any of the more widely accepted systems of natural philosophy. The authority of this type of interpretation lay not in its coherence with a certain natural philosophical system, but in the authority of the Word first spoken by God and written down by Moses in the language of Adam.

Fray Luis de León, Arias Montano's friend from his days at Alcalá, with whom he had discussed natural philosophical questions, as we saw in chapter 1, also ventured into the territory of the hexameral commentary. In fact he wrote two commentaries, the first as a series of lectures titled *De creatione rerum*, and a subsequent one, *Expositio in genesim*, also lectures but delivered almost ten years later during the 1589–90 academic year in Salamanca, that survives in notes taken by one of his students.[54] The commentaries employed two very different hermeneutical approaches. The earlier work closely followed the Thomistic style of questions that typically arose in the context of Genesis commentaries, while in the later one fray Luis was more concerned with exploring questions arising directly from the biblical text in Hebrew.[55] In both works, however, his inclination was to find naturalistic explanations for

52. Honcala, *Commentaria in Genesim*, 9v; Steuco, *Cosmopoeia*, 63; Fuente Adánez, *Exégesis para el siglo XVI*, 126–27. This is a different source for the idea of a celestial fire that challenged the parallel Stoic notion of a celestial fiery *pneuma*: as described in Barker, "Stoic Contributions to Early Modern Science," 138.

53. Honcala, *Commentaria in Genesim*, 9.

54. Thompson, "Lost Works of Luis de León," 199.

55. Ibid., 210.

the events of Creation, pointing out the instances where God could be identified as the sole creative agent and others in which either secondary causes
were at work or they resulted from a creative virtue with which God endowed
matter.

Let us return to the question of the nature of the firmament and the celestial
waters as an example. In the *Expositio in genesim*, fray Luis was well aware
that he was addressing a much-debated topic with little chance of coming
to a resolution. Thus he began by explaining that he would be presenting a
number of conflicting opinions but would leave it to his listeners to form their
own opinions. (He gives a similar summation after discussing the nature of the
luminaries.) When discussing the creation of the firmament he acknowledged
the confusion caused by the Hebrew word *raqiah* (*rāqîaʿ* or firmament). Its
meaning in Hebrew was clearly "extension" (*extensio*), but "our interpreter"
(Saint Jerome) seemed to have followed the Greek definition of *extereoma*,
which means "a support or firmness."[56] Some have argued, he continued, that
this might have been a copyist's error for *extroma*, which, since it derived from
extrando, means "to extend." Regardless of the etymological misunderstand-
ing, the difficulty remained of explaining the nature of this *firmamentum* and
how it divided the waters. He explained that there were two opinions on the
matter: some said it was air, while others said it could not possibly be air, since
it must be capable of holding back the celestial waters from the terrestrial ones.
Those of the first opinion explained that this air was the result of the evaporation of the waters created on the first day, which, once turned into vapor,
diffused and filled all the space above us. Here, dutifully and as expected of a
professor, fray Luis cites as authorities Saint Eucherius, Saint Augustine, Saint
Thomas, and Saint Basil, as well as Rupert of Deutz and Hieronimo Oleastro,
although Steuco was his clear favorite.

To try to settle the matter, fray Luis began with a philological analysis. God
gave the firmament the name *samaiim* (*šāmāyim* or heaven) (Gen. 1:8) immediately after it was created. Therefore the etymology of the Hebrew *samaiim*
should explain the nature of the firmament. However, there seemed to be no
consensus among commentarists on the definition of this word either. For
some it derived from *sa* and *mayn*, suggesting a place of water, while others
took it to derive from *es* and *mayn*, suggesting fire and water. Regardless, fray
Luis surmised, all describe qualities found in heaven: humidity and heat.[57]

---

56. "Quod firmamentum aut firmitatem significant.": León, *Comentario sobre el Génesis*,
80–81.

57. Ibid., 84–85.

He also presented a series of biblical quotations that supported the notion that by "heaven" and "firmament" the Bible meant elemental air. Finally, he presented his personal argument drawn from what he described as "natural reasons" (*naturales rationes*). He argued that the creation of light in Genesis 1:3 also was accompanied by tremendous heat, so much so that the waters of the abyss still covering the earth began to boil and became vapor, releasing the air trapped within the water so that it rose and gave the firmament its airy nature. (Note that fray Luis makes a few Aristotelian assumptions along the way: that the nature of something—in this case "heaven"—is understood as meaning its qualities, and also that the instance of Creation also meant the creation of the four Aristotelian elements, including the air that lay trapped within the waters of the abyss.)

Fray Luis explained that some who objected to the notion that the firmament of Genesis 1:8 was airy argued instead that it was either a distinct region, a distinct element (perhaps watery), or even the sidereal heaven with its multitude of orbs. Paul of Burgos and Alonso Madrigal were of this last opinion and placed the firmament as the eighth sphere, while others made it the ninth and still others conceived of it as an "'I do not know what' kind of body made of water they call crystalline heaven."[58] Cayetan (Tommaso de Vio, OP, Cardenal Gaitanus, 1468-1534), following Saint Augustine, explained that the firmament stood for all the celestial spheres that had existed but folded into themselves (*involuti*) and were unfolded and expanded on the second day. These positions were supported by the Mosaic assertion that the firmament was heaven and thus was a creation of the first day. Fray Luis dutifully presented arguments against all these positions. To dismiss the existence of celestial waters, he relied on a particular interpretation of the Hebrew phrase that was translated in the Vulgate as *super expansionem*. For him this did not mean "above" the firmament, but instead meant "within" a region, thus making the controversial phrase simply indicate that there was water in the airy region.

As expected, fray Luis reserved his own opinion until the very end, presenting what seems to be a compromise between an Aristotelian conception of matter and a Neoplatonic-Augustinian application of the notion of seminal reasons. The heavens, he explained, were created with respect to substance on the first day, but on the second they received a new quality and new duties (*qualitas et officium*). He did not develop the notion further, but we will see below that other natural philosophers saw this as a viable explanation for the unfolding of the world after Creation. It might be a function of the nature of

58. Ibid., 92.

the document as a student's notes, or perhaps a true reflection of fray Luis's exasperation with the topic, but it seems he wrapped up the discussion of the topic rather abruptly that day with a terse "enough on this" (*et hactenus haec*). Indeed.

An exposition of Genesis was also an opportunity to engage in current debates about the nature of the world, something fray Luis de León availed himself of when he touched on the controversial subject of the incorruptible or corruptible nature of the heavens.[59] Following the general format of the disputation, he presented all sides of the debate. For the camp that maintained the incorruptibility of the heavens he enlisted Aristotle, Saint Thomas, and "almost all the Scholastics and Peripatetics." They argued that the heavens have no contraries to oppose them and wear them down and thus remain in a pristine condition; Psalm 148:6 and Job 37:18 also seem to support this.[60] On the side of corruptibility stood the opinions of Hippocrates, Oecumenius of Tricca, Heraclitus, Empedocles, Saint John Chrysostom, Aloysius Lippomanus, and several others who maintained that heaven is immobile and celestial bodies move within it "dividing and cleaving it" (*dividentes eum et scindentes*).[61] While Saint Augustine and Saint Clement seemed to consider the possibility of corruptibility, they certainly maintained that heaven could someday come to dissolution. A number of passages, such as 2 Peter 3:10, Psalm 101:26, and Isaiah 51:6, supported that opinion, while Luke 21:33 and many others testify to the eventual caducity of the heavens. At this point of his exposition, fray Luis mentioned that some were now arguing that the heavens were corruptible after observing the "comet" in 1572. The consensus among astrologers, he noted, was that the "comet" was formed above the sun, which seemed to suggest that this region was susceptible to "foreign impressions" and therefore subject to corruptibility.[62] Fray Luis was referring in this passage to the appearance of a supernova in 1572, a turning point in the history of astronomy. At about this time the notion of corruptibility of the heavens became associated with the question of the reality of the solid and impenetrable celestial orbs described by Ptolemy, largely as a result of Tycho Brahe's publications on this new star. Brahe positioned his argument against the existence of such

59. Ibid., 164–67.

60. "He hath also stablished them for ever and ever: he hath made a decree which shall not pass" (Ps. 148:6 KJV). "Hast thou with him spread out the sky, which is strong, and as a molten looking glass?" (Job 37:18 KJV).

61. León, *Comentario sobre el Génesis*, 162.

62. Ibid., 164.

orbs and argued on behalf of a fluid heaven. His critics argued otherwise, largely based on the notion that fluidity implies corruptibility.[63]

Ultimately fray Luis withheld his opinion, only explaining that the most commonly held opinion was that the heavens were incorruptible, whereas corruptibility was more in concert with Sacred Scriptures. But among Spanish and Portuguese Jesuits questions concerning the structure and composition of the heavens continued unabated. Benito Pereira in his widely read 1590 Genesis commentary, following Steuco's lead, reconciled the Aristotelian notion of celestial incorruptibility with the firmament's fluid nature.[64] The Conimbricenses refused to adjudicate on the matter of corruptibility of the celestial regions and continued to maintain the solidity and existence of celestial spheres.

By the last third of the sixteenth century and in light of consideration of the Copernican heliocentric thesis, natural philosophical speculations like the ones we have seen in the hexameral genre increasingly engaged with the works of astronomers. They began to consider whether the mathematical devices astronomers constructed to explain the motion of the heavens, such as equants, epicycles, and deferents, described the real physical movement of planets and stars or were simply expedient mathematical models useful for predicting the positions of celestial bodies.[65] Complicating matters, however, were powerful epistemic notions defining what it meant to speak authoritatively about natural phenomena. As Peter Barker and Bernard Goldstein have pointed out, the best an astronomer could hope for in the era before telescopic observations (and even afterward) was to offer a *quia* demonstration, one that relied on evidence from observations to show that a mathematical model of the movement of the heavens was plausible. However, in the natural philosophical realm (as in the aspirational realm of physical astronomy) the gold standard was a *propter quid* argument, an explanation that relied on proximate causes to prove an observed effect.[66] These proximate causes had to be axiomatic—unquestionably true—and furthermore had to be derived deductively from established first principles. In early modern natural philosophy, particularly in its Aristotelian embodiments, veracity could be derived from the common

63. Grant, *Planets, Stars, and Orbs*, 348–51.

64. Randles, "Ciel chez les Jésuites espagnols et portugais," 309.

65. In the history of science this debate is framed as one that began in antiquity, pitting realists against instrumentalists, and continued to play out during the sixteenth and seventeenth centuries. For more on the debate see Duhem, *To Save the Phenomena*, and Goldstein, "Saving the Phenomena," 1–12.

66. Barker and Goldstein, "Realism and Instrumentalism."

course of nature or from first principles as defined through philosophical reasoning, but never solely from mathematical arguments. If the common course of nature was deemed unobservable—as was the case for whatever the material substance was that carried planets in the heavens—an axiomatic argument was impossible.

How then could anything be stated categorically about natural phenomena, let alone unobservable phenomena, whether currently occurring in the world or having occurred in the past? For some who felt keenly the natural philosophical disquiet I have been discussing, the inherent lack of certainty offered by the prevailing natural philosophical systems and the epistemic criteria these imposed was seen as a problem. The consequence of not understanding the natural world correctly implied that humanity might not be getting right an important part of what it meant to live properly in the world in order to attain salvation. The epistemic uncertainty that enveloped natural philosophy during the sixteenth century became to them increasingly untenable.

It should come as no surprise, then, that among those interested in resolving the epistemic uncertainty overtaking natural philosophy, the focus turned toward finding an unequivocal foundation for a new natural philosophical system. In this deeply religious era, and despite confessional divisions, all agreed that one fountain of truth remained: the Bible. We will study next some examples of natural philosophers who turned to the Sacred Scriptures in search of epistemic certainty. Whereas medieval theologians/natural philosophers resisted the temptation to create a "Christian science" based on biblical texts and only in rare instances used the Sacred Scriptures to demonstrate scientific truths, the sixteenth century would see the emergence of a number of attempts to derive a natural philosophy from the biblical text.

MOSAIC PHILOSOPHY

In his works, Steuco made repeated references to a "Mosaic philosophy" he thought could be found in the Pentateuch. This notion—albeit largely cleansed of the Hermetic affectations of the Renaissance Neoplatonists such as Steuco—became a cornerstone of Arias Montano's *Magnum opus*, traces of which can be seen as early as the *De arcano sermone*, a treatise included in the appendix of the Antwerp Polyglot. But before Arias Montano could present his own version of this Mosaic philosophy, others took up the idea. Their objectives varied; some wanted to show that a particular flavor of natural philosophy, either Aristotelian, Platonic, Atomistic, or one of their many syncretic variants, was in concert with the revealed Word, while others sought to identify in the Creation account a series of principles of nature on which to fashion a natural philosophical system in concert with nature as God had revealed it to Moses. Nonetheless, they all shared a profound conviction that piety and natural philosophy could be reconciled based on the authority of the Bible and that, furthermore, the revealed Word could serve as a font of knowledge about the natural world. They drew on the hermeneutical practices typical of hexameral commentaries: some preferred allegorical interpretations, but most built their systems from literal interpretations of the Bible. It was a literal exegesis that had to be informed by the careful and systematic observation of the natural world, with the biblical text called on to adjudicate natural philosophical debates.[67]

As Ann Blair has explained, eighteenth-century historians of philosophy looking back at the previous two centuries tended to group these authors into a school: Levinus Lemnius (1505–68), Francisco Valles de Covarrubias (1524–92), Lambert Daneau (1530–95), Otto Casmann (1562–1607), Conrad Aslacus (Kort Aslakssøn, 1564–1624), Johann Heinrich Alsted (1588–1638), and Johann Amos Comenius (1592–1670). Yet, as Blair admits, this grouping suggests a false coherence among their works that falls apart on further scrutiny.[68] I might add that it is also very incomplete, since it largely ignores the work of southern European Catholic theologians and philosophers. Furthermore, I suspect that the only reason Valles (a devout Catholic) was included in the list of Protestant thinkers (compiled by Protestant historians in the eighteenth century) was that all subsequent editions of his book, starting in 1588, were published alongside those of the Protestants Lemnius and Daneau—an association that no doubt has caused the pious Catholic eternal consternation.

A few years before fray Luis de León delivered his lectures on Genesis at Salamanca for the last time, royal physician Francisco Valles set down his own interpretation of the creation story, but with a particular objective that went far beyond the exegetical: to interpret the natural phenomena mentioned in the Bible solely in light of natural philosophy.[69] A prolific author, chief medical officer of Castile (*protomédico*), and personal physician to Philip II—who called him *el Divino Valles*—he wrote extensive commentaries on Aristotle, Galen, and Hippocrates early in his career. The *Sacra philosophia* (1587), however, is the work of a mature thinker who wishes to speak definitively about the

67. Blair, "Mosaic Physics," 34–35, and Crowther, "Sacred Philosophy," 397-98.

68. Blair, "Mosaic Physics," 57.

69. For his biography, see Ortega and Marcos, *Francisco de Valles (el Divino)*, 38–86, and Calero and López Piñero, *Temas polémicos de la medicina renacentista*, 5-9.

natural world. He was convinced that the Sacred Scriptures were written to help humanity attain salvation and not intended to reveal secrets of nature. Yet it was his view that God had given humanity the "heavy burden of philoso-phizing,"[70] so that where the Bible wove into its narrative discussions of nature, it spoke truthfully and thus could be the source of a true doctrine of the natural world. His objective was similar to that of the natural theological tradition—to find knowledge about God in explanations of nature—but his approach was exegetical: to clarify passages in the Sacred Scriptures describing natural phe-nomena. He was particularly interested in instances where he could explain how God acted through natural causes when performing a miracle.[71] (When God does so, Valles maintained, he is making manifest a natural law evident in the nature he has imparted to things.)[72] Valles the empiricist comes through most strongly when he can explain biblical passages using his experience as a medical doctor or can offer rational explanations.

To undertake his project he embraced a literal biblical hermeneutics but set aside the structure of the line-by-line commentary. Throughout the ninety-one chapters of the *Sacra philosophia*, Valles commented on selected scrip-tural passages, particularly those that allowed him to elucidate the subject at hand using natural philosophy and medicine. No Hebraist, his literal interpre-tation was based solely on the text of the Vulgate (with modest references to the Greek Septuagint) and mostly within the bounds of Catholic orthodoxy. In contrast to hexameral authors, he largely sidestepped patristic authors, pre-ferring to compare the opinions of natural philosophers. Here he was careful to survey ancient opinions—from Pythagoras, Democritus, Epicurus, Plato, and Aristotle—along with those of contemporary natural philosophers, juxta-posing their views and finally resolving the issue according to his own opinion and on the authority of the biblical text. He wore his philosophical allegiances lightly, favoring Plato when discussing the role of God as creator (although with the important caveat that the philosopher got the uncreated part of the argument wrong), siding with Aristotle on the eventual caducity of motion and matter, and laughing at Epicurus's version of the subject. While his perspec-tive was informed by a positive skepticism that acknowledged the limitations of sense perception and the variability of the nature of things, he nonetheless maintained that is was possible to make assertions about nature, although he was cognizant that in some instances causal knowledge, or *scientia*, of uni-

70. Valles de Covarrubias, *Sacra philosophia* (1587), 2.
71. Ibid., 291.
72. Ibid., 2.

versals remained beyond humanity's reach.[73] Wherever possible he sought to explain biblical events through natural causes, an approach that irked his religious censors, as we will see below.

For Valles the world was divided into two parts, one constituting a sublime and perfect realm where incorruptible bodies resided, and another realm, lowly and "ignoble," that was created as completely featureless and where change is continually observed. The unfolding of the corruptible world was a divine act that coincided with the creation of the world: "God made everything out of nothing, but also organized the created matter."[74] It was his opinion that Moses had left out these details because he chose to describe only things as they were created in themselves (*per sese*) and not those created only in principle.[75] The unfolding of the creative act, according to Valles, took place in action, but also in potential—as with animals that appear by spontaneous generation or putrefaction, but also with things that generate without seeds, such as metals.[76] Thus, although it went unmentioned in the Bible, some sort of natural force was also put in place that made these things eventually come forth.

Valles's interpretation of Creation makes his notion of a natural creative force clear. On the first day God made the sphere of corruptible bodies with the capacity of generation, which included the elements: earth, water, air, and fire. But they were without order: thus fire remained insensible within the darkness of the watery chaos yet acting to keep it fluid. From this chaos, God first drew out light, which made it possible to differentiate among the corporeal qualities of the other elements. The simple elements were then endowed with matter and form, thereby receiving their essential properties and qualities. On the second day God made the firmament, that is, all the celestial bodies with their many orbs; on the third day he separated the water from the earth. On the fourth day he created the sun, moon, and stars, which Valles believed came "from the same matter as the firmament, just as man has been made from earth."[77] Once the celestial bodies were in place, and through their agency, the rest of the "seedless" things came into being: metals, stones, pitch, and other such bodies that generate accidentally. Valles pointed out that what we call air, water, and so on are not really these simple primordial elements but are mixtures of them. The primordial elements exist only in *potentia* in

73. Ibid., 478, and Valles de Covarrubias, *Libro singular de Francisco Valles*, 518–19.

74. Valles de Covarrubias, *Sacra philosophia* (1587), 22–28, and Valles de Covarrubias, *Libro singular de Francisco Valles*, 27–31.

75. Valles de Covarrubias, *Sacra philosophia* (1587), 25.

76. Ibid., 21–23.

77. Ibid., 25.

bodies and are not to be found in *actus* in the world. Elemental fire, however, because of its tenuous nature, exists everywhere, permeating everything "like a world soul . . . fertilizing them and giving them a certain type of life."[78] (An informed audience would have recognized in this a reference to the Stoic *pneuma*.) Valles then explained in detail how water and fire interacted to cause generation, where the fire acted as the vivifying force or spirit. This was, in his opinion, the real meaning of the biblical phrase *et spiritus Domini ferebatur super aquas* (Gen. 1:2 VUL), which also meant *incubat aquas*, phrases that, Valles noted, are identical in the original Hebrew. (Set forth in the text as a categorical statement, it later earned him an expurgation.)

Valles was just as puzzled as his predecessors about the creation account of supracelestial waters. Yet by resorting to his preceding explanation about the difference between the simple elements created on the first day and those we know as air or water, he was able to reduce the problem to one of semantics. He explained that the waters above the firmament were of a different nature from those on earth; they were celestial or crystalline waters, similar in some aspects to those on earth, but neither cold nor humid.[79] Moses used the term "water" in the passage as an analogy; he was in fact referring to the purest water created on the first day. The firmament itself was a compound of the four purest versions of the elements, including, of course, this celestial water.[80] Heaven was thus brighter, visible and solid, more diaphanous and compact that any body our senses could perceive. It was incorruptible and not dissolvable because this matter does not partake of the principles that cause corruption: hotness, coldness, humidity, and dryness. Stars differed from the firmament only in that they contained different amounts of fire and air.[81]

Valles's discussion of the nova of 1572 is particularly interesting. After noting that some astronomers have come to believe the star that appeared in Cassiopeia was a new creation and that it was located above the sun,[82] he dismissed their arguments on the authority of Ecclesiastes (perhaps Eccles. 3:11–14), which refers to the heavens as perfect and constant. Valles believed the star had always been in the same place but had remained hidden because of some "mutation in the medium" that had persisted until the day it became visible. Not all parts of the heavens were similarly dense, he argued, as can be

78. Ibid., 28, 344–45.

79. Ibid., 36.

80. Ibid., 383.

81. For a comprehensive explanation of Valles's theory see Crowther, "Sacred Philosophy, Secular Theology," 419–22, and Zanier, "'De sacra philosophia'" (1587), 28–30.

82. Valles de Covarrubias, *Sacra philosophia* (1587), 40–42.

seen in the appearance of the Milky Way, lunar spots, and other stars. Furthermore, it could have also been that as this star went about its course it chanced upon a denser part of the heavens, making it appear brighter and then dimmer. In contrast to fray Luis de León, Valles staunchly denied the corruptibility of the celestial regions, a stance that earned him a posthumous reprimand from Tycho Brahe.[83]

Valles sought possible naturalistic explanations for biblical events and phenomena wherever he could. He dismissed magical words and incantations as ineffectual against "beasts or snakes or against illnesses of the body." Any natural effects that might be brought about by "magicians and cabalists" could be deemed the result of their knowledge of demons and some mysterious aspects of nature. Words and incantations, Valles conceded, could cure the vulgar masses because they tended to believe in such things and could be persuaded of their benefit; once convinced, they found solace, and their resulting good spirits could help restore their health. Even the church, he noted, recognized that words heard or uttered as chants and in the repetitions of the rosary were effective against illnesses of the soul.[84] This is not to say that Valles dismissed all the theoretical premises of natural magic. Throughout the *Sacra philosophia* he offered plenty of naturalistic explanations of biblical events that relied on notions of sympathies and antipathies. But he railed against any who interpreted certain biblical passages in support of judiciary astrology, necromancy, quiromancy, or any type of divination.

Valles devoted the third chapter of the *Sacra philosophia* to the nature of human language. As for other commentators, his departure point for the discussion was the passage where Adam named the beasts (Gen. 2:19–20). This passage presented two possibilities that, he noted, had long been debated: whether names represent a natural congruence between the word and thing or whether they are simply human expressions meant for communication.[85] According to the Bible, Adam was said to have "called" animals (and only animals) by their names rather than having "given" them their names. This suggested that the names existed before Adam used them and therefore were based on the animals' natures.[86] For Valles the debate dated from the days of Plato's *Cratylus* and concerned whether words themselves have intrinsic

83. Navarro Brotóns, "Reception of Copernicus in Sixteenth-Century Spain," 60–61. For Tycho Brahe's courteous but stinging criticism of Valles's reliance on ancient philosophers over evidence of the senses, see his *Astronomiae instauratae progymnasmata*, 577.

84. Valles de Covarrubias, *Sacra philosophia* (1587), 79–80.

85. Ibid., 63.

86. Ibid., 64.

power as magi and kabbalists suggested (and, presumably, as did Marsilio Ficino, who although unnamed in this passage clearly informed Valles's work) or whether language was instead a social phenomenon, the creation of use and art (*usu et arte*), and that therefore neither words alone nor language itself could carry such power.[87] Valles sided with the latter opinion, which, as we saw above, let him dismiss incantations and magical words as medically ineffective and patently absurd. But for our purposes this was also a not too veiled criticism of the approach of scholars like Arias Montano. Valles rejected those who sought to find in remnants of Adam's language greater significance than mere appellation. His criticism was aimed at experts in the Hebrew language who thought they knew these words well and had "squeezed from them their properties and nature, derived from the names of the other primitive ones that Adam gave them as it pleased him."[88] He was also making another, somewhat more diffuse criticism of the belief that Adam had had the gift of intuiting the nature of all things and that this knowledge had been communicated to subsequent generations. For Valles, therefore, it was futile to try to recapture any long-forgotten knowledge from the language of the Bible. Instead, it was the natural philosopher's task to investigate nature and describe the events described in the Bible in concert with it, but without forgetting that the ultimate purpose of the task was to better understand the nature of God.

The *Sacra philosophia* was a vehicle designed so Valles could engage with natural philosophical controversies of his times, but also with those of the classical philosophical corpus. Underlying the whole work is a persistently skeptical view of received wisdom, and although Valles does not engage directly with the body of interpretation theologians had poured into hexameral commentaries, he presented his analysis as the veridical interpretation of biblical events. In contrast to Arias Montano, Valles's objective was to construct neither a systematic philosophy nor a wholly new natural philosophy. What he did do was argue consistently in favor of identifying possible naturalistic explanations of confusing or ambiguous biblical events over simply labeling them "miraculous." He proved to be intellectually nimble, drawing from various traditions to support his arguments. For example, he liked Plato's notion of a God creator but thought little of the agency of transcendental forms, preferring Aristotelian notions of potentiality/actuality and seminal principles. Where syncretic explanations fell short, Valles resorted to analogies drawn from his personal experience as a physician.

87. Ibid., 63.
88. Ibid., 81. On this subject see also Zanier, "'De sacra philosophia'" (1587)," 32.

The *Sacra philosophia* was brought to the attention of Roman censors after a denunciation in 1597 by Stefano Guaraldi, the inquisitor of Bologna, primarily because of references Valles made to some censored works. So when the Roman Congregation of the Index in 1599 listed Valles's cover mates, Lemnius and Daneau, as *auctores damnati*, the examination of his work became inevitable. (Lemnius had already been identified as a heretic in the 1583 Spanish *Index* of Quiroga.) In an effort to rehabilitate Valles's book and return it to circulation, it was submitted for examination and expurgation. Among the things the censors found objectionable was Valles's interpretation of the first three days of Creation.[89] One censor, fray Luis Ystella, OP, ordered a few deletions, having noted that Valles was not following rigidly the order of creation—a bit of philosophical nit-picking for the sake of precision, perhaps.[90] But his greater displeasure with the book came from several attempts by Valles to explain miraculous events through natural philosophy, in particular, instances where the power of God only intensifies natural effects. In other cases it was his opinion that Valles had not followed faithfully the patristic exegesis, "because in these [instances] he introduces random new interpretations of little utility."[91] Valles's book underwent extensive censorship in several sections. While some points of censorship simply placed qualifiers on some of the author's pronouncements, many focused on the problematic association Valles made between fire and the spirit of God "that hovered over the abyss" (Gen. 1:2). Also worrisome to the censors were some parallels he set up between the human ability to reason and the immortality of the soul, as well as the rela-

89. "Nam primo die fecit elementorum corpora, et lucem, <add: et coeleum empyreum> secundo fecit firmamentum, hoc est corpus illud totum <delete: totum> coeleste, quod constat multis orbibus: tertio separavit aquam à terra." The words changed are shown in brackets. Valles de Covarrubias, *Sacra philosophia* (1587), 25. On 19 September 1599 the *procurador* of the Dominicans (Paolo Isaresi) requested the examination, which was assigned to fray Ludovico Stella, OP (Luis Ystella Valenciano, 1545–1614, later Master of the Sacred Palace, 1608–14), and to Fray Bonincontro, OP. By 13 November, Stella had his censorship ready, and it was accepted by the Congregation of the Index: ACDF *Diarii* 1, fols. 123v and 125. The extensive original censorship by Ystella is in ACDF, Index, Protocolli T, fols. 213r–14v, while an unsigned different one is at ACDF, Index, Protocolli AA, fols. 848r, 849r–v. This latter one is incomplete and does not seem to have been considered. Both have been edited and published in Baldini and Spruit, *Catholic Church and Modern Science*, 1.3:2435–46. The specific expurgations would not appear in print until the publication of the Roman 1607 *Index librorum expurgandorum*.

90. For Ystella's terse interpretation of the division of the waters, see his *Commentaria in Genesim et Exodum*, 7–9.

91. "*[Q]uòd in illis temerè, et absque ulla utilitate, novas introducat interpretationes.*": Baldini and Spruit, *Catholic Church and Modern Science*, 1.3: 2440.

tion between the regular course of nature and the power of God to intervene at will. Subsequently the Spanish Inquisition also turned its attention to the *Sacra philosophia*, and in the 1612 *Index* of Sandoval it ordered a series of expurgations that coincided on only a few points with the Roman censors.[92] Spanish censors seem not to have been concerned with the correspondence Valles made between fire and the spirit of God. They did, however, object to his naturalistic explanation of Elisha's resuscitation of a young boy (2 Kings 4:34–35), which seems to minimize God's role in the miracle.

*

The emphasis that sixteenth-century authors of hexameral commentaries and Mosaic philosophies placed on reconciling natural philosophy, the Bible, and the world as it was had important repercussions for the objective of the natural philosophical enterprise. They presented the possibility of extracting from the Bible a coherent natural philosophy that could be an alternative to other natural philosophical systems, perhaps banishing all competing schools and setting natural philosophy on its correct path. Although the Bible was the cornerstone of this effort, as we have seen, the project also relied on the empirical tools of observation, experience, and experiment (in its late medieval definition).[93]

Arias Montano was driven by the disquiet that pervaded the natural philosophical inquiry and that we have seen manifested in the Spanish versions of the genres discussed in this chapter. For him the motivation to undertake a project of natural philosophical renovation lay at a far more fundamental level: in the purpose of humanity itself. Human beings were the *very reason for the world*, since everything in the world was made by God to sustain humans, who in turn had to live in constant praise of the God who placed it at their disposal. Along with the power to subject and dominate the earth, God also granted to humanity—or at least to Adam—the gift of knowing the nature of all things on earth: what they were for and their causes, effects, motions, and actions. It was a knowledge—a gift—that had been largely forgotten and that Arias Montano hoped to recover. For someone harboring this belief, the opportunity to create a critical new edition of the Bible in all its original languages must have seemed a godsend.

---

92. Sandoval y Rojas, *Index librorum prohibitorum et expurgatorum* (1612), 342.

93. As Gianna Pomata has explained, during this time in medicine, the *observatio* became an epistemic genre: Pomata, "Observation Rising," 48.

CHAPTER 4

The Antwerp Polyglot:
Hints of a New Natural Philosophy

Very little is known for certain about Arias Montano's disquiet concerning natural philosophy in the years before the Antwerp Polyglot; the letter to Luis de León is the sole tantalizing clue. Yet it is evident throughout his many interventions in the project that during the sojourn in Antwerp he continued developing earlier ideas about the relation between the Sacred Scriptures and natural philosophy. What is beyond doubt is that his involvement in the polyglot project was nothing short of a scholarly explosion. The catalyst for this work might have been the stimulating intellectual climate he found in Antwerp and the vitality aroused by collaborating with the world's foremost scholars of biblical languages. But it would be naive to presume that the ideas Arias Montano was so intent on stamping onto the project were the sudden product of his departure from Spain; it is likely that the opposite is true. His contributions to the *Apparatus*, in particular, result from an approach to scripture that he had nurtured in Spain.

This chapter considers Arias Montano's involvement in the Antwerp Polyglot Bible, focusing on his intellectual contributions to the project and how these reveal his earliest ideas about the Bible as a source of natural philosophy. These concerns appeared in three textual forms: the prefaces for the project, his editing of Santes Pagnino's Latin translation of the Hebrew Old Testament, and several treatises in the explicatory *Apparatus sacer* or scholarly appendix to the Bible. Of these treatises, the *De arcano sermone* is of particular importance; it illustrates Arias Montano's first attempts at developing an approach to natural philosophy derived from the philological analysis of the Hebrew Bible and shows where he rehearsed the methodologies that served as the foundation for the natural philosophy he would develop some twenty years later in the *Magnum opus*. Its publication would bring him significant problems, and it came close to jeopardizing the whole of the polyglot project. But Arias Montano remained steadfast about its inclusion. As Luis Gómez Canseco, one of his principal scholars, notes, "It seems that in the *De arcano sermone* were hidden—and still remain hidden—some of the essential keys to his thoughts, and that the treatise was not a casual result, born from the royal project, but rather the consequence of a larger, more ambitious plan whose roots we must search for not in Flanders, but in the immediately preceding years."[1]

As Arias Montano stepped onto the international stage afforded by his role in the Antwerp Polyglot, he had to carefully balance the interests of the many factions with vested interests in the project. First came those of his patron, Philip II, and those of the Catholic Church. He also had to "manage" the interests and concerns of a number of contributors, reviewers, and censors and, not least, those of Antwerp publisher Christophe Plantin, who had originated the enterprise and shouldered a significant portion of the risk. Perhaps a person of more modest or pragmatic convictions would have played it safe and steered the ship close to the shores already navigated by the Complutensian Polyglot Bible, but Arias Montano did not choose to do this. He saw a unique opportunity to improve, and even change, the way the Bible was studied and interpreted, using what he understood to be the best scholarship and approaches available.

The challenges this posture brought during the post-Tridentine era soon made themselves apparent. But as we will see, the objections to the Antwerp Polyglot did not stem from fear concerning the possible infiltration of Protestant ideas into the project or a general condemnation against providing material that would endorse personal study of the Bible. The bulk of the objections were against the very exegetical methodology and the unusual conclusions it supplied. This chapter explores the methods Arias Montano advocated, most apparent in his treatise *Liber Ioseph, sive De arcano sermone*.

THE POLYGLOT PROJECT

Since the mid-1560s, Christophe Plantin had been nurturing the idea of publishing a polyglot Bible and had spent considerable effort and treasure in the enterprise before seeking Philip II's patronage in 1568.[2] In a 1565 letter to

1. Gómez Canseco, "Los sentidos del lenguaje arcano," 47.
2. Rooses, *Christophe Plantin*, 121–23.

Andreas Masius he refers to the project as an expanded edition of the trilingual *Biblia Complutensis*, a reference to the grand project of Cardinal Cisneros and the University of Alcalá.[3] But though Plantin acknowledged the Spanish Complutensian Bible as a precedent, he sought to bring to fruition a far more comprehensive project. The earliest source that documents Christophe Plantin's interest in having Philip II sponsor the polyglot project dates from 19 December 1566 and makes no reference to its being a new edition of the Complutensian Bible.[4] This, however, was not Plantin's first contact on the subject with the cleric and royal secretary Gabriel de Zayas, since the letter suggests they had already communicated about it. Their acquaintance dated to 1554 or 1555, when the secretary lived in Flanders.[5] By way of the powerful Cardinal Antoine Perrenot de Granvelle (1517–86), Plantin sent proofs showing the typography and general scheme of the Bible he was offering to make. These had been printed during spring 1566, and it is reasonable to assume that the contact with the Spanish court occurred soon thereafter or on Plantin's return from the Frankfurt Book Fair.[6]

The surviving documentation makes it clear, as historians have noted, that Plantin sought the protection of the Catholic king to shield himself against suspicion of having Protestant sympathies.[7] The letters frame the project as

3. C. Plantin to A. Masius, Antwerp, 26 February 1565. In the letter Plantin also acknowledges Masius's offer to provide him with a Targum in Latin that he had acquired in Rome. Plantin thanks him and mentions that Bishop Lindanus had provided a New Testament in Syriac: Rooses and Denucé, *Correspondance de Christophe Plantin*, 3:1–3. Léon Voet notes that "Plantin developed his initially modest concept of a slightly revised edition of the Alcala polyglot Bible into a work that was as original as it was monumental.": Voet, *Golden Compasses*, 1:60. The Complutensian Polyglot (1514–17) was the first printed polyglot Bible. The Old Testament appeared in side-by-side columns in Hebrew, Latin (Vulgate text), and Greek (Septuagint) with an interlineal Latin translation. The Targum of Onkelos in Chaldaean (Aramaic) with a Latin translation was placed at the bottom of the page for the first five books of the Old Testament (Pentateuch). The New Testament appeared in side-by-side versions of Greek and Latin text of the Vulgate. The fifth and sixth volumes also had two dictionaries (Greek–Latin, Hebrew–Chaldaean–Latin) and an index of proper names in Latin cross-referenced to their Hebrew or Greek equivalents, as well as a Hebrew grammar.

4. The originals are at AGS, Estado 583, fols. 58–59, 44–45. Zayas's translation that was forwarded to Arias Montano is MS EST, fols. 179–82, as noted in Macías Rosendo, *Biblia Políglota*, 65–70.

5. Voet, *Golden Compasses*, 1:56.

6. Rooses, *Christophe Plantin*, 118.

7. Voet, *Golden Compasses*, 1:50–56, and Macías Rosendo, *Biblia Políglota*, xxiv, citing Rekers, *Arias Montano* (1972), 62.

necessary in the face of what he portrays as similar projects imminent under the sponsorship of Duke Augustus, elector of Saxony. Plantin omitted the fact that he was a Protestant, simply noting that other parties, both Protestant and Catholic, had shown interest in the project. Plantin accompanied his petition with an impassioned profession of faith and a call for studying the Sacred Scriptures under the guidance of "prelates and ministers of the Holy Roman Catholic Church" to forestall the errors "new ministers" might bring to the interpretation of the Word, "seeing themselves inflated with some measure of science in the letters and languages" and leading others to believe that their interpretations were inspired by the Holy Spirit.[8]

In his original proposal, Plantin favored placing the Hebrew text of the Old Testament side by side with the Latin translation by Santes Pagnino, with the facing page showing the Septuagint in Greek with a new Latin translation.[9] (Such was Plantin's commitment to this that the page proofs that made their way to Spain via Zayas had the Pagnino text alongside the Hebrew Old Testament.) At this early stage Plantin also suggested including a Hebrew dictionary as well as the New Testament in Syriac (an ancient dialect of Aramaic), something he reiterated when he referred to the project as a "quintilingual" Bible. The bottom of the Old Testament page would include the corresponding passage from the Targum, a paraphrase of the first five books of the Bible in "Chaldaean" (that is, in Aramaic written in Hebrew characters), and alongside it its paraphrase in Latin. The New Testament would appear in Greek, Latin, and Syriac with new Latin and Hebrew translations. Furthermore, Plantin sought not only the financial backing of the Most Catholic Monarch Philip II but also his protection against possible objections to the project by ecclesiastical authorities.[10]

As Plantin's proposal made its way through the consultation that characterized Philip's court, a curious change took place. The project was reconceptualized, most probably by Zayas; rather than being a new quintilingual Bible, it was presented as a new edition of the Complutensian Bible. In a com-

---

8. Plantin to Zayas, Antwerp, 19 December 1566, in Macías Rosendo, *Biblia Políglota*, 65–70.

9. Santes Pagnino, OP (c. 1470–1541), was a humanist biblical scholar who in 1528 published a new Latin version of the Bible directly from the Hebrew and Greek texts, the *Veteris at novi Testamenti nova translatio* (Lyon, 1528), followed by the very influential *Thesaurus linguae sanctae* (1529). His translations were characterized by his commitment to literal translation and philological exactitude. Vanderjagt, *"Ad Fontes!,"* 2:185–89.

10. For a detailed study on the documents that constitute the Antwerp Polyglot, see Sánchez Salor, "Contenido de la Biblia Políglota," 279–301.

munication Zayas sent to Arias Montano describing Plantin's project, Macías Rosendo has noted the careful additions and redactions Zayas made as he conflated Plantin's letters and translated them into Spanish. Whereas Plantin had made no references to the Complutensian Bible in his original letters to the king (or in the surviving communications with Zayas), Zayas's paraphrase begins with a statement establishing a parallel between the two Bibles and seeking to ensure that this project would measure up to the Complutensian Bible.[11] Furthermore, the supposed missive from Plantin now included a comment about the scarcity and cost of the Complutensian Bible and a promise to make the Antwerp edition cheaper. (Unfortunately, most of the printed Complutensian volume sets went down with the ship that was transporting them to Rome, making this monument of humanist scholarship very scarce and sought-after by the 1560s.) Clearly there was a concerted effort to make the Antwerp project a legacy of the Complutensian Polyglot Bible. Zayas, perhaps with Plantin's acquiescence, had anticipated how effective positioning the project in this matter could be and how many obstacles this approach would obviate. In the eyes of the Spanish court, Spanish faculties of theology, and the Inquisition, it became a Spanish product, the *Biblia Regia*, rather than a Flemish exercise in biblical scholarship.

Once Plantin's proposal had been taken up at court, the king ordered the project studied by three theologians: the Hebraist Martín Martínez Cantalapiedra (1515–79) of the University of Salamanca; Juan de Regla (1500–1574), a Hieronymite friar, delegate to the second session of the Council of Trent and former confessor of Charles V; and Arias Montano, who by then served as one of the king's chaplains. Of the three, only Arias Montano's opinion survives. It was an unconditional endorsement of the project, which he justified on several grounds.[12] First, it was the duty of a Christian monarch to safeguard and promote the study of the Word of God—a duty, the Bible states, that God will reward with many blessings. To lend some urgency to the matter, he pointed out that Francis I of France, archrival of Philip II's father, had sponsored the publication of Bibles in Latin, Hebrew, and Greek. (Their publisher was the Protestant Robert Estienne, and his "correction" of the Vulgate had come under severe criticism by ecclesiastical authorities.) The message was clear: If Philip did not underwrite this venture, other monarchs—perhaps less inclined to preserve the integrity of the Bible—would undertake the project. Philip's support not only would be a service to all Catholics, but would im-

11. Macías Rosendo, *Biblia Políglota*, 68.
12. Arias Montano presumably to Philip II, n.d., in Macías Rosendo, *Biblia Políglota*, 71–73.

pede heretics and be a great service to the universal Christian church of non-Latin-speaking peoples, maybe even to Hebrews. Arias Montano echoed the concern about the scarcity of the Complutensian Bible and how sought-after copies had become; he also framed the project as an enhanced version of the Complutensian Bible because it would include other Targumim to the Old Testament that had been omitted in the earlier work. Tellingly, he was mute on whether Pagnino's Latin translation of the Pentateuch should supersede the Vulgate. Plantin's proposal to use Santes Pagnino's Latin translation of the Hebrew Pentateuch was not without controversy. Just a few years earlier, the Council of Trent had declared that the authorized translation of the Bible was Saint Jerome's Vulgate.[13] This decree had been a blow to Christian Hebraists who recognized Pagnino's translation as a more accurate and philologically polished version.

The king then consulted with the General Council of the Inquisition in Spain. In the communications with this body Plantin's project was presented as a reprint of the Complutensian Bible, albeit with some additions.[14] They in turn instructed Arias Montano to consult with the Faculty of Theology at the University of Alcalá. Having done so during spring 1567, Arias Montano returned to court with the endorsement of both bodies. The council then recommended to the king that Arias Montano be put in charge of the project because of his "condition as a priest and a scholarly theologian, expert in Sacred Scripture and knowledgeable in biblical languages."[15] His reputation must have preceded him, since he had yet to publish any theological works.

The royal directives given to Arias Montano, however, made some significant changes to Plantin's original project.[16] Rather than using Santes Pagnino's Latin translation of the Hebrew text in the Pentateuch, the Latin text was to be that of the Vulgate, just as it had appeared in the Complutensian Polyglot. The argument for this was threefold: it was considered the most "authoritative" translation by the church; it was the most important translation and therefore should not be absent from such a project; and, for these two reasons, it deserved the prominent place alongside the Hebrew text. In addition, the project should extend the Old Testament past the Pentateuch

13. On the relevant Council of Trent decree, see chapter 5, note 11 below.

14. Philip II to the Inquisitor General (?), n.d., in Macías Rosendo, *Biblia Políglota*, 74.

15. Philip II to Arias Montano, "Instrucción a Arias Montano sobre la Biblia regia," 15 March 1568. Published as document 19 in González Carvajal, "Elogio histórico," 140–44, and in Macías Rosendo, *Biblia Políglota*, 79.

16. Macías Rosendo, *Biblia Políglota*, 76–83; Macías's excellent introduction synthesizes the complicated publication history of the Antwerp Polyglot.

(where the Complutensian Bible ended) and include the other books of the Old Testament following the Chaldean text as it had been printed in Rome and Venice. (This meant that in addition to the Targum of Onkelos for the Pentateuch, the rest of the canonical books of the Old Testament would appear with their Targumim, except for the books of Chronicles, Esdras, Nehemiah, and Daniel, where none were available.) The New Testament should also include the Syriac text as published in Vienna.[17] The Gospel of Matthew should also appear in Hebrew and the rest in Syriac, and alongside them their Latin translations. The instructions also insisted on including four dictionaries (Hebrew, Greek, Chaldean, and Syriac) to aid in the study of the Sacred Scriptures. The instructions specified that the Hebrew dictionary should be "the best available" but should not include examples, only cite where the words appear in the Bible. Finally, the project should also include the Gospel Canon Tables of Eusebius of Caesarea to aid in the study of the New Testament.

Beyond the editorial guidelines, the instructions placed on Arias Montano the duty of conveying the importance of these changes to Plantin and seeing them through to the final product. Arias Montano was to ensure that the project moved along with the utmost "diligence, care, and attention" (*diligentia, cura et attentione*) and that Plantin dedicated as many printing presses as possible to its production so it would quickly come to fruition. Furthermore, Arias Montano was to be aware of the competence and fidelity to the church of those involved in the project. The instructions ensured his close participation in the project, instructing him to correct "by your own hand [*de tu puño y letra*] the proof pages and, once corrected, to sign each of them, as he dutifully did[18] (figs. 4.1 and 4.2). In addition, Arias Montano was to send sections of the Bible to Spain as they were completed. The preface also received special mention. Arias Montano was to make sure it reflected the reasons for the project and to make note especially that the matter had been discussed and considered by "men of great prudence, knowledge and virtue."[19] Additionally, he was to send drafts of the prefaces so they could be reviewed before publication.

Arias Montano could have interpreted the role of editor as analogous to a supervisor with fiduciary duties over the monarch's investment and the theological orthodoxy of the project. But he viewed his role in a far more expansive

---

17. This is the Peshitta, or "simple" version of the Syriac New Testament, specifically *Liber sacrosancti evangelii de Iesu Christo . . . Caspar Craphtus Elvangensis suevus characteres Syros ex norici ferri acie sculpebat*. Michael Cymbermannus . . . (Vienna, 1555).

18. Macías Rosendo, *Biblia Políglota*, 83.

19. Ibid., 82.

Interp.ex Græc.lxx. GENESIS. ΓΕΝΕΣΙΣ. 47

* Et statuam pactum meũ inter me & inter te, & inter semen tuum post te in generationes suas, in fœdus sempiternum, vt sim tuus Deus, & seminis tui post te. * Et dabo tibi & semini tuo post te terrã quam habitas, omnem terram Chanaan in possessionem æternam. & ero illis in Deum. * Et dixit Deus ad Abraam: Tu autem pactum meum custodies, & semen tuum post te in generationes suas. * Et hoc pactum quod obseruabis inter me & vos, & inter semen tuum post te; circumcidetur vobis omne masculinum: * Et circuncidetis carnem præputij vestri: & erit in signũ fœderis inter me & vos. * Et infans octo dierum circuncidetur vobis, omne masculinum in generationes vestras. * Vernaculus domus tuæ, & emptitius ex omni filio alieno, qui non est de semine tuo circuncisione circuncidetur. et erit pactum meũ in carne vestra in fœdus æternum. * Et incircun-

CHALDAICAE PARAPHRASIS TRANSLATIO.

Et statuam pactum meum inter verbum meum, & inter te, & inter filios tuos post te in generationibus suis in pactum sempiternum: vt sim tibi in Deum, & filiis tuis post te. Et dabo tibi & filiis tuis post te, terram habitationis tuæ, vniuersam terram Chanaan in hæreditatem ęternã: & ero eis in Deum. Et dixit Deus Abrahæ: Tu autem pactum meum seruabis, tu & filij tui post te in generationibus suis. Hoc est pactum meum quod seruabitis inter verbum meum & inter vos, & inter filios tuos post te: vt circuncidatur in vobis omnis masculus. Et circuncidetis carnem pręputij vestri: & erit in signum fœderis inter verbum meum, & inter vos. Et filius octo dierum circuncidetur in vobis: omnis masculus in generationibus vestris; tam natus domi quàm emptus argento ab omnibus filiis populorum qui non sunt de filiis tuis. Circuncisione circuncidetur tam natus domi tuæ quàm emptus argento tuo, & erit pactum meum in carne vestra in pactum æternum. Et masculus habens præputium qui non circunciderit carnem præputij sui, peribit homo ille de populo suo: quoniam pactum meum irritum fecit. Et dixit Deus ad Abraham: Sarai vxorem tuam non vocabis nomen eius Sarai: sed Sara est nomen eius. Et benedicam ei, & insuper dabo tibi ex ea filium, & benedicam ei, & erit in gentes, & reges qui dominabuntur populis, ex ea erunt. Et cecidit Abraham super faciem suam, & gauisus est. Et dixit in corde suo: Nunquid homini centum annorum nascetur filius? & si Sara nonagenaria pariet? Et dixit Abraham coram Deo: Vtinam Ismael permaneat coram te. Et dixit deus: In veritate Sara vxor tua pariet tibi filium, & vocabis nomen eius Isaac; & statuam pactum meum cum eo in pactum æternum, & filiis suis post eum. Et super Ismael suscepi orationem tuam: ecce benedixi ei, & crescere eum faciam, & multiplicabo eum valde nimis: duodecim principes generabit, & ponam eum in gentem magnam. Et pactum meum statuam cum Isaac; quem pariet tibi Sara in tempore isto anno altero. Et cùm consummasset loqui cum eo; ascendit gloria Domini desuper Abraham.

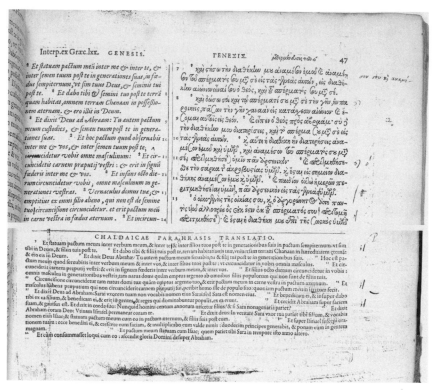

FIGURE 4.1. Details from a proof sheet of the *Antwerp Polyglot*. Plantin-Moretus Museum (B 948, 1:47).

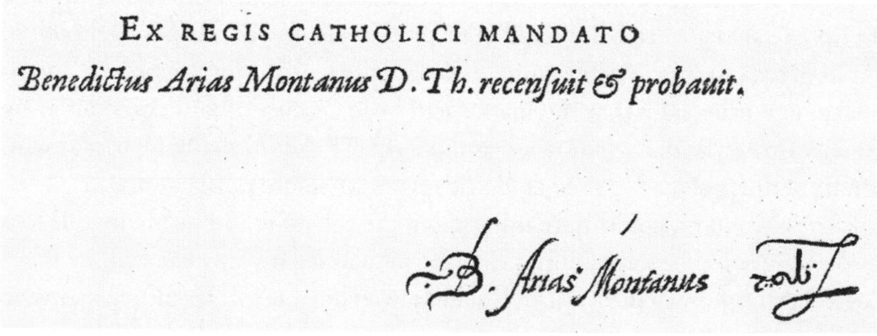

EX REGIS CATHOLICI MANDATO
Benedictus Arias Montanus D. Th. recensuit & probauit.
B. Arias Montanus

FIGURE 4.2. Arias Montano's signature at the end of the book of Genesis. *Antwerp Polyglot*, 1:183. Special Collections, Sheridan Libraries, Johns Hopkins University.

way. He soon saw himself as also a contributor to scholarly aspects of the enterprise. His writing the prefaces ensured that it would be his vision, his premises, and his interpretation that would frame the project. By writing the interpretative essays in the *Apparatus*, he would convey a particular approach to biblical interpretation. And by editing Pagnino's Latin version of the Hebrew Bible, he would provide the project with an alternative version of the Sacred Scriptures of the Old Testament that reflected his own understanding of them.

PLANTIN'S TEAM AND THEIR BIBLICAL TRADITIONS

By the time Arias Montano arrived in Antwerp in May 1568, Plantin had been working on the project for some time with several scholars he had recruited as contributors. To carry out such a vast project of Orientalist erudition, Plantin turned to the leading experts of the time. It was a varied group of theologians and humanist scholars representing the spectrum of heterodoxy, ranging from wholly committed Protestants and other sectarians to messianic kabbalists and orthodox Catholics.[20] Among the principal contributors was Andreas Masius (ca. 1514–73), an early proponent of the project. He wrote the Syriac grammar and dictionary that appear in volume 6 of the Antwerp Polyglot. He also lent the project a manuscript of the Chaldean version of the books of Joseph, Judges, Samuel, and Kings he had purchased in Rome. These had only been partially translated into Latin by Alfonso de Zamora, so Arias Montano had to translate the rest.[21] Guy Lefèvre de la Boderie (1541–98), a student of Guillaume Postel (1510–81), acted as editor of the Syriac sections. He traveled to Antwerp at the insistence of Pope Pius IV and spent three years there copyediting the Syriac sections of the Bible and preparing the Syriac-Chaldean dictionary.[22] Specifically, he edited the Latin translation and a Hebrew transliteration of the Syriac New Testament supplied by Postel. With the help of his brother Nicholas, he edited the Hebrew, Chaldean, and Syriac texts and contributed to the editing of Santes Pagnino's Old Testament translation. Their association with Postel was a problem, something Arias Montano and Plantin noted and Lefèvre de la Boderie acknowledged.[23] (In 1555 Postel had been imprisoned as a lunatic by the Venetian Inquisition for his messianic and apocalyptic ramblings. Once

20. For a detailed survey of the religious commitments of the contributors, see Wilkinson, *Kabbalistic Scholars of the Antwerp Polyglot Bible*.

21. Macías Rosendo, *Biblia Políglota*, 96, 150.

22. Ibid., 99n63.

23. Ibid., 164–67.

back in Paris in 1562 he was arrested again and imprisoned at the monastery of Saint Martin des Champs until his death.)[24] Working hand in hand with Arias Montano on the copyediting was Plantin's son-in-law Franciscus Raphelengius (1539–97), a scholar of Greek and Hebrew who had studied in Paris with Jean Mercier. He also contributed a treatise on the Chaldean paraphrase to the *Apparatus*, the result of his textual corrections to the Chaldean paraphrases. The bulk of his contributions consisted of copyediting Pagnino's Hebrew dictionary and grammar as well as Pagnino's translation of the Hebrew Old Testament. In these tasks he had the assistance of Cornelius Kiel, Antonius Spitaels, and Theodore Kemp.

The wide range of confessions and exegetical traditions of the contributors to the Antwerp Polyglot has given rise to very varied historical interpretations of how they shaped the project. A particular preoccupation of the project's historiography has been pinpointing Arias Montano's position in the spectrum of heterodoxy this group represented. For many of his biographers it is, in fact, the driving historical question. Ben Rekers found the key to Arias Montano's attitude toward Protestantism in the Plantin family's affiliation with the Family of Love.[25] Rekers presented Arias Montano as a "secret" follower of the sect, to the point of alleging that he installed a cell of the sect among the Hieronymites at the monastery of El Escorial, the very home of Philip II and the bastion of the Counter-Reformation. He also identified the Orientalist Postel and his former students, the Lefèvre de la Boderie brothers, as followers of the Family of Love.[26] Contemporary Spanish historiography dismisses these allegations, largely because of the tenuous evidence Rekers offered in establishing Arias Montano's membership in the sect. Furthermore, contrary to Rekers's intimations, the careful studies of Arias Montano's work undertaken in conjunction with their recent translation reveal his profound commitment to Catholicism. Rekers seems to have interpreted Arias Montano's quietist attitude and inclination toward confessional reconciliation as signs of a simmering heterodoxy.[27]

Having moved beyond Rekers's thesis, recent scholarship has placed the

24. On Postel, see Bouwsma, *Concordia Mundi*; Secret, "Emithologie de Guillaume Postel," 381–437; Secret, "Guy Le Fèvre de la Boderie," 245–57; and Kuntz, "Guillaume Postel."

25. On the Family of Love and its spiritual leader, Henry Jansen Barrefelt, or Hiël, see Hamilton, "Hiël and the Hiëlists"; Hamilton, "From Familism to Pietism"; Hamilton, *Family of Love*; and Rekers, *Arias Montano* (1972), 70–104.

26. Rekers, *Arias Montano* (1972), 45.

27. Bécares Botas, *Arias Montano y Plantino*, 30, 57–59. See also Wilkinson, *Kabbalistic Scholars*, 28–31.

Antwerp Polyglot in a sort of academic tug-of-war. One faction, mostly Spanish, sees the *Biblia Regia* as the culmination of the tradition of Spanish humanism with its emphasis on multilingual and philologically driven exegesis. The other camp interprets the project as stemming from northern European scholars, both in deed and in spirit, while acknowledging the Spanish contribution as mostly financial and fiduciary. Robert Wilkinson, for example, sees the Antwerp Polyglot as the production of a group of Christian kabbalists and Orientalists from northern Europe—led by Postel and Masius—whom he considers the true drivers behind the project.[28] For him the contributions of the Orientalists Masius and Guy Lefèvre de la Boderie are clear indications that the Antwerp Polyglot project was informed by scholars heavily influenced by Kabbalah and its associated mystical interpretation of scripture. These scholars placed particular emphasis on the Syriac New Testament, which they thought contained providentialist and eschatological messages that had to be interpreted using kabbalist methodology. By extension, Wilkinson diminishes Arias Montano's intellectual contributions to the project but portrays him as sympathetic with their approach, if not complicit.

A third perspective, more in keeping with Arias Montano's attitude, sees the project as unified by a common sense of purpose and methodology. The purpose of the enterprise was to bring to light a Bible reflecting the best that humanist scholarship had to offer. The steps those involved in the project took testify to their commitment to a humanistic approach to biblical scholarship. They sought the oldest and most reputed texts of the Sacred Scriptures and brought them from the Vatican and the University of Alcalá. All the reference material necessary for the careful, informed study of the whole Bible was provided in the *Apparatus*. Grammars and lexicons were made available so that a curious student could engage with the Bible directly. With the treatises in the *Apparatus*, Arias Montano made sure modern readers understood the cultural and historical context from which the Bible had sprung. The team Plantin assembled shared a commitment to the linguistic depuration of ancient texts and to studying the Bible from philological, cultural, and historical perspectives. Where they differed radically was on how they interpreted the results of their philological investigations. Regardless of their methodological coherence, it would have been impossible for this group to produce a unitary exegesis,

28. This author claims that the kabbalistic material appears in the Syriac New Testament while Masius's contributions, a Syriac grammar and a lexicon, are "free from mystical or kabbalistic content." In contrast, he describes Lefèvre de la Boderie's Dictionarium syro–chaldaicum as full of kabbalistic content: Wilkinson, *Kabbalistic Scholars*, 28–32, 81–82, 82n15.

something they never attempted and that was far beyond the scope of what Plantin proposed, Philip II agreed to, and Arias Montano saw come to fruition. For whereas some might see irreparable fissures in the spectrum of confessions that came together at Plantin's press, Arias Montano prudently saw only an unsurpassable concatenation of skills and agreement in philological method.

The Prefaces

Arias Montano's balancing act as editor of the Antwerp Polyglot is most apparent in the prefaces he wrote for most of the sections of the project. The inquisitorial officials involved in endorsing the project—whether in Spain, Paris, or Louvain—realized it had to be represented to the public in a way that allayed any concerns about the integrity of the biblical text and, furthermore, attested that the interpretive apparatus supplied was within Catholic orthodoxy. It was up to Arias Montano to convey this in the prefaces; yet, as we will see, he still managed to transmit his personal vision not only of the polyglot project, but also of the future avenues of investigation made possible by the sources and tools at hand. The prefaces as a whole reveal that since the days of the Antwerp Polyglot Arias Montano was committed to some the central tenets underpinning the natural philosophy he developed years later in the *Magnum opus*.

All the prefaces, as stipulated in the guidelines presented to Arias Montano at the outset, were carefully reviewed and amended before publication, with the first general preface getting special attention.[29] Although the final censorial responsibility for the project was in the hands of the Faculty of Theology at the University of Louvain, the king instructed three professors from Alcalá to supervise the project: Ambrosio de Morales, fray Luis de Estrada, and Pedro Serrano, the last a former teacher of Arias Montano at Alcalá.[30] The last two were thorough in their critiques of his prefaces.[31] Beginning in March 1569, a litany of concerns arose as they reviewed several versions of the general preface: the reasons the king sponsored a multilingual Bible had to be (more) clearly defined; the Bible should not be called the *Biblia Philippica*; for political and theological reasons, the pope's support should be sought and noted in the preface; the final determination on the preface should be delayed until the end

29. Macías Rosendo, *Biblia Políglota*, documents 48–52; Arias Montano, *Prefacios*, lxxviii–lxxxv.

30. Domínguez Domínguez, *Arias Montano y sus maestros*, 42–49.

31. Their comments and those of a number of other influential individuals asked to comment on the prefaces can be read in documents 14, 17, 18, 32–46, and 50 in Macías Rosendo, *Biblia Políglota*.

of the project, and it needed to state clearly that the purpose of the enterprise was to correct the corruptions that Jews and heretics had introduced into the Sacred Scriptures; finally, the preface was too long and florid and should be revised. Arias Montano patiently addressed the concerns, eventually gathering a number of positive opinions and remitting a new version of the general preface to the king so that the final decision rested with the royal authority.

The long version of the general preface prevailed. The reasons Arias Montano refused to pare it down become clear when we consider the arguments that he presented in light of the exegetical ambitions he nurtured and that would serve as the framework on which he would construct the reform of natural philosophy. The very first idea a reader of the Antwerp Polyglot encountered in the general preface was the importance of knowing yourself and being cognizant of the purpose for which human beings were created: in God's design everything in the world was given a purpose (*finis*), a distinct genus, and a way of being that was suitable to its nature.[32] Arias Montano followed this with an elegant statement of the argument for design: We can survey all creation, from the simplest elements to the most complex being without finding a single thing in nature that does not manifestly argue and prove that what happens in them is not alien to the reason, order, and will (*ratio, ordo, et consilium*) of a mind that governs all, which we call God. Acknowledging this purposefulness of nature, he continued, is simply a preamble to discovering the ultimate purpose of humanity: attaining knowledge of God and partaking in the promised salvation. As a consequence of the Fall, he explained, man emerged from Paradise without the gifts God had given him and that had raised his awareness beyond his natural abilities. Worse yet, human beings had lost their divine guide and could rely only on reason, experience, and hard work to understand the world around them. The subsequent "forgetting" of prelapsarian knowledge of nature led to the invention of philosophy and the concomitant confusion caused by philosophers who thought they could understand nature without the guidance of God. The resulting religious and philosophical sects left humanity in great confusion and enveloped in a fog of errors, fables, and idolatries. The Bible was a conduit, a light of God if you will, that could—if studied carefully—guide humanity out of this confusion and error.

Not before laying out the historical narrative of the dire straits in which humanity finds itself—almost halfway through the preface—did Arias Montano arrive at the points the censors had insisted must be front and center: the need

32. Arias Montano, *Prefacios*, 2.

to study the Sacred Scriptures using ancient languages (Hebrew, Chaldean, Greek, and Syriac); the way a renewed emphasis on the study of scripture could counteract the errors of their time (a veiled reference to Protestantism); the way these enemies had corrupted the Bible (perhaps an ambiguous reference to Protestants and Jews); how this edition sought to redress this situation; and finally, the reasons the king had undertaken the project. Arias Montano relegated to a second preface the acknowledgments to the contributors and a survey of the work's contents in which he emphasized that the project was meant, in its design, intent, and completion, to encourage the study of the Sacred Scriptures from the purest original sources (*purissimae fontes*) and that every tool had been provided to make this possible.

In the general preface Arias Montano put his intellectual commitments on full display. His argument was structured as a logical sequence of what a scholarly audience would consider universal truths, followed by a description of a historical event that allowed the argument to pivot to specific instances that had led to the current state of religion (and philosophy). Building on the argument from design for the existence of God, taken from the natural theological tradition, he singled out the purposefulness of nature, and particularly that of humanity. Adam's fall was the historical event that set the course of humanity astray, a course that could be righted only by rediscovering from the Sacred Scriptures how God intended humans to live in concert with their own nature and that of the world. With the implicit authority of the monarch and, he hoped, the pope, he had presented the argument for the study of the Bible in its original languages as a viable way of returning theology and philosophy to the single stream flowing from the biblical fountain.

Arias Montano's Biblical Translations

Once the general preface had been approved, the censors' concerns, some of which Arias Montano managed to defuse early on, shifted to the rest of the project. Serrano in particular did not want non-Vulgate Latin translations to be part of the project if they were in any way presented as sanctioned versions of the Bible; should they be included, they should be framed as as only commentaries or reference material.[33] Despite his concern, two new Latin versions formed part of the Antwerp Polyglot: Pagnino's Latin translation of the Hebrew Old Testament, with corrections by Arias Montano and his collabora-

---

33. Pedro Serrano to Gabriel de Zayas, Alcalá, 6 November 1569, in, Macías Rosendo, *Biblia Políglota*, 154–55.

tors, and Arias Montano's Latin translation of the Greek New Testament.[34] In their prefaces Arias Montano presented them, especially the edited Pagnino interlineal Hebrew-Latin translation, as essential to studying the Old Testament from its original sources. He knew this decision was controversial, and he secured two sets of approvals. In addition to approvals from the censors of Louvain—Augustin Hunnæus (or Huneo), Cornelius Reyneri Goudanus, and Johannes Harlemius[35]—the edition was also approved in 1569 by theologians from the University of Paris—H. Nebrisius, Jean Faber, I. Benedictus, Dequictebeufius (Jean Cinqarbres), Genebrardus (Gilbert Génébrard), and none other than Guillaume Postel.

Arias Montano, with the assistance of Raphelengius and Guy Lefèvre de la Boderie, extensively edited Pagnino's Latin translation, indicating changes in italics and showing the superseded text in the margins along with the Hebrew roots of key words. Although his Spanish censors noticed the absence of the Hebrew roots in the main version of the Old Testament and expressed their wish that these be included as they had been in the margins of the *Biblia complutensis*, Arias Montano included roots only in the Pagnino version, and then only those relevant to his editing.[36] By doing this he effectively shifted the scholarly focus of the translation exercise from the Vulgate to the volume containing the Pagnino/Arias Montano version. For most scholars aspiring to become Christian Hebraists, wanting to engage directly with the *hebraica veritas*, or even just willing to make the effort to read the Bible in Hebrew, the absence of the roots in the main text could be cumbersome. They would inevitably have to turn to the seventh volume and the Pagnino/Arias Montano translation (fig. 4.3). In fact, the demand for this volume was so great that Plantin reprinted it several times.

The primacy of the Hebrew language was the thrust of Arias Montano's

34. The Old Testament of the first four volumes of the Antwerp Polyglot used the same Hebrew version as the Complutensian Bible, but now fully vocalized and accented. It was accompanied by the Latin Vulgate translation and the Greek Septuagint (or version of the LXX), with Arias Montano's Latin translation. The seventh volume contained the same Hebrew text, but now with Pagnino's interlineal Latin version.

35. Lanoye, "Benito Arias Montano (1527–1598)," 23–44.

36. The marginalia in the Pagnino Old Testament included in the Antwerp Polyglot obeyed the following convention: plain Latin fonts indicate Pagnino's original version, which was superseded by Arias Montano in the interlineal translation; italic Latin fonts indicate an alternative version of the Pagnino text as suggested by Arias Montano; Hebrew fonts in capital letters indicate patterns or roots of verbs; and Hebrew fonts in lowercase indicate variants that exist among Hebrew Bibles.

preface to the edited version of Pagnino's interlineal translation of the Old Testament into Latin because of Hebrew's unrivaled position as the language instituted by God (*a Deo ipso instituta*) and its status as the archetype of all human languages.[37] Hebrew was therefore immensely fecund, even containing within itself a double meaning (*duplex sensus*) that, when enriched with literary adornments and allegories, resulted in its having an "almost infinite" number of meanings. Throughout history, Arias Montano argued, Hebrew had been placed in a sort of "sacred citadel" (*sacra arx*) where it had been preserved and venerated for its antiquity. It was as old as the world itself, Moses having used it to name the most noble parts of the world: heaven, earth, light. It was the vehicle through which God communicated his ineffable name. Even the Hebrew characters used to express that name, Arias Montano explained (following the Jewish tradition), held so much significance that their true meaning could not be expressed in the letters of other languages.

To convey the profound semiotic value of the Hebrew language, Arias Montano described the relationship between it and the first human (and thus with Moses as the first interpreter) as similar to the relationship people have with their native tongues. It is the language in which people can best express their thoughts; thus readers are led to surmise that the Hebrew text is closest to what Moses meant to convey. Furthermore, Arias Montano argued, some languages lack the proper vocabulary or terms for certain arts and disciplines that were invented and initially described in a different language. This often forces those interested in mastering those arts and disciplines to learn that other language, since the best way to learn about foreign things is to understand the language that contains this knowledge. At this point Arias Montano made one of his rare calls on authority in support of the study of Hebrew; Saints Augustine and Jerome turned to Hebrew to recover the true meaning of scripture through the *hebraica veritas*.

The preface to Pagnino's interlineal translation was also an opportunity for Arias Montano to discuss the premises that guided his revisions of the Hebrew-to-Latin translation. Hebrew, he explained, is peculiar in that many words and phrases are ambiguous but cannot always be translated to convey the same level of ambiguity. And yet, because the Holy Spirit had chosen these ambiguous words, it was the exegete's responsibility to take into account all possible meanings and include them in a proper translation. The inverse was also true; something expressed clearly in Hebrew might unwittingly be translated ambiguously into another language, with subsequent translations

37. AntPoly, 7:†1r-3v, and Arias Montano, *Prefacios*, 82–101.

only increasing the confusion. The best way around these issues was to make a Latin translation as "Hebrew" as possible by setting aside Latin style and conventions and translating word by word. This was precisely what Pagnino had tried to do.[38] Yet even Pagnino's laudable effort was not without fault. Critics noted that in some instances he had not been accurate or had offered translations that deviated from the mysteries of the Catholic faith, corrections Pagnino had humbly accepted. Arias Montano presented the current edition as a continuation of this effort. The editors examined each phrase, each verb with its tenses, and all parts of the sentences, seeking the most suitable translation "in light of the property of the Latin language." Arias Montano explained further that this edition should be studied using the other reference material included in the *Apparatus*, particularly the *De arcano sermone* and the treatise on idiomatic phrases in Hebrew, the *Communes et familiares hebraicae lingua idiotismi* in volume 7 of the Antwerp Polyglot.

Including edited versions of Pagnino's Old Testament was very important to Arias Montano; among other things, it was an opportunity to define a biblical text on which to build his natural philosophy. The editing was a tremendous undertaking requiring meticulous effort by him and his collaborators. Some examples from the book of Genesis might illustrate what the translation achieved and show how he navigated the lexical maze between the Hebrew *Beresith*, the Vulgate, and Pagnino's edited interlineal translation. The stated intention was to achieve linguistic precision, whether this precision was found in preserving the ambiguity of the Hebrew original or in finding a word or phrase in Latin that exactly conveyed the Hebrew meaning. The first instance occurs in the second sentence of the Bible (fig. 4.3). The interlineal Latin translation reads, "In principio creavit DEUS caelos & terram. Et terra erat solitudo & inanitas: & *caligo super facies abyssi*: & spiritus DEI *motabat* super facies aquarum."[39] Arias Montano indicated his changes in italics in the interlineal text and moved Pagnino's original to the margin. In this case, "*caligo super facies*" superseded "tenebrae in superficie," and "*motabat*"[40] replaced "flabat." The passage was also annotated with three Hebrew roots, none of which explains his change.[41]

38. Arias Montano, *Prefacios*, 93.

39. The Vulgate translated Genesis 1:1–2 as, "In principio creavit Deus caelum et terram. Terra autem erat inanis et vacua: et tenebrae super faciem abyssi: et spiritus Dei ferebatur super aquas."

40. *Motabat* is a frequentative of moto in the third-person singular imperfect.

41. הָיָה (*hāyâ*): *erat* or *esse*; b תהה (*thh*): *solitudo* or *desolation*; c בהה (*bhh*): *inanitas*, vacuous or void.

The first change hinged on adding the word *caligo*[42] to what had been a purely descriptive clause indicating simply that there was darkness over the surface of the abyss. To understand the reasons for adding *caligo*, readers would have had to identify the root of the Hebrew word חֹשֶׁךְ (*hōšek*) as חָשַׁךְ (*hāšak*), then look up this primitive root in the thesaurus that accompanied the Polyglot.[43] There they would have found that the root means to shroud or darken something, and also that the dictionary entry lists חֹשֶׁךְ (*hšek*) as darkness (*obscuritas, vel tenebrae*). It seems that Arias Montano located the correct interpretation by going with the meaning suggested by the root as something (e.g., fog or mist) that could darken the region above waters. In Latin *caligo* can be both a noun and a verb, but here it can grammatically be only a noun meaning a thick air, mist, vapor, or fog. Consequently the phrase was presented as meaning that some sort of cloudy atmosphere was above (*super*) the shape (*facies*) of the abyss.

The exercise would have been similar in explaining the use of *motabat*. In this case the word to be interpreted was רָחַף (*rāhap*), itself a primitive root. The Polyglot's thesaurus defined the particular usage in Genesis 1:2 as "to move itself" but also noted that Saint Jerome had preferred *ferebatur*, while the Targum preferred *sufflabat* and Rabbi Salomon had chosen *incubabat*.[44] Here again, Arias Montano turned to the meaning supplied by the root instead of adjudicating the meaning based on authorities' interpretations. Replacing *flabat* with *motabat* changed the Pagnino translation from "the spirit of God blew over the surface of the water" to "the spirit of God continuously moved over the surface of the water." The change was subtle, but it suggested a desire to make the sentence describe a turbulent natural process or agitation. Having earlier replaced a word that referred to a quality of darkness (*tenebrae*) with a word that indicated something with a material nature, such as fog or mist changed the meaning of the verse significantly. In Arias Montano's translation the space above the waters now contained matter in the form of some cloudy substance. This matter was operated upon continuously by the moving agency of the Spirit of God or Elohim. We will see in chapter 8 that in the *History of*

---

42. *Cālīgo*, ĭnis, f.: a thick air, mist, vapor, fog.

43. "חָשַׁךְ Tenebrescere, obscurari: *Isa. 13, 10. Hiph. intransitivè et transitivè accipitur iuxta exigentiam loci: ut Ham. 5, 8. et 8, 9. Ier. 13, 15. etc.* חֹשֶׁךְ Obscuritas, *vel* tenebrae: *Gen. 1, 2. Exod. 10, 21.*": Raphelengius, "Thesauri hebraicae linguae," in AntPoly, 6:34.

44. "רָחַף Movere, *aut* moveri. *Ut Ierem. 23, 9.* רָחֲפוּ Commota sunt, *vel* moverunt se. *Sic in Piel, Gen. 1, 2.* מְרַחֶפֶת Movebat se. *Hieron.* Ferebatur. *Al.* Sufflabat. *R. Salo.* Incubabat. ut Deut. 32,11. יְרַחֵף Incubabit, *pro* incubat. *Tharg.* Tegmen facit.": Raphelengius, "Thesauri hebraicae linguae," 6:119.

Nature Arias Montano interpreted this as an agent (a moving Spirit of God) acting upon matter (fog, mist) to initiate the process of creation.

Arias Montano also did significant editing of the biblical passage on the separation of the waters (Gen. 1:6–7). His text read, "Sit *expansio* in medio aquarum, & sit *dividens inter aquas ad aquas*. Et fecit Deus *expansionem*, & divisit inter aquas quae (erant) subter expansionem &, inter aquas quae (erant) super expansionem" (italics show Arias Montano's changes). He made three significant changes: replacing "firmamentum" (used in the Vulgate and in Pagnino) with *expansio*, replacing Pagnino's "dividat aquas ab aquis" with *dividens inter aquas ad aquas*, and adding the verb *erant* ("were") as needed. The choice of *expansio* was determined by the root of רָקִיעַ (*rāqîaʿ*), that is, רָקַע (*rāqaʿ*). The Polyglot's Hebrew thesaurus gives its meaning as "to expand" or "to extend," so that "heaven" or "firmament" is the name given to something "because it is expansive and stretched out over the land."[45] In Arias Montano's translation God's command thus became: "Let there be expanse in the middle of the water and let there be dividing the waters from the waters. And God made the expanse and divided between the waters which were above the expansion and the waters which were under the expansion." The presence of this *expansio* was now divorced from the act of dividing (which in this version was done by divine command), although it still could be understood as something interposed between the two types of waters. It simply became something that carries out the action of separating but no longer bears the suggestion of being something firm, let alone solid, a characteristic that in his estimation was not suggested by the root רָקַע (*rāqaʿ*). In this case Arias Montano chose to retain the ambiguity of the original as suggested by the Hebrew root of the word while increasing the descriptive precision of what this *expansio* was doing during the Creation: dividing waters.

From these brief examples it is apparent that Arias Montano's exegetical approach relied extensively on returning the meaning of words to the definitions most closely associated with their Hebrew roots, summarily dismissing definitions accrued over many editions of the Santes Pagnino/David Kimhi *Thesaurus*, especially if these definitions had been contributed by later editors of this very popular Hebrew-Latin dictionary. By doing so, Arias Montano

---

45. "רָקַע Expandere, extendere. *unde in Benom*, רֹקַע Extendens: *Isa. 42, 5. et Psal. 44, 24. Sic Ezech. 6, 11. in Imperat.* וּרְקַע Et extende pede tuo. *Hieron*. Et allide. *Al.* Percute pede tuo. *Thargum*, Conculca . . . רָקִיע Caelum *sive* firmamentum *vocatur* Expansio, *eo quòd sit expansum et extensum super terram: Genes. 1, 6. Et in regimine* רְקִיע: *versu 14.*": Raphelengius, "Thesauri hebraicae linguae," 6:123.

was adhering to the belief that the true meaning of Hebrew, and therefore its most ancient meaning, was encoded in the root of a word. This would be one of his guiding methodological premises for interpreting the events of the book of Genesis in his *History of Nature*. It was a message that was not lost on attentive readers of the *Apparatus* of the Antwerp Polyglot. He was not trying to disguise his approach at all; in fact, he made it crystal clear in his treatise *De arcano sermone*.

THE *APPARATUS* AND THE *DE ARCANO SERMONE*

By late 1569 Arias Montano was well aware that the versions of the Bible and the reference material he was assembling would have plenty of critics, and he tried to keep the particulars of the *Apparatus* from his Spanish overseers. He explained to the royal secretary Zayas that the *Apparatus* had to be examined only in its totality, so that none of those who wanted to cross out things solely from jealousy would find reason to do so.[46] In contrast, the official censors from the University of Louvain, Augustin Hunnæus and Cornelius Reyneri Goudanus, endorsed the inclusion of the Pagnino/Arias Montano Latin translation as a separate volume, even suggesting that those who could not afford the first five volumes could buy this one volume and have access to the Hebrew and Greek versions of the Old and New Testaments.[47] They were careful, however, to make it clear they were not endorsing Arias Montano's translation, but only certifying that nothing appeared in the translation against faith or good customs.[48] Meanwhile, in Spain the rumor circulated that the Vulgate had been superseded by Pagnino's version, and as fray Luis de León reported to Arias Montano, some theologians in Salamanca, led by León de Castro, a professor of Greek, had begun to criticize him and his stewardship of the project.[49] Arias Montano persisted, and as these letters reached him, he hurried to put the finishing touches on the most controversial treatise in the *Apparatus*.

The *De arcano sermone* poses as a glossary of biblical terms, but it is much more than that. It is informed by the notion that biblical Hebrew—properly deciphered—reveals the arcane and occult properties of things and their natures and forces. As Arias Montano explained in the prefatory letter to the

46. Arias Montano to Gabriel de Zayas, Antwerp, 23 December 1569, in Macías Rosendo, *Biblia Políglota*, 169.

47. Censors of Louvain to Arias Montano, Leuven, 22 February 1570. In ibid., 176–79.

48. Censors of Louvain to C. Plantin, Leuven, 20 May 1571, in ibid., 252–55.

49. Fray Luis de León to Arias Montano, Salamanca, 28 October 1570, in ibid., 242–44.

"studious reader," the publication of the Antwerp Polyglot had forced him to accelerate a portion of another large project he had been working on in order to prepare the *De arcano sermone*.[50] In this other "greater work" (*maius opus*)—the first reference to a plan to write the *Anima*, *Corpus*, and *Vestis* trilogy—he hoped to elucidate the nature of all things using the Sacred Scriptures. The treatise at hand was meant to provide an additional tool for the personal interpretation of scripture. Arias Montano explained that the book/s title, *Liber Ioseph, sive De arcano sermone* (The book of Joseph, or On the arcane language), recalled the figure of Joseph, son of Jacob, who had been granted the gift of interpreting dreams and explaining them to others (Genesis 37–50). The cover page also included a note from the publisher (Plantin) explaining that, in addition to eleven thousand biblical references, with this treatise (and the following one, *De actione*) readers would be furnished with a "complete commentary of the Sacred Scriptures."

Arias Montano also made it clear in the prefatory letter that interpreting the Sacred Scriptures requires some command of the Hebrew language. Thankfully, he explained, this language has been transmitted through the ages with great care, so the precepts of its grammar and other properties can be acquired by someone knowledgeable in Latin. Furthermore, he noted, the Hebrew of the Bible preserves the true meaning of words, especially the names of things (*nomina rerum*), because these were bestowed "by the diligent and accurate observation of nature" (*ex diligenti atque accurata naturae observatione*).[51] Understanding the true meaning of these Hebrew words was the key to understanding the meaning of divine oracles, expressed as they had been in a language that represents the thing itself, its nature and properties. Far from only capturing the literal meaning of something as defined by its Hebrew etymology, the *De arcano sermone* recognized the symbolic capacity of language to construct layers of meaning by means of associations beyond the purely literal. These are also part of the definitions Arias Montano provided in the treatise.[52]

Arias Montano approached this project first as an exercise aiming to recover a forgotten lexicon from the darkest reaches of history. He compared his work with that of other antiquarians, such as Horapollo and subsequent authors who had tried to decipher Egyptian hieroglyphs; those efforts, although

50. For a synthetic discussion, see Fernández Marcos, "Lenguaje arcano y lenguaje del cuerpo," 69–74, and Fernández López, "Exégesis, erudición y fuentes en el *Apparatus* de la *Biblia Regia*.": Arias Montano, *Libro de José*, 91, 403, [fol. 1].

51. Arias Montano, *Libro de José*, 91, 403, [fol. 1].

52. Perea Siller, "Capacidad referencial," 32–37.

valuable, had contributed little to the study of the Bible.[53] In contrast to other dictionaries or reference works, the *De arcano sermone* sought not so much to give translations of specific words as to furnish a description of a thing's properties, nature, and essence (*proprietates, natura, ac vis*). From these he then drew the arcane and latent (*arcana et latens*) meaning of words as embedded in the biblical Hebrew lexicon. He made it clear that he had not relied on other authors but had determined these meanings solely from his own careful study of how words are used in different passages of scripture, and also by carefully observing things and how they are ordered in nature. This first level of interpretation led to other levels that relied on the referential capacity of language to recall a broad range of associations, some obvious and others far more mysterious and arcane. To attentive readers, as Fernández Marcos has eloquently described it, these levels of meanings gradually reveal themselves like the unfolding of a pleated fan.[54]

How was the *De arcano sermone* to be used? Arias Montano explained: Upon encountering a word in the Sacred Scriptures that is used in an unusual context—say, "lion," where the context suggests that the word is not signifying the animal—he advised readers to consult this treatise and exercise their judgment when choosing one of the meanings offered, so that the selection fits the context as suggested by the passages preceding and following it. His seemingly straightforward approach hid tremendous interpretive fluidity, especially since it was conceivable that the very meaning of a Hebrew word was imprecise and ultimately contingent on context. Biblical interpretations resulting from this approach could therefore be equally varied. Seemingly advocating an astonishing level of interpretive independence, Arias Montano placed the agency of exegesis on what he presumed were well-prepared and diligent readers willing to do the hard work of studying the Bible in its original languages.

The *De arcano sermone* opens with three short essays explaining the basis for the lexical study that follows. In the first, *De divisione rerum ex quibus arcanus instituitur sermo, ac symbola petuntur* (On the division of things from which the arcane language is constructed and its symbols are sought), Arias Montano established an ontological division of all the things that are

---

53. The reference is to the work reputedly written by Horapollo Niliacus and translated into Greek in 1505. Subsequent Latin translations during the sixteenth century made Egyptian emblematics popular, with editions by Trebazio and later interpretations by Jean–Pierre Valerian. Henricus Glareanus (Heinrich Loris, 1488–1563) is also mentioned, but it is not clear that he ever published a book on hieroglyphs. He did publish books on some of the topics featured in the *Apparatus*, such as Roman weights and measures, biblical chronology, and geography.

54. Fernández Marcos, "Lenguaje arcano y lenguaje del cuerpo," 70.

(or might be) in the world. He set forth a logical structure for how material and spiritual entities were organized and related to one another. He explained that the human mind, given its limitations, perceives everything in the world as belonging to one of two kinds, either "things" or "actions." The words that refer to "things" are the subject of the *De arcano sermone*, while "actions" are the subject of the following treatise, *Liber Ieremiae, sive De actione*. Things, in turn, can be of two kinds: those that have no author—God sits alone in this category—and those created by God. Within the realm of created things Arias Montano identifies five further divisions that together encompass the lexical dimensions of the sacred language.[55] These are place, time, duration, things that are "in themselves" (*sunt per se*), and things that are "in others or appear in others" (*insunt vel adsunt*) and can be considered ornaments of the things that are "in themselves." The relation between the last two categories is established by how things that fall within the class "in others" can be of utility or used positively or negatively by the things that are "in themselves." He further subdivided the "*sunt*" category into two categories according to the incorporeity or corporality of each: *spiritus* and *corpus* (fig. 4.4 shows my schematic interpretation).

It was this section of the *De arcano sermone* that the censors at Louvain found most puzzling. It was indeed an unconventional classification of how things are signified in the world, particularly as it pertained to the category "created things." A reader familiar with Aristotle's *Categories* might have tried to map Arias Montano's scheme onto the Stagerite's but would have been puzzled first by the division that placed "actions" at the same genus level as "things."[56] (In the Aristotelian scheme, the four categories indicating actions ["being in a position," "having," "acting," and "being acted upon"] share the same level as "substance.") The category of "created things" presented some familiar ground: "time," "duration," and "place" mapped neatly onto three of the ten Aristotelian logical categories, but *sunt per se* and *insunt* do not so easily match up with Aristotelian categories. The Aristotelian category "substance" might correspond to *sunt per se*, but this would require placing the Aristotelian categories "quality" and "relatives" into the *insunt* basket. Historians of philosophy have long noted the debate surrounding the *Categories*: Did Aristotle intend them to have metaphysical valence, or were they related to his

55. Arias Montano, *Libro de José*, 93, 404, [IV].

56. The ten Aristotelian categories are generally considered to be substance, quantity, quality, relatives, somewhere, sometime, being in a position, having, acting, and being acted upon (*Categories*, 1b25–2a4): Gracía and Newton, "Medieval Theories of the Categories."

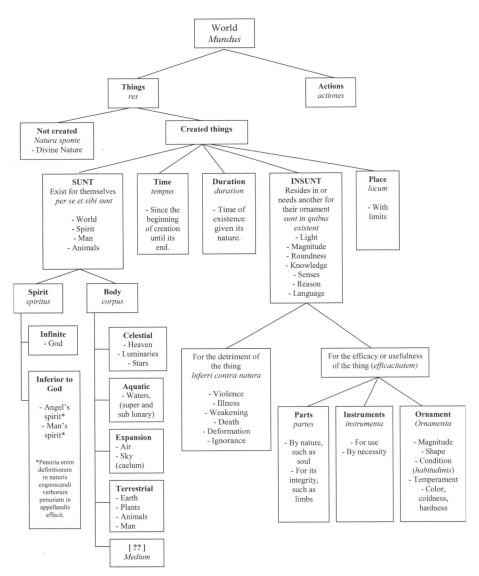

FIGURE 4.4. Interpretation of "On the Division of the Things on Which the Arcane Language Is Established and Its Meaning Sought" (*De divisione rerum quibus arcanus instituitur sermo ac symbola petuntur*), in *De arcano sermone. Antwerp Polyglot*, 8:†1-†5. Special Collections, Sheridan Libraries, Johns Hopkins University.

semantic theory and therefore pertained to logic rather than ontology?[57] If the former, the distinction made in the *Categories* between "substance" and "quality" poses serious difficulties with Aristotelian hylomorphism as described in the *Physics* and *Metaphysics*. However, we must recall that in this treatise Arias Montano was not discussing the division of real things as they occur in nature or their metaphysical ontology, but rather how language sometimes creates arcane associations between symbols in the form of words based on human perception. Thus, rather than an ontological division, he was explaining the semantic relation between things that "are in themselves" (*sunt per se*) and "time," "duration," "place," and "qualities." Yet undoubtedly Arias Montano's conception of language rested on the notion that there was—particularly in the language of the Sacred Scriptures—a strong isomorphic relation between words and things.

In the next section, *De symbolorum, sive Arcani sermonis tractatione et usu, brevis observatio* (Brief observation about the handling and use of the symbols of the arcane language), Arias Montano explained how to extract the meaning of symbols from the sacred language, or how signifiers and their meanings should be recovered from the text.[58] He suggested fourteen ways of examining word/thing relations. Foremost was considering both the genus and the species of things, since this is how they are organized in nature. The genus should be considered first, since it reveals the nature of the thing, followed by the species (which reveals the different types within the genus), and then the thing's relation within a species (which determines its relation to class). Meanings could also be derived by examining the relations between efficient causes and their effects (and vice versa) and by examining the temporal relations (precedents, concomitance, consequences) between things. Careful attention should also be paid to rhetorical devices commonly used in the Bible: metaphors, allegories, comparisons, examples, archetypes, metonyms, synecdoches, and periphrases.

In the third section, *Animadversiones quaedam, ad arcani sermonis rationem tractandam, opportunae* (Certain useful observations about handling the plan of the arcane language), Arias Montano reminded readers that the meaning of a word must also coincide with the nature of the thing itself, or with its virtues, efficacies, or actions. Otherwise there was danger of coming up with meanings that are inappropriate or contrary to the nature of the thing itself. According to Arias Montano, although there were many ways of learning the nature of things, such as by reading philosophical books, by practice, or through experience, the most certain method of all was studying the Sacred Scriptures. The investigator will find the Bible's arcane language very consistent and accurate, but only so long as determined meanings are given to determined things (*certae res*) based on their nature.[59] For example, milk (a nondubious thing) in the Bible is always portrayed as being good, while wine can either be good or cause some problems; thus "milk" and "wine" must be interpreted accordingly. Furthermore, when a thing is used as a symbol, it will always convey a meaning as determined by its true nature, but it may also convey a different one given its context, as with Judas's kiss.

It was only after these preliminaries that Arias Montano placed the formal preface to the *De arcano sermone*. Whereas the preliminary material discussed in the letter to the reader and the introductory essays addressed historical, lexical, and methodological issues, the preface itself focused on purpose: Why attempt to understand the nature of things encoded in the Bible's arcane language? Here again Arias Montano reprised the arguments he set forth in the general preface, but now he related them to language: human beings are the *cause* of Creation, since everything in the world was made by God so that humans could preside over it and live in constant praise of the God who placed it at their disposal. Along with the power to subject and dominate the earth, God granted humanity the gift of knowing nature: the utility, causes, effects, motions, and actions of all things that have been made under the heavens. For evidence of this design, Arias Montano referred to the famous passage in which Adam names the beasts (Gen. 2:19–20) and the notion of an Adamic language in which this knowledge was encoded. He cast the *De arcano sermone* as an exercise in recovering the relation between word and thing that existed when the Bible was written. This required not only the study of Hebrew but also the ability to identify and describe accurately the natural object being referred to, so that what one learned while studying the language could be used to elucidate the meaning of the biblical passage that made reference to the natural thing.

At the end of the preface we run into a Montanian notion central to his later natural philosophical work: the immutability of the essential nature of things. For him the beginning of the second chapter of Genesis, "Thus the heavens and the earth were finished, and all the host of them," meant that the limits of

59. Ibid., 99, 407, [3].

the world had been defined by number and delimited in time at the conclusion of the seven days of Creation[60] (Gen. 2:1 KJV). In the Montanian notion of Creation, God made a fecund nature, but the divine creative process took place only during the six days and not beyond. His interpretation suggests a Creation that was definitive and finite but that could perpetuate perfectly the genus as originally created. The creative act was not continuous, nor did things have the potential to change their genus.[61] Establishing the works of Creation as immutable was an essential component of Arias Montano's exegetical approach, for the things contained in the world *must* remain unchanged if their study is to tell us anything about the relation Adam observed between sign and signifier and Hebrew, the archetype of all language.

Consider some examples that illustrate this perspective and approach. Take the entry for the metal lead (עֹפָרֶת, *'ōpāret*); it reads in part, "The word stands for the metal, its weight, and its lowly nature, since it means 'dust turning to dust.'"[62] It is made by the concretion of the heaviest particles of dust, Arias Montano continued. Lead has a vile and feminine nature and thus is good only for weighing things down or obstruction; in the fire it returns to its dusty nature faster than other metals. Thus Arias Montano concluded that when lead is mentioned in the Bible it means something useless, something that obstructs. Yet it can also indicate the severity of a punishment intended to wipe away the stain of sin, because lead is used to purify silver. Arias Montano found some of lead's characteristics evident in the root of עָפָר (*'āpār*), a feminine word meaning dust. But he also drew from his own experience with the metal itself, noting its weight, its dullness, how it behaves in the fire, and how it is used in metallurgy. In the Montanian scheme anything feminine suggested a vile, lowly, and weak nature, as he explained later in the definition of "feminine": in this language this genus means the weak and is a mark of lesser value.[63]

This combination of etymological analysis and experience also informed his explanation of the many ways of referencing fruits in the Bible. First, the usage of "fruit" is determined by the purposes for which fruits and living things were created: to be useful to humanity. In Hebrew the word for fruit, פְּרִי (*pĕrî*), may mean simply the fruit itself; but because it is derived from the verb פָּרָה (*pārâ*) it can also signify "to propagate," "to generate," "to greatly amplify,"

60. "[T]emporísque rationes constitutae omnes, certísque numeris et terminis definitae sunt.": ibid., 101, 409, [1].

61. "Cuius admirabiles partus unum tantum Dei imperium sine ullo alio seminis conceptu, neque graviditatis taedio facit.": ibid., 102, 410, [2].

62. Ibid., 239, 469–70, [61–62].

63. Ibid., 317, 502, [94].

and, interestingly, "to bring forth its virtue."[64] Therefore the word can also be used to signify descendants and propagation not only of humans, but also of plants and animals. It is also used to signify the results of one's labor, whether good or bad according to the virtue of the "tree" that yielded the fruit. Arias Montano illustrated each potential meaning with one or more quotations from the Bible.

Often he did not stop to explain the origin of his interpretations. In a later section, while discussing the meaning of *mustum* (wine, but also fresh, unfermented grape juice), he explained that it is used to indicate "a vehement force and efficacy, and a spirit of great impetus,"[65] citing Job 32:19, where a diligent reader would have found the word יַיִן (*yayin*), whose root means "effervescence."[66] In fact, of the more than 530 entries in the *De arcano sermone*, only three dozen explain the meaning of the word by referring to a specific Hebrew word or root. (This tally does not include the section on the meaning of the Hebrew names of God, where Hebrew roots feature prominently.) The rest of the entries either omit this information, though they are clearly informed by the etymology of the words, or establish meanings using biblical intertextual references and a good dose of experience and observation.

The overall arrangement of the subjects in the treatise, as Gómez Canseco observes, follows the order of creation or the first mention of the natural phenomenon (or thing) according to the Genesis account.[67] The definitions in the *De arcano sermone* start with a discussion of things that have a spiritual nature, beginning with God and the interpretation of his arcane names and the subtle difference they suggest about the nature of God. This is followed by discussions of spirits, angels, and demons. Then Arias Montano turns to things with corporeal natures—though not before reminding readers that his focus

64. Ibid., 256, 479, [71]. It is interesting to note how far he went beyond the definitions provided by the thesaurus included in the Antwerp Polyglot: "פָּרָה Crescere, fructificare: *Ier. 3, 16. Sic Deut. 29, 18.* Radix פָּרָה ראֹשׁ fructificans venenum. *Et Genes. 49, 20.* Ramus פֹּרַת crescens. et פֹּרִיָּה Fructificans: *Psal. 128, 3. Et in Hiphil,* הִפְרָה Crescere facere: *Genes. 41, 51. etc.* פְּרִי *et cum accentu* פֶּרִי Fructus: *1, 12 et Ierem. 12, 20. Et cum affix. sic mutatur* פִּרְיוֹ Fructus eius, *etc.* פֶּרְיְךָ Fructus tuus: *Hos. 14, 9.* [*Et aliud nomen, Cantic. 3, 9.* אַפִּרְיוֹן Lectus quo deferuntur sponsae. *Al.* Carruca. *Hieron.* Ferculum. i. gestamen.]": Raphelengius, "Thesauri hebraicae linguae," 6:98–99.

65. Arias Montano, *Libro de José*, 257, 478, [70].

66. The definition given in Raphelengius's *Thesaurus* does not include the discussion of the root, but only gives the translation, "*vinum*": Raphelengius, "Thesauri hebraicae linguae," 6:38.

67. Gómez Canseco, "Los sentidos del lenguaje arcano," 52.

here is on how the language of the Bible refers to them—while postponing for later in the text the discussion of their parts, natures, and utility. The things in the world that resulted from the creative act of God are discussed first: earth, water, light/darkness, and day. From the day followed the creation of time, and from time the measure of all things and their limits (*terminus*), which are defined by "number." These are followed by the creation of "place" as defined by its limits, and finally air. He further divides the discussion of each of these fundamental aspects into two parts, one examining the meaning of something relative to its place and the other the thing itself. Thus we find, for example, one section describing water as a thing (chapter 8) and another describing water as a place (chapter 18). The rest of the treatise discusses "composite things," including man, as well as things that result from the arts (architecture, vestments, war) and social conventions (accounting terms, laughter).

By dividing the definitions between meanings related to the thing itself and the thing in a place, Arias Montano was able to further distinguish between something's essential qualities and those determined by its situation relative to place. In the case of water as a thing, we learn that its nature and significance can be derived from how it was described in the order of Creation, first as "abyss" and later as "waters" and "sea."[68] When water is described as the "abyss," it is presented as being "rather burdensome and unmanageable" (*graviora et molesta*); Genesis 1:2 describes it as covering the land and making it dark, barren, and empty. However, by the sixth and ninth verses of Genesis 1 the abyss has become ordered, so that "waters" now suggests a nature that is mild, useful, and fecund. An interpreter of the Sacred Scriptures should understand that whenever "waters" is used in the text the resulting interpretation should take into consideration the waters' capacity to support life, cleanse, cool, and oppose fire. Water as "sea," described in chapter 18, takes on added characteristics when considered with respect to place. For Arias Montano the separation of the waters that occurs in Genesis 1:6–7 further determines the arcane meaning of "water." At this point, however, he did not have much to say about the celestial waters, though this would change in the *History of Nature*. Here he limited his description of the celestial waters to their being retained by the convex dome (*tantùm conuexa concameratae [sic] complectuntur*) of the heavens, adding that "what these contain or nourish is uncertain to us."[69] The rest of the section describes the many roles and characteristics of water on earth, as a place of fecundity and abundance as well as a place from which

68. Arias Montano, *Libro de José*, 124–25, 420–21, [12–13].

69. "[Q]uid praetereà contineant vel alant, incertum nobis est.": ibid. 154, 435, [27].

humanity derives great utility (e.g., fisheries, navigation); yet water can some-times overwhelm humanity and serve as an instrument of divine punishment.

Throughout the treatise Arias Montano made repeated references to its being a preliminary exploration of the capacity of the arcane language of the Bible to explain the *rationes naturarum* of all created things, something he hoped, with the grace of God, to explain at length at a later time. What is patently obvious after comparing the *De arcano sermone* to the *Anima* and *Corpus* is that those later works were based on ideas Arias Montano had dis-tilled years earlier. The readers of both exercises would find them unconven-tional, unexpected (particularly in the context of an orthodox post-Tridentine Bible), and very confusing.

Early Objections to the *De arcano sermone*

The official censors assigned to the Antwerp Polyglot from the University of Louvain, Augustin Hunnæus and Cornelius Reyneri Goudanus, were the first to encounter Arias Montano's *De arcano sermone*, and they—and a number of other theologians they consulted—strongly advised against including it in the *Apparatus*.[70] They had several objections. First, they argued that neither the learned nor the ignorant (*indocti*) would derive much benefit from what they deemed an "obscurely written" treatise: the learned because they had the commentaries of the church fathers and other reference material at hand to help them interpret the Sacred Scriptures, and the *indocti* because they would find it confusing. In fact, the censors continued, no one they consulted could understand its purpose or utility! Furthermore, the proem was particularly confusing because of the obscure ideas expressed and the novel phraseology used.[71] Finally, they explained to Arias Montano that because he had chosen not to call on any authorities to support his conclusions, relying only on collat-ing references within the Bible to determine the meaning of certain words and passages, many readers would simply dismiss the treatise, since it is not always possible to find the true meaning of a word by collating meanings drawn from different parts of scripture. The censors were clearly uncomfortable with the interpretive fluidity and independence from authority that were being advo-cated. To mollify Arias Montano they asked him to consult other erudite men

70. Their objections are also discussed in Gómez Canseco, "Sentidos del lenguaje arcano," 45. The *De arcano sermone* was also examined and approved by Sebastian Baer of the College of Censors of Antwerp and Franciscus Sonnius, bishop of that city, but their opinions do not survive.

71. Censors of Louvain to Arias Montano, Louvain, 20 August 1570, in Macías Rosendo, *Biblia Políglota*, 230.

who would be candid with him (and not inclined to flattery) and to base his decision on their judgment. Later they even suggested he circulate the manuscript anonymously to gauge opinions. In his reply, Arias Montano said he intended to show the treatise to others, but he also wanted to know if there was anything in it that might seem to offend the purity of the Catholic religion, which he did not want to do.[72] There was not, they replied. Arias Montano had drawn a theological line in the sand. The interpretive tools contained in the *Apparatus* would go forth as planned. These letters must have been very difficult to write for Hunnæus and Goudanus, two champions of the Antwerp Polyglot. Yet they thought the matter so important that they could not dissimulate their bafflement over what Arias Montano was up to in the *De arcano sermone*, and they feared a negative reception of his work.

This was the first but hardly the last time theologians criticized the treatise. From this point on the criticism would only get more acrimonious and personal. While in Rome in 1572, Arias Montano heard firsthand a number of other objections to the *Apparatus* and rushed to make some changes before more copies were printed. These included deleting some apparently offensive names from Lefèvre de la Boderie's Syriac-Chaldean lexicon. A second printing of the *De arcano sermone* appeared in August 1573, but any changes seem to have pertained only to its typesetting.[73] In 1575, however, during Arias Montano's second stay in Rome, the correspondence between the exegete and Plantin reveals that Arias Montano requested additional changes to the *De arcano sermone*, which were then made; such requests continued until 1587.[74]

THE *APPARATUS* OF THE ANTWERP POLYGLOT AS A HISTORICAL/CULTURAL PROJECT

The Antwerp Polyglot was shaped in fundamental ways by the historical and cultural sensitivity of the contributors to the project, particularly as it pertained to their skills as philologists and humanists. But it also reflected a new sensitiv-

72. Arias Montano to the censors of Louvain, Antwerp, 23 August 1570, in Macías Rosendo, *Biblia Políglota*, 234–37.

73. Rooses noted the differences but did not explain them in detail: Rooses, *Christophe Plantin*, 126. See the letter from Plantin to Arias Montano, Antwerp, 1–7 August 1572, in Dávila Pérez, *Correspondencia conservada en el Museo Plantin–Moretus*, 1:103–8.

74. Letter from Plantin to Arias Montano, Antwerp, 15 October 1575, in Dávila Pérez, *Correspondencia en el Museo Plantin–Moretus*, 1:239–44, and Macías Rosendo, "*De arcano sermone*," 40–41. Plantin reprinted the volumes that contained the *Apparatus sacer* on several occasions owing to the great demand for them. See Arias Montano, *Antigüedades hebráicas*, 40.

ity of Renaissance scholars to the passage of time and to history. Nowhere is this attitude more apparent than in the cultural treatises that Arias Montano wrote and included in the eighth volume of the Antwerp Polyglot and that were subsequently reprinted and published as the *Antiquitatum judaicarum libri IX* (1593). There we find the histories and cultural studies that historicized the Bible and presented it as a cultural product of a particular historical moment of the Jewish people of the Bible. From the histories we learn how the Promised Land was partitioned (*Chaleb*), about the dispersal of the descendants of Moses, and about the time and ages (*Daniel*). The descriptions included the land of Canaan (*Chanaan*), the priestly vestments and ornaments (*Aaron*), the city of Jerusalem (*Nehemiah*), Noah's ark, the tabernacle, and temple (*Exemplar*), and weights and measures in ancient Israel (*Thubal-Cain*).[75] In many ways these treatises follow in the tradition of Titus Flavius Josephus's *Antiquitates Judaicae* (ca. first century AD), whose work Arias Montano cited in his biblical commentaries.[76] In a subsequent project, the translation of the *Itinerary* (1575) of Benjamin of Tudela—a twelfth-century Jewish explorer from Navarre who traveled to the Holy Land—Arias Montano stated his views on the importance of understanding the cultural and geographical dimensions of the Bible.[77] As Zur Shalev has noted, "Montano saw the true significance of Benjamin's travels not in his precise distances and acute observations. It was Benjamin's ability to disclose facts and insights that may enable an eager and devout reader to approach Scripture in an informed way."[78]

The historical sensitivity Arias Montano brought to the study of the Bible was informed by a notion of historical distance that was central to the work of Christian Hebraists.[79] Yet bridging this historical distance was essential to

75. For a careful study of these treatises, particularly as they relate to sacred geography, see Shalev, *Sacred Words and Worlds*, 23–71. For Spanish translations and comprehensive introductory studies, see Macías Rosendo and Fernández López, in Arias Montano, *Antigüedades hebráicas*, 13–85.

76. Perea Siller and Pozuelo Calero, "*Phaleg* en su entorno," 1:337–39.

77. For an English translation, see Tudela, *Itinerary of Benjamin of Tudela*. The book was prohibited in the 1583 *Index* of Quiroga because its publication violated the fourth rule of the Index about publishing works by Jews that contained arguments against the Catholic faith, its customs, and its ceremonies, or against the biblical interpretations of church fathers and doctors: Quiroga, *Index et catalogus librorum prohibitorum* (1583), 2–3, 12.

78. Shalev, "Benjamin of Tudela, Spanish Explorer," 23.

79. Historical distance is a function of the lapse of time and geographical displacement that implies a definitive cultural difference between the past and the present. It suggests that the past is somehow "foreign" to our times. For some reflections of how this notion informed the sensitivities of Renaissance humanists, see Lowenthal, *Past Is a Foreign Country*, 74–93.

producing a proper literal interpretation of the Bible. Applying the best practices of humanist philology and the developing discipline of Jewish antiquarianism meant not only understanding Hebrew grammar and lexicon, but also bringing to the text the historical contextualizing sensibility of Renaissance humanism. This type of contextualizing historicism became characteristic of sixteenth-century Christian Hebraists.[80] It was one of several raisons d'être for Arias Montano's contributions to the *Apparatus*. The goal was to re-create the cultural, material, and linguistic context of biblical protagonists of the Old Testament in order to better understand the way it was delivered to this specific cultural group, with the ultimate aim of further elucidating the divine mysteries contained in the Bible.[81] Curiously, the strongest argument in favor of this approach was an important cultural marker among Jews, based in tradition dating from the days of antiquity. Even when certain Jewish communities in Europe spoke derivative dialects of Hebrew, or no Hebrew at all, biblical Hebrew had persisted through the ages as the language of worship; the Torah was always read in Masoretic, or biblical Hebrew, and its writing was routinely taught at rabbinical schools.[82] Within this historical imaginary of Christian Hebraists, the text of the Pentateuch had been meticulously copied, correlated, and preserved ever since Moses wrote it. This tradition of pristine biblical descent provided Christian Hebraists with something other Orientalists-or even classicists-could only dream of: a carefully transmitted and uncorrupted ancient source.

Yet the notion of historical distance required a balancing act, resulting in what Amnon Raz-Krakotzkin describes as a discourse of integration and separation.[83] While cultivating a philological and exegetical approach that sought

80. On the interest in all antiquities during the Renaissance, see Momigliano, *Classical Foundations of Modern Historiography*, 54–73. On Christian Hebraists in particular, see Coudert and Shoulson, *Hebraica Veritas?*, and Heide, *Hebraica Veritas*, 44–66. On Jews' interest in their own antiquities, see Malkiel, "Artifact and Humanism in Medieval Jewish Thought," 21–40.

81. As I noted in chapter 1, the strand of Orientalist scholarship employed in the elaboration of the Antwerp Polyglot—at least as far as it concerned Arias Montano—descended from a specific school of medieval Iberian rabbis associated with David Kimhi, who taught a literal interpretation of scripture and whose tradition of scholarship continued at the University of Alcalá through the agency of Alfonso de Zamora and Cipriano de la Huerga: Fernández López, *Alfonso de Zamora y Benito Arias Montano*, 23–26. On Kimhi's influence on medieval and early modern Hebraic studies, see Talmage, "David Kimhi, the Man and the Commentaries," and Friedman, "Sixteenth-Century Christian-Hebraica."

82. Burnett, *Christian Hebraism*, 95.

83. Raz-Krakotzkin, *Censor, the Editor, and the Text*, 24–25.

to bridge historical distance was in keeping with humanistic scholarship, it could also present difficulties to Christian Hebraists, who often sought to distance themselves from contemporary Jews.[84] Although the antiquarian fashion gave Arias Montano some intellectual latitude to pursue exegetical strategies that would otherwise have been seen as blatant Judaizing, it did not always convince his critics, particularly at a time when the Spanish Inquisition was concerned with the existence of crypto-Jews in postexpulsion Spain.[85] Furthermore, accusations that the Talmud expressed anti-Christian sentiments—an issue extensively discussed at the Council of Trent—and that Jews had corrupted the biblical text in order to undermine Christian tenets were widespread throughout Christendom and were routinely hurled against the Jews of the day. (Some Christian Hebraists could argue that the ancient copies of the Old Testament they worked with had escaped the Jewish "perfidy.")[86] So for Christian Hebraists who did not want to run afoul of ecclesiastical authorities—of any Christian confession—maintaining a historical distance between the people of the Bible and the Jews of their own time required great care.

*

For Arias Montano the opportunity to edit the Bible and prepare a new interpretive apparatus presented a unique opportunity to correct what he understood to be fundamental flaws in the way the Bible had previously been studied and interpreted. Yet as we will see in the next chapter, the challenges this posture entailed during the post-Tridentine era soon made themselves apparent. The objections Arias Montano and the Antwerp Polyglot faced did not stem from fear that Protestant ideas would infiltrate the project or from a general condemnation of providing material that would endorse personal study of the Bible, let alone the intimation that a new natural philosophy could be derived from its study. The initial objections were against the very exegetical methodology that Arias Montano featured in the *Apparatus* and the unusual conclusions he drew from this approach.

84. Grafton, "Christian Hebraism and the Rediscovery of Hellenistic Judaism."
85. Alpert, *Crypto-Judaism*, 17, 38–39, and Kagan and Dyer, *Inquisitorial Inquiries*, 11–16.
86. Roth, *Medieval Jewish Civilization*, 167–68.

CHAPTER 5

Arias Montano Castigated

Despite the sheltering mantle of the Western world's most powerful monarch, things did not go smoothly for Arias Montano when it came time to unveil the Antwerp Polyglot. Since settling into his tasks in Antwerp he had been receiving a constant series of communications that portended a storm. Most acknowledged the merits of the project yet were concerned enough with certain aspects to make their misgivings known. Others objected vociferously to the whole enterprise. Arias Montano at first dismissed them by invoking those favorite bogeymen of Renaissance scholars, *los envidiosos* (the jealous ones). But as 1571 arrived and the project was nearing completion, he realized he needed to shift his efforts toward developing a strategy for unveiling the project in a favorable light and preparing for its defense.

In what follows, we will study Arias Montano's critics' objections and see how he successfully addressed their concerns. Yet to explore the many and varied political positions and motivations behind the criticisms would go beyond the scope of this book. Those were fraught times to say the least. The completion of the project coincided with the hardening of Spanish Habsburg rule over the Low Countries, the arrival of the Duke of Alba as the king's representative, and the unleashing of the Spanish Fury. In Rome the papacy was negotiating a new power balance with the Spanish; it wanted defense against the Turkish threat and at the same time sought to stem Spanish meddling in the affairs of the Vatican, particularly on matters of ecclesiastical jurisdiction. In Spain the Antwerp Polyglot project coincided with the incarceration of a group of professors from the University of Salamanca, including fray Luis de León, who had been accused before the Spanish Inquisition of holding heretical ideas. Furthermore, along with sponsoring the Antwerp Polyglot, Philip II had granted Christophe Plantin the privilege of printing all the new editions of prayer and liturgical books, infuriating many members of the Spanish clergy, who saw the revenue they derived from their sale vanish.[1] The many implications of this backdrop for the Antwerp Polyglot and Arias Montano's involvement with it have already been studied at length, and here I will say only that Arias Montano, as both contributor and editor of the project, had to carefully consider the wider implications of the positions he took in defense of his work. This chapter will therefore focus on those issues and criticisms that have a direct bearing on his exegetical methodology and on the natural philosophical conclusions he drew from it. The bulk of the concerns originated first in Spain and Louvain, playing out later in Rome, where Arias Montano journeyed in 1572 to present the Antwerp Polyglot to the pope and again in 1575–76 to defend the project's orthodoxy. The final chapter took place in Spain and took the form of a castigation solicited by the Spanish Inquisition and written by the Jesuit Juan de Mariana.

OBJECTIONS TO THE ANTWERP POLYGLOT

The earliest and loudest of the critics of the Antwerp Polyglot was the professor of Greek at the University of Salamanca, León de Castro, OP.[2] What began as whispers in Arias Montano's ears in 1571 would by 1575 become an unavoidable and threatening situation that not only had the attention of Philip II, but was under serious consideration by the Spanish inquisitor general as well as by the pope. During the polemic's later stages it caused Arias Montano enough concern that he hesitated to return to Spain, fearing he might be charged and brought before the Inquisition. It is clear that León de Castro's criticism was not just a personal crusade; behind him stood powerful members of the Spanish church concerned with the king's decision to have the breviaries, prayer books, and liturgical material printed in Antwerp. These critics might have also

1. Morocho Gayo, "Felipe II: Las ediciones litúrgicas," 857.

2. León de Castro's crusade against the Antwerp Polyglot has been well studied over the years, since the epistolary exchanges and archival materials have been made available to historians. In addition to the prefatory material in the collections of Arias Montano's correspondence by Macias Rosendo and Dávila Pérez, curious readers can read well-informed syntheses of the polemic in Ortega Sánchez, "El enfrentamiento entre Arias Montano y León de Castro"; Domínguez Domínguez, "Correspondencia de Pedro Chacón (III)," 379–420; Arias Montano, *Prefacios*, lii–lxxvii; and Macías Rosendo, "*Apparatus sacer* en la *Biblia Regia de Amberes*," 27–39.

shared León de Castro's concerns with the methodology used in preparing the Antwerp Polyglot, but little historical evidence remains, so maestro León is left to stand alone as the malicious instigator. There is no doubt that León de Castro tried to recruit a broad spectrum of influential figures: his fellow professors at Salamanca, members of the royal court in Madrid, the Spanish and Roman Inquisitions, Bishop William Lindanus of Ruremonde (1525-88), and, through members of his Dominican order, the very Chair of Saint Peter. What made his allegations about the Antwerp Polyglot so persuasive?

There is a fundamental difficulty in answering this question, for León de Castro's formal complaint to the Inquisition has not survived. We have an early complaint from him, dozens of letters by third parties describing the shenanigans of the scholar from Salamanca, and some summaries of his principal objections. When we look at his objections as a whole, what emerges is a case against the Antwerp Polyglot based solely on doctrinal and methodological grounds, but from a person who had convinced himself that the editor of the project was a Judaizer whom he would gladly burn at the stake, or at the very least see humiliated in an *auto da fe*. In a memorandum addressed to the king that dates from the early years of the polemic (ca. 1569), León de Castro raised concerns about replacing the Vulgate with Santes Pagnino's translation of the Bible. Pagnino, he alleged, had added words in many places to make the meaning Jewish (*hacer sentido de los judios*) and to "follow the Jews" while undermining the interpretation of the apostles and Saint Jerome.[3] He alleged that there were at least a dozen faulty sections where—most tellingly, in his view—the text had been interpreted in such a way that it ceased to refer to the coming of the Messiah or to the Resurrection. (If asked or ordered, he promised to make these known in a report.) Pagnino, according to León de Castro, "does not translate according what is written in Hebrew, but according to how the Hebrews explain it."[4] He also objected to the publication of the Bible in Syrian, "because all those Syriacs and the Orientals were Arians," and thus it is likely that the Arian heresy had infected the text. Only a council of many learned theologians could be charged with cleansing it. He also suspected that printing the Gospel of Matthew in Hebrew, no matter how old the manuscript, could be a hoax perpetrated by Jews against the Christians. By 1574, after going over a copy of the Antwerp Polyglot that he had urged the University of Salamanca to purchase, León de Castro went to work on his unsolicited censorship. He now made charges against Arias Montano personally as well,

3. AGS, Estado 143, in L. Gil, "Advertimiento del maestro León de Castro," 50.
4. Ibid.

accusing him of being aligned with "the rabbis." It was a scathing and even libelous critique. Others stepped up to Arias Montano's defense, among them Pedro Chacón (ca. 1525–81), a priest and longtime friend. Chacón had moved to Rome in 1570 to continue his philological studies as a member of the Roman Curia. He directed a letter to León de Castro after reading the complaint he had submitted to the Roman Holy Office.[5] Not mincing words, Chacón wrote: "I dislike greatly as well the way you have carried out [the criticism], because I take it to be prejudicial to the good opinion I had of your Christianity, zeal, doctrine, and judgment, because from what I see in these papers you have let yourself be driven by passion, and you lunge to wound your opponent with abandon."[6]

Others who listened to León de Castro's complaints or read his letters could not fail to note that the objections stemmed from a deep-seated animosity, almost an obsessive hatred, of anything associated in the least with Judaism. Fray Luis de León, responding to a 1570 letter from Arias Montano in which he had asked fray Luis's opinion about the tenor of León de Castro's allegation, characterized his Salamancan colleague as someone for whom "all that is written, or has a taint of having been written by rabbis, is something excommunicated."[7] León de Castro also denounced Pagnino and François Vatable and even, in fray Luis's opinion, seemed to "not forgive Saint Jerome" (presumably for having translated the Bible from Hebrew). Fray Luis knew León de Castro very well. They had served in a series of meetings in Salamanca to correct the Latin translation of the Hebrew Bible published by Robert Estienne in 1545 that included controversial notes by François Vatable (d. 1547), a Royal Lecturer of Hebrew in Paris.[8]

If we set aside the more vitriolic aspects of León de Castro's statements and focus on the doctrinal and methodological grounds he used to support his charges, it is evident that the main opposition to the Polyglot project focused on the proper way to study the Sacred Scriptures. As a Dominican, he firmly opposed the humanistic and philological approaches advocated by Christian

5. Chacón also wrote the *Historia de la universidad de Salamanca* (1569), which remained unpublished, and while in Rome he participated in the reform of the Julian calendar: Domínguez Domínguez, "Pedro Chacón," 193–94.

6. Pedro Chacón to León de Castro, Rome (est. 1575 to May 1576), in Domínguez Domínguez, "Correspondencia de Pedro Chacón (II)," 204. This article publishes BNE MS 1946, fols. 15r–43r. There is also a copy of the letter at HSA, B1352, box 5.

7. Fray Luis de León to Arias Montano, Salamanca, 28 October 1570, in Macías Rosendo, *Biblia Políglota*, 242–43.

8. León, *Comentario sobre el Génesis*, xxxiv–xxxv.

Hebraists, Augustinians, and others who found the Scholastic approach of Saint Thomas limiting, if not wholly misguided. His arguments were based on his conviction that the Old Testament had been intentionally and maliciously altered by "the rabbis" and thus could not serve as a credible source. Chacón's letter cited a number of León de Castro's arguments to this effect. Jews, León de Castro maintained, had altered the Bible to persuade their own not to abandon the faith. By changing a period here or a letter there, they carried out this treachery, particularly in the arguments from scripture used against them by Christians.[9] His opinion of the Old Testament was categorical: "The text we have now is more Judaic perfidy than Hebraic truth."[10] León de Castro's opposition stemmed from his objecting in principle to any new edition of the Sacred Scriptures using primary sources, and specifically to relying on the Hebrew Bible as a credible source from which to correct the Vulgate. Thus he considered including the Pagnino Bible in both the Hebrew and Greek translations into Latin an affront to the authority of the Vulgate.

It is worth noting that during the years of gestation of the Antwerp Polyglot the question of the Vulgate's authority was far from resolved. When the fourth session of the Council of Trent adjourned in 1546, the council had agreed only that the Vulgate would be "held as authentic; and that no one is to dare, or presume to reject it under any pretext whatever."[11] Exactly what this meant,

---

9. Domínguez Domínguez, "Correspondencia de Pedro Chacón (II)," 205.

10. Ibid.

11. The *Decree concerning the Canonical Scriptures* (or "*Insuper*") issued during the fourth session listed the canonical books of the Old and New Testament and declared "anathema" anyone who did not accept these books as sacred and canonical as contained "in the old Latin Vulgate edition." The decree stated, "Moreover, the same sacred and holy Synod,—considering that no small utility may accrue to the Church of God, if it be made known which out of all the Latin editions, now in circulation, of the sacred books, is to be held as authentic,—ordains and declares, that the said old and Vulgate edition, which, by the lengthened usage of so many years, has been approved of in the Church, be, in public lectures, disputations, sermons and expositions, held as authentic; and that no one is to dare, or presume to reject it under any pretext whatever. Furthermore, in order to restrain petulant spirits, it decrees, that no one, relying on his own skill, shall,—in matters of faith, and of morals pertaining to the edification of Christian doctrine,—wresting the Sacred Scripture to his own senses, presume to interpret the said Sacred Scripture contrary to that sense which holy mother Church,—whose it is to judge of the true sense and interpretation of the Holy Scriptures,—hath held and doth hold; or even contrary to the unanimous consent of the Fathers; even though such interpretations were never (intended) to be at any time published. Contraveners shall be made known by their Ordinaries, and be punished with the penalties by law established.": Waterworth, *Council of Trent*, 19–20. The text of the decree masked an unease acknowledged in the council members about error in the Vul-

especially where it concerned the "original" versions of the Bible in Hebrew and Greek, remained for years open to interpretation. A faction of Hellenists (to which León de Castro belonged) granted authority to the Greek version of the Old Testament, or Septuagint, made in the third century BC by the seventy interpreters during the reign of Ptolemy II. They reasoned that this was the version of the Bible cited by the apostles in the New Testament and by the early Christian church; therefore some believed it had escaped the "perfidy of the Jews." In fact, Chacón pointed out that León de Castro seemed to vacillate on whether even the Septuagint was uncorrupted and whether the corruption by the Jews had taken place before or after Saint Jerome's translation. (If before, even the Vulgate was to be taken as suspect, a point Chacón used as the coup de grâce in his devastating critique of the lion from Salamanca's arguments.)

During the early days of the Polyglot project fray Luis de Estrada—who had been at the fourth session of the Council of Trent—had explained to Gabriel de Zayas that the council members had not condemned versions of the Bible in its original languages; rather, they should be used to correct later versions and enrich interpretations, as long as these were in keeping with the intent of the Holy Spirit. But, he continued, this had to be done by garnering agreements (*con grande acuerdo*), so as not to inflame some "stubborn and ignorant men" who, hiding behind the little Latin they know, cry "Lutherans!" whenever they encounter versions of the Sacred Scriptures in Hebrew or Greek, or those who studied only Greek and think that only the Septuagint is correct and anyone who dares study the Bible in Hebrew is a Judaizer.[12] As early as 1569 Estrada alluded to "a certain person" who had persuaded some at court that to follow the Hebrew Bible is to Judaize; this person had even slandered the memory of Pagnino. Although unnamed, the reference was clearly to León de Castro, whose opposition to the Polyglot project was visceral and unrelenting.

León de Castro had honed his arguments a few years earlier in the public debates that surrounded a petition by a Salamanca printer to publish the Vatable Bible, and he often repeated these arguments in his published works.[13] Especially after the Council of Trent's decree in support of the Vulgate and its interpretation by church fathers, advocates of the literal interpretation us-

---

gate and a lack of clarity concerning the status of the Hebrew and Greek versions of the Bible, issues that would come to the fore with the Antwerp Polyglot and as efforts to correct the Vulgate continued in Rome during the latter half of the sixteenth century, culminating with the Sixto-Clementine edition of 1592: Wicks, "Catholic Old Testament Interpretation," 2:627–35.

12. Letter from fray Luis de Estrada to Gabriel de Zayas, Alcalá, 11 July 1569, in Macías Rosendo, *Biblia Políglota*, 132–36.

13. Domínguez Domínguez, "Correspondencia de Pedro Chacón (II)," 214n17.

ing the original Hebrew and Greek versions of the Bible had come under increased criticism by a camp that sought to cleanse exegeses of what they perceived as humanistic innovations. The dispute with Arias Montano was part of a second charge by León de Castro. In 1572, with the collaboration of Bartolomé de Medina, he accused a number of professors from the University of Salamanca of holding theologically suspect ideas, which—to put it simply—stemmed from their reliance on Jewish sources to explain the Bible, their rejection of patristic authority, and their failure to uphold the authority of the Vulgate. Three professors were incarcerated by the Inquisition: Arias Montano's friend fray Luis de León; Martín Martínez de Cantalapiedra (1510–79), a professor of Hebrew; and Gaspar de Grajal (1530–75), a professor of Sacred Scriptures. (All, not incidentally, were of *converso* origins.) The Salamancans had opposed León de Castro and Medina at the meetings convened to discuss the Vatable Bible. Joining them in prison later was Alonso Gudiel (1526–76), a professor of Sacred Scriptures at the University of Osuna.[14] All the accused were advocates of the literal interpretation of the Bible using Near Eastern languages to derive their exegesis. Historian Angel Alcalá has identified the "hermeneutical laxity" this approach implied, and the multiplicity of literal meanings derived from it, as the fundamental reason for persecuting the Salamantine professors.[15] The case was resolved in 1576–77, when fray Luis de León and Martín Martínez de Cantalapiedra were absolved and returned to their chairs at Salamanca, though not before Grajal and Gudiel died in prison. As Arias Montano learned of the increasingly strident allegations against him and the Antwerp Polyglot, he must have had in mind the trials his brothers in exegesis were enduring in Spain.

Given this situation, securing the Vatican's endorsement for the Antwerp Polyglot gained urgency. It did not come easily. Some factions at the Vatican did not look kindly on the autonomy the Spanish monarch exercised in the production of the new polyglot Bible; for political as well as theological reasons, they thought the project should have been overseen by Rome. It had always been the intention that the Antwerp Polyglot would receive the pope's approbation and blessing, as had the Complutensian Bible, but Pius V proved

14. The cases have an extensive bibliography. For a summary, see Alcalá Galve, "Peculiaridad de las acusaciones de fray Luis," 65–80. Transcriptions of the cases are available in Alcalá Galve, *Proceso inquisitorial de fray Luis de León*; Pinta Llorente, *Proceso criminal contra el hebraísta salmantino, Martin Martínez de Cantalapiedra*; and Pinta Llorente, *Procesos inquisitoriales contra los catedráticos hebraístas de Salamanca: Gaspar de Grajal, Martínez de Cantalapiedra y fray Luis de León.*

15. Alcalá Galve, "Peculiaridad de las acusaciones," 78.

reluctant to grant it. By January 1572 Juan de Zuñiga, the Spanish ambassador in Rome, was writing to tell Arias Montano the reasons for the the the pope's reluctance.[16] Although the king had hoped the description of the project and a testimonial from the Faculty of Theology of the University of Louvain would suffice to get the endorsement, no rubber stamp was forthcoming. Some of the pope's reluctance stemmed from a previous failed attempt to edit the Bible in Rome; because of the problems on that occasion, he resolved that the project could be endorsed only by a general council composed of cardinals and many theologians.[17]

As to the project for which his approval was now being sought, the pope asked Cardinal Guglielmo Sirleto (1514–85), librarian of the Vatican Library, and Cardinal Archangelo de Bianchi, OP (earlier Bishop of Teano, 1516–80) to look into the matter and study the papers Arias Montano sent. Sirleto had contributed a short treatise to the *Apparatus* of the Antwerp Polyglot comparing variants of the text of the Psalms. After reading a description of the project[18]—but without seeing the final product—the cardinals raised several objections, and the pope even threatened to have the Polyglot "closely examined" and maybe "banned" (*la hará ver muy particularmente y podría ser que la vedase*).[19] Their reasons, as relayed by Juan de Zuñiga, made it clear that the cardinals had not gotten a very specific description of the contents of all the volumes of the Antwerp Polyglot. They raised concerns about the versions of the Syriac and Latin translations used and specifically noted that, because there had been changes (*mudanzas*) to certain places in Santes Pagnino's version, the pope should not approve it until it was duly examined. The fourth point they raised concerned the treatises of the *Apparatus*, motivated

16. Macías Rosendo, *Biblia Políglota*, 273.

17. This is in reference to the second commission (1569) to amend the Vulgate, which included Cardinals Sirleto and Carafa but had failed to reach a definitive version. See Höpfl, *Beiträge zur Geschichte der Sixto-Klementinischen Vulgata*, 77–101, and Denzler, *Kardinal Guglielmo Sirleto*, 130–34. The commission followed the guidelines set by Sirleto: as many biblical manuscript versions as possible needed to be compared to determine existing typographical errors; the Vulgate was to be checked for consistency against the original Hebrew text and the text of the Septuagint; and the Septuagint had to be held in the highest regard because it has already been quoted by the apostles: ibid., 131.

18. Because of the references made in its description of the project, the document is likely BAV Vat. Lat. 6149, fols. 34–38: "Operis Sacrorum Bibliorum Regis Catholici consilio et sumptu Aria Montano procurante ad Catholicae Ecclesiae usum instructi brevis enarratio," Rome, n.d.

19. Don Juan de Zuñiga to Philip II, Rome, 4 February 1572, in González Carvajal, "Elogio histórico," 159–60.

by the concern that only "very unquestionable things" should accompany the Sacred Scriptures. Of the treatises *De arcano sermone* and *De ponderibus et mesuris*, they said they might contain things that were "very uncertain and not investigated" (*muy inciertas y no averiguadas*) and that the *De arcano sermone* might even be kabbalistic.

Arias Montano set off for Rome in May 1572 with a copy of the recently completed Bible. He hoped to address these concerns with the pope in person, but Pius V died before he arrived. His meeting with Sirleto was preceded by a flurry of letters to the librarian from Cardinal Granvelle, Pedro de Fuentidueña—then in Rome and involved in refuting the *Magdeburg Centuries*[20]—and others. It did not take long for Arias Montano to win the learned man over to his cause. The achievement was significant, since only a year earlier Sirleto had been appointed to lead the Roman Congregation of the Index, the body at least nominally responsible for censoring all the books in Christendom. A few months later Sirleto wrote to Philip II singing Arias Montano's praises and expressing his opinion that the Polyglot was an "opera certe molto digna e excellente."[21] The new pope, Gregory XIII, who was sympathetic to the pro-Spanish faction of the Curia, quickly granted the *breve*, and by September 1572 had granted the rest of the necessary privileges.

When reporting on all these matters to the king, Arias Montano downplayed the particulars that had slowed down papal approval and explained that the problem had stemmed from the Roman Curia's jealousy at seeing such a "notable and universal work" (*tan insigne y tan universal obra*) coming out of Spain.[22] This jealousy might have been ameliorated, Arias Montano noted candidly, had the Spaniards made the petition "with a bit more humility" (*con un poco de sumisión*). (The request had been written by the Duke of Alba, not known for diplomatic subtlety.) The rest of the problems originated from the misinformation that had preceded the project. Luckily Pius V had recognized the importance of the project and referred the matter to Sirleto, who was, in Arias Montano's estimation, the most learned man in the College of Cardinals and the person the pontiff relied on for advice on such matters. Of Cardinal Tiani (Bianchi), Arias Montano said only that he was a follower of the way of Saint Thomas and had no other type of studies. Yet the 1572 approval proved to be only a temporary reprieve from the Polyglot's difficulties in Rome.

20. Macías Rosendo, *Biblia Políglota*, 489.

21. BAV Vat. Lat. 6946, fol. 180, as cited in Denzler, *Kardinal Guglielmo Sirleto*, 136, and Höpfl, *Beiträge zur Geschichte der Sixto-Klementinischen Vulgata*, 311.

22. DIE, 41:275–80.

León de Castro's crusade grew fiercer after the Antwerp Polyglot received papal endorsement. In 1574 he felt emboldened to raise his formal complaint against the project, and the members of the Spanish Council of the Inquisition asked him to put the details of his complaint in writing.[23] The complaints ranged from general concerns to specific places where he alleged that the editor had inserted rabbinical interpretations. Though León de Castro was more than happy to discuss the contents of his complaint with anyone who would listen, defenders of the project—Zayas, Salinas, and Fuentidueña—could not even get their hands on the report. As maestro Francisco Salinas told Arias Montano, the main problem seemed to be the editing of the Pagnino translation and, more specifically, the reliance on Hebrew lexicons (*de Rabinos*) in editing the interlineal translation. León de Castro argued that in effect Arias Montano had made a new translation of the Bible and added a rabbinical interpretation in the margins.[24] As reported by Chacón, León de Castro complained that Arias Montano had placed in the Pagnino Bible "all manner of Hebrew teachings that truly exist, or that some little Judaizer feels like coming up with, which he places on the margins without mentioning the ones Jerome noted."[25] On the same point, Pedro de Fuentidueña understood León de Castro's argument to suggest that, given that rabbinical dictionaries also contained the definitions used in the Vulgate, Arias Montano had intentionally followed the meanings of the rabbis and not those dictated by the Catholic Church.[26] The accusation was that he had fished for "novelties" in the lexicons to bring forth an unauthorized version of the Bible.

Fuentidueña summarized the principal accusations León de Castro had made through 1575 in a letter to Cardinal Stanislaus Hosius (1504–79), prince-bishop of Warmia and papal legate. He objected on the following points:

That the version of Santes Pagnino was included.

That it describes [Pagnino's] as a very proper translation.

That the Vulgate is not conferred great authority.

That treatises are added to the Bible that are from rabbis hostile to the Christian religion.

23. As Fuentidueña reported to Zayas on 4 May 1574, in Macías Rosendo, *Biblia Políglota*, 322.

24. Francisco Salinas to Arias Montano, Salamanca, 13 July 1574, in Macías Rosendo, *Biblia Políglota*, 329.

25. Domínguez Domínguez, "Correspondencia de Pedro Chacón (II)," 216.

26. Macías Rosendo, *Biblia Políglota*, 330–31.

That it puts other versions in certain passages of the Holy Scriptures used in the Vulgate edition to establish some tenets of the faith.

That later versions of the Vulgate having been confirmed by decree of the Sacred Council of Trent are not allowed to take refuge in Hebrew or Greek sources.

And various other matters that, to shorten this exposition, I pass over.[27]

Cardinal Hosius responded favorably to Arias Montano's cause but referred the matter to Cardinal Sirleto. Sirleto refused to raise the issue with ecclesiastical authorities unless a formal accusation was presented against the Antwerp Polyglot or Arias Montano's person.[28] The lion retreated from his Roman offensive.

Over the years Arias Montano had cultivated what proved to be an important and influential friend in Cardinal Sirleto. After meeting in Rome in 1572, the two men started collaborating on various important projects involving revisions or expurgations of biblical texts. Sirleto asked Arias Montano to collaborate on the ongoing revision of the Vulgate, but the king cut that prospect short when he ordered Arias Montano to return to Antwerp. In 1573 Arias Montano informed Sirleto that he was seeking copies of the Talmud so he could take part in its ongoing expurgation. Sirleto was leading an effort to censor Hebrew works and the Talmud, an endeavor that continued until the publication of the Clementine Index.[29] Regardless of Arias Montano's standing among influential members of the Roman Curia, Pedro de Fuentidueña strongly advised him to seek clarification from the pope concerning the Trent Decree on the Vulgate and the situation relative to Pagnino's version. We have testimony that Arias Montano requested the clarification from Joannes Harlemius at the University of Louvain, but no evidence survives on whether he sought clarification from Rome.[30]

---

27. Pedro Fuentidueña to Cardinal Hosius, Salamanca, 10 August 1574, in González Carvajal, "Elogio histórico," 169–71.

28. Macías Rosendo, *Biblia Políglota*, 360–62n351.

29. On the expurgation of the Talmud, see Raz-Krakotzkin, *Censor*, 68–76. Clement VIII's actions after Sirleto's death show a hardening stance against these works. He issued two bulls, one of which, *Quum Hebraeorum malitia* (28 February 1593), concerned the Talmud, kabbalist works, and Jewish commentaries and mandated that they be burned rather than expunged as Sirleto and Arias Montano had hoped: Parente, "Index, the Holy Office," 183.

30. Dávila Pérez, *Correspondencia conservada en el Museo Plantin–Moretus*, 1:173–78. It is unclear whether Arias Montano raised the question before the Congregation of the Council of Trent, the pertinent Vatican body adjudicating such matters.

In the Low Countries, León de Castro had found a kindred spirit in Bishop Lindanus. Although Lindanus had initially supported the Polyglot project, his relationship with Arias Montano soured when the *Apparatus* included a scathing critique of Lindanus's earlier work on a Hebrew Psalter from England that the bishop believed proved some versions of the Psalms had been intentionally corrupted by the Jews.[31] The offended bishop—far from satisfied by Arias Montano's tepid retraction—then directed most of his criticism against Arias Montano's editing of Pagnino's translation of the Bible. He also objected to the Hebrew dictionary included in the *Apparatus*, since it followed the Pagnino/Kimhi version and not that of Philo of Alexandria, which Saint Jerome had used.[32] More fundamentally, he objected to claims concerning the integrity of the Hebrew Bible as the source of the *hebraica veritas* because he believed it had been corrupted by "the perfidy of the rabbis." In 1575 he also exchanged a series of charged letters with two supporters of the Polyglot project: Gilbert Génébrard, professor of Hebrew and Sacred Scriptures at the Collège de France, and Joannes Harlemius, a Jesuit professor of Sacred Scriptures and Hebrew at the University of Louvain. In letters to Génébrard and in the extensive correspondence with Harlemius, he pointed out specific instances where, in his opinion, the Jews had corrupted the Sacred Scriptures. Like León de Castro's criticisms, most of these concerned passages of the Old Testament that Christians had read as prefiguring the Messiah.[33]

Having been summoned back to Spain in 1575 to take charge of other royal commissions, including organizing the new Royal Library at El Escorial, during a stop in Italy on his journey home Arias Montano realized that the complaints from the lion from Salamanca were also getting heard at the Vatican and could no longer be dismissed as "jealousy." Arias Montano wrote to the inquisitor general of Spain, the bishop of Cuenca, for clarification on the matter.[34] The temperate tone of this letter barely hides the anguish he must have felt. Not only was the Polyglot project threatened, his personal reputation was being slandered by accusations of heterodoxy and being a Judaizer. On

31. Antonio Dávila Pérez has studied the exchanges between Arias Montano and Bishop Lindanus extensively in a series of important articles: see "'Regnavit a ligno Deus, affirmat Arias Montano, negat Lindanus'"; "Retractación o pertinacia"; and "Dos versiones de la '*De Psalterii Anglicani exemplari animadversio*.'"

32. Macías Rosendo, *Biblia Políglota*, 388.

33. Ibid., 337–45, and Dávila Pérez, "'Pro Hebraicis exemplaribus et lingua.'"

34. Arias Montano to the inquisitor general, bishop of Cuenca, 12 August 1575, in DIE, 41:316–20.

the same day he also wrote asking Philip II's permission to stay in Rome to continue challenging León de Castro's allegations and to further his studies with the learned company and resources the city provided.[35] Despite persistent rumors that León de Castro was going to pursue the matter officially in Rome, it seems a formal accusation was never made. I could find no vestiges of such an accusation at the Archive of the Congregation of the Doctrine of the Faith (ACDF) or among the papers of Cardinal Sirleto, who headed the Congregation of the Index during those years.

In 1913 historian Hildebrand Höpfl identified and reproduced a document from the Vatican Library that he described as a judgment against the Antwerp Polyglot made by the Sacra Congregatio Cardinalium Concilii Tridentini Interpretum, or the Congregation of the Council of Trent.[36] In an important historiographical correction concerning the 1575–76 interlude of the Antwerp Polyglot in Rome, Morocho Gayo pointed out that Höpfl had misidentified the material.[37] On closer examination the document reveals itself as a *censura* made by an individual about the Greek version used in the project.[38] Nor is there any evidence that this opinion was adopted and forwarded as a policy position by any of the pertinent Vatican councils that would have issued a statement on this delicate matter.[39] Furthermore, the historiography of the Antwerp Polyglot's Roman episode often conflates this *censura* with a 12 January 1576 *declaratio* by Cardinal Antonio Carafa (1538–91) on the authority of the Vulgate.[40] This declaration contains the often-quoted statement about not permitting, under penalty, any alterations to the Vulgate text, "of neither a

35. Arias Montano to Zayas, 19 August 1575, in DIE, 41:321–24. See also Arias Montano to Philip II, Rome, 12 August 1575, in Macías Rosendo, *Biblia Políglota*, 432–36.

36. Höpfl, Beiträge zur Geschichte der Sixto-Klementinischen Vulgata, 108–9, 315–17.

37. Morocho Gayo, "Trayectoria humanística, 2," 264.

38. BAV Vat. Lat. 6207, fols. 176–77. Its title is "Judicium ratione corrigendi Biblia Graeca," unsigned and undated. Höpfl might have used the description on the back of the document, "Judicium de Biblias Regias et de ratione corrigendi eas," as the title he gives the document. (Copy examined at the Vatican Film Library at Saint Louis University, Missouri, microfilm 727.)

39. Batiffol, *Vaticane de Paul III à Paul V*," 73–74n1. Rekers further mischaracterized the document by attributing it to Roberto Bellarmino (who was not head of the Congregatio Concilii): Rekers, *Arias Montano* (1972), 61–62. Wilkinson repeated Höpfl's mischaracterization in Wilkinson, *Kabbalistic Scholars*, 95.

40. These declarations are found as annotations made by Cardinal Carafa in BAV Vat. Lat. 6326, fols. 2–227v. Declarationes Sacrae Congregationis Illustrissimorum Cardinalium super Concilium Tridentinum in eiusdem Concilii decretis.

single period, clause, part, diction, syllable, or iota."[41] Cardinal Carafa—then the first precept of the Congregation of the Council of Trent—later succeeded Cardinal Sirleto as librarian of the Vatican. For our purposes, it is essential to disaggregate the two documents and note that Cardinal Carafa's statement was not made in relation to the Antwerp Polyglot. Nor was it ever meant to be an official indictment against creating a new version of the biblical text using the Hebrew Bible or the Septuagint. At the time he wrote this document Carafa clearly thought the definitive version of the Vulgate still lay in the future, but he argued that any corrections introduced should be derived from neither the Greek nor the Hebrew text, but should be based only on earlier manuscripts of the Vulgate itself.[42] A few years later he headed a commission to create the Sixtine Bible, on which Arias Montano was invited to participate. So, rather than finding a hostile environment in Rome, as much of the historiography suggests, Arias Montano found a very welcoming place where he felt understood and protected. He was surrounded by influential individuals who shared his conviction that it was possible to create a new, more accurate version of the Sacred Scriptures; it was where he wished to continue the work he had begun to publish in Antwerp.

MARIANA'S *CENSURAE*

That the epilogue of the charges against the Antwerp Polyglot played out in Spain should come as no surprise. It was where León de Castro filed his formal accusation, and the matter thus came entirely under the jurisdiction of the Spanish Inquisition. Following the standard procedure and given the nature of the *denuncia* as against a book and not against Arias Montano's person, it had to be handed over to a *calificador*, or examiner. The Jesuit Juan de Mariana (1536–1624) was selected to write the censorial report on the Antwerp Polyglot, intended to be the final word on the whole affair.[43] By selecting a Jesuit, the Inquisition might have been trying to strike a middle ground. The order advocated the study of Hebrew, and although it maintained the primacy of the

41. "[A] propter huiusmodi verba S. C. C. censuit incurri in poenas vel si solo periodus, clausula, membrum, dictio, syllaba, iotave unum quod repugnat vulgatae editioni immutatur.": BAV Vat. Lat. 6326, fol. 18v.

42. Batiffol, *Vaticane de Paul III à Paul V*, 76.

43. For Mariana's biography, see Noguera Ramón, "Historia de la vida y escritos de p. Juan de Mariana," 1:i–lxxxiv; Mariana, *Obras del padre Juan de Mariana*, 1:iv–xlix, and Jiménez Guijarro, *Juan de Mariana (1535–1624)*.

Vulgate on doctrinal matters, it taught an exegetical approach that included consulting the Hebrew Old Testament and Greek Septuagint as resources.[44] The lengthy report was ready by August 1577, as Mariana informed the inquisitor general in a missive that includes a chapter-by-chapter summary of the report.[45] Once thought lost, a fair copy of the whole *censura* survives at the Vatican Library.[46] Mariana clearly had before him not only León de Castro's charges, but also Arias Montano's response. Mariana divided the report into two parts. In the first he addressed eight general premises that undergirded León de Castro's accusations, the last of them containing twenty-six specific allegations. The second part was devoted to nine aspects of the Antwerp Polyglot that Mariana himself found wanting. Because these did not touch on matters of faith they did not merit a new denunciation, and because they were not part of León de Castro's denunciation they were not relevant to that case. In Mariana's *censurae* Arias Montano stood as the sole party responsible for the Antwerp Polyglot; even when he singled out a certain contributor, such as Lefèvre de la Boderie, it was Arias Montano who had to answer. In what follows I will focus on aspects of the *censurae* that relate to Arias Montano's exegetical methodology, since these are the most relevant to the development of his natural philosophy.

Mariana began by first justifying the study of Hebrew and other biblical languages because of their utility in the study of Sacred Scriptures and then defending the integrity of the Hebrew Bible against León de Castro's accusations that they had been corrupted by the Jews. He also corrected León de Castro's assertion that among the ancient biblical texts only the Septuagint survived

44. Wicks, "Catholic Old Testament Interpretation," 636–39.

45. Juan de Mariana to the inquisitor general, Cardinal Quiroga, Toledo, 16 August 1577, in Macías Rosendo, *Biblia Políglota*, 464–68.

46. BAV Barb. Lat. 674. fols. 14–66. Juan de Mariana, *Jo. Marianae censurae in Biblia Regia, quae nuper diligentia et industria D. Benedicti Arias Montani in lucem editae sunt* (consulted as microfilm 9789 at the VFL, St. Louis University). The contents of this document correspond to the sections and chapters described in the August 1577 letter cited above. A comprehensive summary was published in Noguera Ramón, "Historia de la vida y escritos de p. Juan de Mariana," xx–xxxi. For more on the *censura* when it was thought lost, see Asencio, "Juan de Mariana y la Políglota de Amberes," 50–80; Rey, "Censura inédita del P. J. de Mariana de la *Políglota Regia de Amberes*," 523–48. A fragment from fols. 62v–63v appears translated in Macías Rosendo, *"Apparatus sacer,"* 37. Wilkinson's analysis of the section on the Syriac New Testament relies on Mariana's *Pro editione Vulgata* of 1609, where the Jesuit reproduces many of his earlier arguments: Wilkinson, *Kabbalistic Scholars*, 96–99.

uncorrupted, because it had been cited by the apostles. This was not a valid argument, alleged Mariana, since the apostles also cited the Hebrew Bible. On the question of exegesis, Mariana maintained that there could be more than one way to read the Bible and therefore more than one literal interpretation, so that even if the church preferred a particular interpretation or translation, it did not mean others were to be condemned. Furthermore, the object of translating the Bible was not to correct the Hebrew or Septuagint versions—as León de Castro thought should have been done—but to faithfully translate the text in question. Errors and variations were to be expected in such ancient texts, but these were not necessarily injurious to religious tenets. With these arguments Mariana was coming out in full support of the literal interpretation of Sacred Scriptures directly from the Hebrew, not only undermining León de Castro's principal arguments, but fully endorsing the general methodological approach behind the Antwerp Polyglot.

Despite his commitment to studying the Sacred Scriptures in Hebrew, however, Mariana drew the line at using the Masoretic Text to correct the unvocalized Hebrew Bible; it was not a valid approach because not only was the Masorah a later text, but he considered it had been altered by the Jews to the detriment of the Christian faith.

Mariana also proved to be no friend of the Pagnino Bible. Nevertheless, he corrected one of León de Castro's more virulent accusations: that Arias Montano had included rabbinic commentaries in the marginalia of the Pagnino Old Testament (volume 7 of the Antwerp Polyglot). Mariana explained—better than Arias Montano had, in fact—that the marginal annotations followed a four-part convention depending on the combination of languages and fonts, and that neither of these indicated rabbinic commentaries.[47] This led Mariana to León de Castro's frequent accusations against Judaizers. He first drew distinctions between Christian Hebraists. Clearly, Mariana asserted, León de Castro was not referring to persons who had abandoned the Christian faith owing to "Jewish perfidy" or he would be raising an injurious lie against someone as devout as Vatable (and, though unnamed, Arias Montano). Mariana then identified a suspicious tier among certain Christian Hebraists—Pagnino among them—whom he considered to have become too passionate about the study of the Hebrew language and the teachings of the rabbis, to the point of adopting some of their opinions. The number of Christian Hebraists of this type was high, especially in Spain, he added, but many such scholars could

47. See chapter 4, note 36 above.

also be found where the Hebrew language was commonly known or where there was greater freedom of opinion.[48] After criticizing several instances in Pagnino's Bible where he detected anti-Christian translations, Mariana suggested that perhaps it would have been best to omit the Pagnino Bible entirely rather than taint the project by association. Neither was the censor convinced that the approvals the Antwerp Polyglot had garnered from the professors of Louvain and Paris were authoritative enough to exempt the project from further inquiry as Arias Montano had alleged. After all, he explained, the review by the theologians in Louvain had been cursory and did not include a close examination of the treatises of the *Apparatus*, while the papal approval was granted as a *motus proprius*, "which is the same usually granted to common books."[49] The suggestion here was that including the Pagnino translation with Arias Montano's editing in a project of this magnitude should have been subjected to far more scrutiny, for it appeared to grant to a particular interpretation the status of canonical text that it certainly did not have.

Although Mariana did not relate any of these problems directly to Arias Montano, the message he sent to the inquisitorial authorities who read the *censurae* was clear: the theological supervision of the project had been lax, and Arias Montano, by having followed Pagnino so closely, could be suspected of belonging to the suspicious tier of Christian Hebraists. Therefore the Antwerp Polyglot merited closer examination. Mariana, however, reaffirmed the merits of studying the Bible in its original languages and made it clear that doing so should not imply or carry the charge—a capital charge in Spain, we should remember—that the scholar was a Judaizer.

Like Chacón before him, Mariana found León de Castro's specific accusations against Arias Montano's orthodoxy—twenty-six instances in all—either incorrect, doubtful, or contradictory, particularly where León de Castro's claims rested on his flawed version of the history and integrity of the Vulgate and the Septuagint. Mariana largely dismissed León de Castro's objections, but he backed three minor points and added a few admonitions to Arias Montano's rather sparse response to the accusations. First, Mariana pointed out, there were only three instances in Arias Montano's edition of Pagnino that should be corrected (Job 19, Psalm 15, and Psalm 21 VUL). Second, Arias Montano should have chosen a Hebrew dictionary other than the Pagnino/ Kimhi one, which relied too heavily on rabbinical scholarship. This had led him into some errors, for example, repeating the notorious mistake of inter-

---

48. Mariana, *Censurae*, fol. 20v.

49. "[Q]uae quem admodum aliis vulgo libris solet.": Mariana, *Censurae*, fols. 21v–22r.

preting the word גֶּבֶר (*gāber*) in Isaiah 22:17 as "rooster" instead of "man."[50] And finally, Mariana found that the project could have spoken more strongly in defense of the authority of the Vulgate by including patristic commentaries in the *Apparatus*.

In studying Arias Montano's responses to León de Castro's accusations, Mariana found that they were too brief and that he had not sufficiently explained the reasons behind the changes and commutations of letters that appear in the Hebrew Bible.[51] (Not wanting to leave this matter unresolved, Mariana cited several authorities who testified that these alterations were common in Hebrew, including Saint Jerome and Rabbi David Kimhi.)[52] In fact, in most of the twenty-six resolutions Mariana showed that his knowledge and access to Hebrew and rabbinic sources were ample to defend Arias Montano's editorial choices, finding explanations for particular translations, admitting variations, and for the most part exonerating the Antwerp Polyglot from any theological heterodoxy. Yet in what is largely a defense, he expresses displeasure with Arias Montano's insistence on following definitions indicated by the roots of Hebrew words and consequently changing the meanings of certain words in translating instead of following the Vulgate's translation where definitions were close enough.

50. Mariana, *Censurae*, fol. 26v. Indeed, under the definition of *gallus* in the *De arcano sermone* Arias Montano included the reference to Isaiah 22:17. Arias Montano, *Libro de José*, 276, 485, [77]. Yet this was an unfair criticism, because in his editing Arias Montano had modified Pagnino's "Ecce Dominus migrando migrare facti, o fortis" to "Ecce Dominus transportans te transportatione vir."

51. Arias Montano's point-by-point response to León de Castro's accusations that Mariana saw might in fact survive in a manuscript written in a sixteenth-century hand at HSA B1351, "Defension y respuesta, de la traslacion latina ad v[er]bum del Hebreo, q[ue] esta en el segundo tomo del sacro apparato de la biblia real." It is in Spanish, undated and unsigned, but internal evidence suggests it was written by Arias Montano, responding to accusations made by "m. l." (maestro León?) against the *biblia real* ("De lo q. m. l. ha observado y alborrotado en este particular"). The document recounts the gestation of the Antwerp Polyglot, dwelling on the numerous consultations and approbations it underwent, including those that took place in Rome in 1572 and in 1575–76. It also describes León de Castro's intervention at court with the king and later in Rome concerning the inclusion of the Pagnino ad verbum Bible, which he despised and continued to disparage in his books and apparently in some printed broadsheets. The document then makes some clarifications concerning the Hebrew Bible, the use of punctuation in Hebrew, and the allegation concerning the intentional corruption of the text by Jews. It also addresses—indeed quite briefly—the same twenty-six specific objections made by León de Castro and commented on by Mariana in the *Censurae*.

52. Mariana, *Censurae*, fol. 30.

In the second part of the report, Mariana turned to the *Apparatus*, which León de Castro had not examined as carefully as the biblical translations. Mariana found the Chaldean paraphrase a useful addition but reprimanded Arias Montano for not being as careful as he could have been in identifying and pointing out all the places where the text had been corrupted by the Jews. He also noted the references to Guillaume Postel (who, he pointed out, was of *converso* origin [*conversus esse*]) and Jean Mercier, both of whom contributed to the preface, grammar, and translation of the Syriac New Testament by Guy Lefèvre de la Boderie. He castigated the Syriac version because he considered it an improper translation with omissions that affected matters of faith, aspects the translator should have noted in the preface—and that the editor should have caught.[53] For the same reasons he found particularly offensive the Chaldean grammar by Andreas Masius and the dictionary by Lefèvre de la Boderie, which openly relied on Jewish Kabbalah to identify the roots of several words and, furthermore, praised the Zohar as prophetic and even divine because of the mysteries revealed from the Sacred Scriptures.[54] It was inexplicable to him why authors with such an affinity for Kabbalah were included in a Bible sponsored by a Catholic king! To dismiss the merits of any kabbalistic intepretation and by way of example, Mariana briefly synthesized an unattributed kabbalistic interpretation of the first three letters of Genesis 1 (ברא, *br'*) as "In the son, the father created the heaven and the earth" and of the six occurrences of א in the first verse as an indication of how long the world would exist until Judgment Day. (This explanation does not appear anywhere in the Antwerp Polyglot.) He pointed out that kabbalistic texts circulated in Spain

53. This issue concerned 1 John 5:7 ("For there are three that bear record in heaven, the Father, the Word, and the Holy Ghost: and these three are one," KJV), which was not to be found in the Syriac Bible, and which the church maintained the Arians had removed so as to support their heresy. Mariana explains this issue further in Mariana, "Tractatus II—Pro editione Vulgata," 106–18.

54. See, for example, the definition of the word *alius* in Lefèvre de la Boderie, *Dictionarium syro-chaldaicum*: "אִידְךָ Alius, vel potius ordo fructorum lapidum . . . [R]adix est יד quae in Tikun Hazohar est dictio mystica. Iod enim decem nomina divina, quae vocantur decem Sephiroth Belimah innuit, et quicquid denario numero comprehenditur, ut novem ordines Angelorum cum ordine Animastico, ut ita dicam, id est cum ordine mentium humanarum. 10. Sphaeras caelorum in mundo visibili magno, 10. cortinas Tabernaculi in typicis, decem praecipua membra corporis humani in Microcosmo, et cetera id genus; quae fusius explicavimus in Encyclico Poëmate nostro, quod octo Circulis, Librisve absolvimus. Daleth verò ejusdem nominis יד in numeris valet 4. et 4. literas magni et sacrosancti nominis illius יהוה iuxta illud יהוח הויח כי יד Quoniam manus Domini Tetragrammaton est הויה.": AntPoly, 6:6.

and noted that works by Johannes Reuchlin and Francesco Giorgi had been banned.[55]

Of the other treatises of the *Apparatus*, Mariana complained that in some instances the definition used in the translation did not correspond with the one given in the accompanying lexicon, whether the Chaldean or the Hebrew. This was the case with רָחַף (*rāḥap*) in Genesis 1:2, which, as I mentioned in chapter 4, Arias Montano had changed to *motabat*, whose precise definition the Hebrew lexicon did not include; furthermore, it was translated as *insufflabat* in the Chaldean paraphrase.[56] Among the handful of examples he studied, Mariana cited Arias Montano's inconsistency in translating קְשִׂיטָה (*qěśîṭâ*), changing Pagnino's "money" to "lamb" in Genesis 33:19 and Joshua 24:32 but changing Pagnino's "cattle" to "money" in Job 42:11. Arias Montano ignored the Vulgate translation, Mariana admonished, where the word is consistently translated as "sheep" (in Latin, *agnus*). Mariana noticed that when one turned to the treatises in the *Apparatus* where Arias Montano offered additional explanations, he sometimes contradicted the very translations that appeared elsewhere in the Antwerp Polyglot.

Mariana noticed the extent of the changes made to the Pagnino Bible and pointed out instances where he thought Arias Montano had made hasty decisions or had not considered all possible meanings of a word when making a change.[57] For example, Mariana found the translation of *maledictus* in Genesis 4:10 [*sic*] and Genesis 3:14 to be inconsistent.[58] In Genesis 7:18 Arias Montano replaced *roboraverunt se* (they strengthened themselves) with *invaluerunt* (they grew strong), and in Genesis 14:4 he replaced *defecerunt*, perhaps too vague a term, with *rebellaverunt*, a word that more clearly im-

55. Mariana, *Censurae*, fol. 57v. On the banning of the Kabbalah, see Burnett, *Christian Hebraism*, 104 5, 227. On negative sentiment toward the Kabbalah in Spain, see Secret, *Kabbalistes chrétiens de la Renaissance*, 218–27.

56. Mariana, *Censurae*, fol. 48.

57. "Gen. 1 dictionis *expansiones* et *motabat* ab illo desumpsit et c. 7 *invaluerunt*, et 14 *revellaverunt*, multaque alia, qua recensere longum esset ab illo est mutatur. Deut. c. 29 v. 18 ubi nos *fel et absynthium* Pagnino et Vatablus professe *venenum* reddiderunt, qua dictio ראש aliquando intoxici significatione invenitur. Arias relictis omnibus maluit eam dictionem in *capitis* significatione interpretari, quo videlicet nec pes, nec caput uni reddatur forma, quemadmodum ille ait, qua enim sententia excellis. Ariae verbi constare possit et *sit inter vos radix fructificans caput, et absynthium* mitto alia; quoniam, ut saepe dixi: non vacavit omnia considerare et fortassi ad rem de qua agitur hac ipsa nimis multa fuerunt.": Mariana, *Censurae*, fol. 59.

58. Mariana, *Censurae*, fol. 58v. Pagnino's version of Genesis 4:11 reads, "Et nunc maledictus tu à terra, quae aperuit os suum, ut susciperet sanguines fratris tui de manu tua."

plies rebellion. Another translation puzzled Mariana because it deviated not just from the Vulgate, but also from the translations of Pagnino and Vatable. Mariana deduced that Arias Montano had consulted the meaning of the root of ראש (rōʾš) in Deuteronomy 29:18 and, finding that "head" and "principal" were possible meanings, had changed the sentence from "lest there should be among you a root that beareth gall and wormwood" to "lest there should be among you a root that beareth head and wormwood." Mariana could not resist quipping that he could not make heads or tails of the substitution.[59] In other instances, Mariana complained that Arias Montano had followed the literal translation too rigidly and had put forth a translation that favored the Protestant interpretation.[60] For example, in the treatise *Idiotismus*, he noted that Arias Montano based his work closely on that of Bartholomeus Westheimer's *Phrases in Sacrae Scripturae*, which might have let some (unspecified) Protestant ideas from that banned author creep in.

As for the treatises of the *Apparatus*, Mariana mentioned that they had been found not worthy (*indigna*) of the project by all the learned men who had seen them so far. His objections, which mostly concerned the *De arcano sermone*, were succinct and on point. He noted that the prefatory essays of the treatise were written in a way that made familiar concepts seem remote by deviating from generally accepted precepts of physics, though he does not specify what aspects. And if he was not mistaken in his opinion, in these sections Arias Montano seemed to be following the doctrines of Ramón Llull. This was a curious claim, because the divisions Arias Montano put forth in the *De arcano sermone* (discussed in the previous chapter) do not map onto the ontological divisions of Llullian art.[61] However, if we interpret Mariana's

59. In fact, the Hebrew lexicon in the Antwerp Polyglot listed *caput* as the first definition and mentions that others favored *venenum*, following Rabbi David Kimhi: Raphelengius, "Thesauri hebraicae linguae," 6:114.

60. The issue concerned Psalm 15:10 VUL ("Non derelinques animam meam in inferno, nec dabis sanctum tuum videre corruptionem.") which Arias Montano amended in the Pagnino edition as follows: "Quoniam non relinques animam mean in sepulchro: *non dabis* [nec permittes] misericordem tuum *videre foveam*. [, ut videat corruptionem]." (Pagnino's version is shown in square brackets. It appears in that Bible as Psalm 16:10.) In the *Idiotismus* Arias Montano went further and replaced *animam* with *corpus*. Mariana, *Censurae*, fol. 60. For a detailed explanation of what was objectionable about the translation, see Mariana, "Tractatus II–Pro editione Vulgata," 73.

61. "Et initio quidem remittit multiplices dictiones, quas afferre ad schola physica consuetudine prorsus remota sunt, et ni ego multum fallor ex Raymundi Lulli schola.": Mariana, *Censurae*, fol. 62v. On the Lullian art, see Rossi, "Logic and the Art of Memory," 38.

claim more broadly, we can find a few reasons for the objections. Mariana might have seen Arias Montano's work as a reference to the pansophic ideal of a universal science that sought to find a single language—that of the Creator— from which to uncover, interpret, and articulate ideas about the natural world. Or perhaps Mariana's objections stemmed from the fact that Arias Montano, like Llull, studiously avoided using Scholastic terminology that might have made the section more intelligible to an audience accustomed to discussing ontology or categories using Scholastic terms. Or further yet, that in including a long exposition on the divine names (*Arcanorum nominum interpretatio*),[62] Arias Montano was following the lead of the *De divinis nomibus* of Pseudo-Dionysius the Areopagite (d. ca. 500), from whose work Llull had drawn inspiration for his list of divine attributes.

Throughout the second part of the *censurae*, and particularly when he commented on Arias Montano's treatises in the *Apparatus*, Mariana comes across as being irked as much by Arias Montano's lack of citations as by his manner of stating his etymological speculations as definitive translations without engaging with other opinions, even those that supported his own interpretations. An exasperated Mariana complained, Why didn't Arias Montano cite at least Jewish sources, such as Maimonides's *De siclis*, when he discussed coinage or the meaning of the Ark of the Covenant?[63] Yet it was also troublesome when Arias Montano did mention his sources. Mariana complained that the extensive discussion of the divine names of God in the *De arcano sermone* borrowed much from the Kabbalah and did not follow the teachings of the patristic fathers or of "our" theologians; Arias Montano seemed to have been led by the teaching of the rabbis, citing the Mishnayoth more often than Saint Jerome and even saying as much in the preface of the treatise *De mensuris*.[64] After explaining the relation between the Talmud and the Mishnayoth—which, he pointed out, Saint Jerome had criticized and the Catholic Church had prohibited since 1553—he admonished Arias Montano for mentioning its use and relying on it for the treatises in the *Apparatus*.

Interestingly, there is only one instance in the whole *censurae* where Mariana criticizes Arias Montano for a potential natural philosophical error. As Mariana read it, Arias Montano had incorrectly placed the creation of plants

62. Arias Montano, *Libro de José*, 105, 412, [4].

63. Mariana, *Censurae*, fol. 63v. For Arias Montano's sources in this and the other treatises in volume 8 of the *Apparatus*, see the excellent annotations in Arias Montano, *Antigüedades hebráicas*.

64. Ibid., fol. 63.

and tress on the fourth day and furthermore considered plants and trees "inanimate."[65] On the first point Mariana was incorrect. In the sentence where the reference occurs Arias Montano was referring not to plants but to inanimate bodies created on the fourth day, the sun and the moon, which indeed exhibit two different natures.[66] (Arias Montano knew the order of Creation perfectly well, since later in the text he clearly indicated that plants were created on the third day.)[67] About the second point, Mariana observed that the notion of plants as inanimate was against the more commonly accepted opinion of Aristotle and the Peripatetics, following instead the opinions of the Stoics and Epicurus as described in Plutarch's (46–120 AD) *Placita philosophorum* (book 5, chapter 24), Clement of Alexandria's (ca. 150 to ca. 215 AD) *Stromata*, (book 8), and Theodoret of Cyrus's (d. 457) *Graecarum affectionum curatio*, discourse 5, chapters 24–25; these, he suggested, were the sources of the introductory section of the *De arcano sermone*. Yet Mariana's assertion was off target; indeed, all three authors mention the Stoic notion of plants as inanimate and note that this position contradicted Plato and Aristotle, but they do not set forth the same metaphysical ontology that serves as the treatise's foundation; thus they could have not been Arias Montano's sources. Yet Mariana's larger point, that Arias Montano considered plants inanimate along Aristotelian lines (or even Platonic ones), is correct. Arias Montano did not consider plants as having a soul and life (*anima et vita*), but they did share the earth's capacity to sustain growth and, through the agency of ELOHIM working on its seminal forces, the capacity to germinate, grow, and reproduce.[68] Arias Montano's language here suggests that God had instilled in the earthy matter he referred to as ARESZ (אֶרֶץ, *'ereṣ*) distinct natures and forces capable of begetting and sustaining life. Earth, once possessing these, had furnished itself with the plants

65. "Profecto quo loco, herbas et arbores quarto die factas esse affirmatur contra expressa Gen. verba c. 1 v. 12 et 13. easdem ponit inter inanimata contra Aristot. Peripateticorum communemque scholae oppinionem ex sententia videlicet Stoicorum atque Epicuri, ut Plutarchus lib. 5 De placitis philosophorum c. 26, Clem. Alex. Stroma[ta]. 8, Theodor. contra Graecos Serm. 5 auctores sunt eius libri initio, cui titulum facere placuit Joseph sive de arcano sermone.": Mariana, *Censurae*, fol. 62v.

66. "Toto hoc genere, maiora, digniora atque specie ipsa pulchriora sunt, quaecunque quarta die sunt edita, duplici etiam naturae gradu distincta: quorum altera pars duobus etiam terminis distinguitur; herbam enim virentem et herbam facientem semen exhibuit illa dies.": Arias Montano, *Libro de José*, 95, 405, [2].

67. Arias Montano, *Libro de José*, 101, 410, [2].

68. *Corpus*, 353; *HistNat*, 244.

and trees that adorn it and bequeaths this nature to them.[69] As we will see in chapter 10, in the *History of Nature* he completely avoided discussing the animate/inanimate distinction between plants and animals.

Despite his sometimes scathing criticism and, at best, measured praise, Mariana did not recommend banning any part of the *Apparatus*, having found nothing theologically wrong in it. He insisted, however, that it could be improved and corrected by its author. In sum, Mariana considered the *Apparatus* a work that opened the door to endless controversies and thus had no place in a royal project. Mariana reached this conclusion for entirely different reasons than León de Castro had raised, which he found largely baseless. Instead he identified problems with the project that were far subtler, but potentially no less subversive. The problem was not that the whole Antwerp Polyglot bore obvious hallmarks of Judaizers. After all, there was nothing wrong with citing Jewish sources on matters that did not concern faith. The problem was that some contributors (including Andreas Masius and Guy Lefèvre de la Boderie) had acknowledged the work of others (such as Guillaume Postel and Sebastian Münster) who were known to have held heretical ideas. This was enough to taint the whole project.

It is evident in the *censurae* that Mariana's unease with the project stemmed largely from Arias Montano's philosophy of translation as evinced by his editing of the Pagnino Bible. In his view Arias Montano had included in the project a very idiosyncratic translation of the Bible based on a hyperliteral interpretation of the Hebrew text, a translation that Mariana himself could barely explain and that Arias Montano had refused to explain by citing conventional sources and methods. In his reluctance to cite sources, especially patristic authorities or even Saint Jerome, Mariana found a measure of disrespect toward the entire Catholic biblical tradition, including the Vulgate itself. In the few instances where Arias Montano had cited sources, such as his reference to the Mishnayoth, Mariana found them improper and unconvincing. And in the *De arcano sermone*, his explanations of the division of things in the world were so unconventional and so against accepted Aristotelian wisdom that Mariana had to reach for the Llullians, Stoics, and Epicureans to find anything like them.

Largely because Mariana had not found anything against the Catholic faith in the Antwerp Polyglot, the Spanish Inquisition was under no obligation to pursue León de Castro's accusations further, so his opinion put the matter to

69. Arias Montano, *Libro de José*, 101, 410, [2].

rest. It is unclear to what extent Philip II influenced this outcome, because Mariana's report certainly could have been read as a call for further scrutiny. But by the time the report was finished, fray Luis de León and Martín Martínez de Cantalapiedra had been cleared by the inquisitorial tribunal and were back in the lecture halls of the University of Salamanca, and the adversaries of Spanish Christian Hebraists were retrenching.

The public polemic surrounding the Antwerp Polyglot was not over, however. In the next phase the three Spanish actors took their opinions to print.[70] León de Castro published his objections in a book that took almost six years to work its way through the censors. The *Apologeticus pro lectione apostolica* (1585) offers an extended argument in defense of the Vulgate and patristic commentaries, as well as making charges about how first the Jews and then heretics had corrupted the Sacred Scriptures. The first book revisits the original twenty-six objections plus a critique of the *Idiotismus*, yet without a single mention of Arias Montano or the Antwerp Polyglot.[71] Mariana also continued studying the matter and would publish his more polished views in his 1609 "Tractatus II," in the second treatise, *Pro editione vulgata*.[72]

Perhaps aware that León de Castro was preparing to present his criticism in a book, Arias Montano took advantage of Plantin's publication of a new, standalone edition of the Pagnino Bible in 1584 to make widely known his response to the firestorm of some years back. In its second preface, titled *De varia Hebraicorum librorum scriptione et lectione commentatio* (On the study of the diverse writing and reading of Hebrew books) Arias Montano began with a brief history of the Bible. He explained the existence and validity of two types of Sacred Scriptures written in Hebrew: the purely consonantal one of Moses

---

70. The controversies surrounding the Antwerp Polyglot did not suddenly cease with Mariana's pronouncement. They continued into January 1579 when the inquisitor general, Cardinal Quiroga, learned of complaints by Dominican friars in Seville, but more importantly when he received a letter from Cardinal Giacomo Savelli (1523–87), then inquisitor mayor of Rome. He ordered Savelli's letter reviewed by "personas graves" at the University of Salamanca. The king, noting that Savelli's objections concerned only the Targum, ordered that Arias Montano not be notified, out of concern that he would once again have to defend himself. The king's comment, written in his own hand, reads, "he will not hear it from me" (*de mi no lo sabrá*). AHN, Inq. L. 284, fols. 204–5. For a synthesis, see Pizarro Llorente, *Gran patrón en la corte de Felipe II*, 269–71.

71. Castro, *Apologeticus pro lectione apostolica*, 93–104.

72. Mariana, "Tractatus II—Pro editione Vulgata," 33–126. The book was later censored in Spain, and the political treatise *De monetae mutatione* was banned.

and one fully vocalized during the time of Ezra known as the Masoretic Text. He considered the second essential for understanding the first, a position he maintained despite Mariana's admonition. As Saint Jerome instructed, "vowel and points are to be used for inference and judgment."[73] In other words, the vocalization points should be used to clarify meanings, and Saint Jerome said as much. The consonants in Hebrew scriptures also varied between copies, for a number of reasons: copyists might have confused similar-looking letters, such as the notoriously troublesome ר (*resh*) and ד (*dalet*), or the sounds of the letters might be too similar to differentiate. Further complicating the matter was the variety of pronunciations within Hebrew, as biblical reference to the ancient use of the word *shibboleth* testified. These lexical issues are compounded by the fact that many words in Hebrew have multiple, sometimes contradictory meanings.[74] The editors of the Pagnino edition tried to correct these discrepancies, Arias Montano wrote, by using texts that are vocalized and always following the rules of Hebrew grammar, as is clearly noted in the marginalia. In what amounted to a defense of his approach to translation, Arias Montano stood his ground, reaffirming the merits of consulting the Masorah as a source for biblical exegesis and displaying a nuanced understanding of the problems associated with oral and textual transmission.

Arias Montano then responded point by point to León de Castro's accusations. Although he never mentions him by name, he alludes to a certain critic as an Erostratus, someone willing to burn down the temple for the sake of personal fame. He had been wrongly accused by this critic of allowing corruptions introduced into the Bible by the Jews to persist in his edition and also of changing the Sacred Scriptures for nefarious purposes. After maintaining the integrity of the Hebrew and Greek Bibles from which the current translations were derived, he tackled most of León de Castro's accusations by giving extended and far more detailed replies than those he had supplied Mariana. Arias Montano explained the grammatical and lexical reasons for the offending changes, often defending Pagnino's choices; elsewhere he highlighted instances of metathesis (changes in word order), as in the title of Psalm 9, or expounded on the reasons a *verbo ad verbum*, word-for-word, translation may lead to meanings different from those rendered in the "elegant" Latin of the Vulgate. Once again, Arias Montano's defense of his approach was sys-

---

73. "[P]uncta et vocalis adhibentur pro consequentia et arbitrio.": Arias Montano, "De varia hebraicorum librorum scriptione et lectione commentatio," in *Biblia hebraica* (1584), 1.

74. Ibid., 1–3.

tematic and equable, and by the same token it still avoided denouncing the "Jewish perfidy" or the Kabbalah as early modern Catholic readers had come to expect.

WAS THE *DE ARCANO SERMONE* KABBALISTIC?

Arias Montano's equanimity in the face of such charged accusations lends some credence to Mariana's claims that our biblist was no enemy of the Kabbalah and that the *De arcano sermone* drew from the teachings of its rabbis. But can the treatise be described as kabbalistic?[75] Arias Montano—and most Christian Hebraists—clearly shared some of kabbalists' fundamental premises, such as the primacy of Hebrew and the importance of understanding Genesis as a way to gain knowledge about the nature of God. Yet these were also essential components of the long tradition of natural theological thought. Like the kabbalists, Arias Montano believed the Hebrew Bible held the key to a deeper understanding of the Creation and of humanity's place in it, but he did not see knowledge gained this way as esoteric, accessible to only the initiated. Instead, he stated repeatedly in the prefatory material of the Antwerp Polyglot that the objective of the *Apparatus*, especially the *De arcano sermone*, was to provide the tools necessary for making this deeper understanding attainable to anyone willing to make the effort. Indeed, the approach presented in the *Apparatus* required a sort of initiation, but only if we consider mastering Hebrew an initiation rite, as would be following the philological approach to literal interpretation and immersing oneself in the culture of the ancient Jews. When we consider that Kabbalah arose in the thirteenth century as Jewish mystics' response to the rationalism of Moses Maimonides (d. 1204) and his followers, we see that Mariana's other accusation about Arias Montano siding with the rabbis might have been more on target. In his prefaces to the Antwerp Polyglot one hears echoes of Maimonides's rational arguments in the *The Guide for the Perplexed*,

> The object of this treatise is to enlighten a religious man who has been trained to believe in the truth of our holy Law, who conscientiously fulfils his moral and religious duties, and at the same time has been successful in his philosoph-

---

75. For a general introduction to the history and tenets of Kabbalah, see Tirosh–Samuelson, "Kabbalah," and Tirosh–Samuelson, "Kabbalah and Science in the Middle Ages." On early modern Jewish magic and kabbalistic practices, see Idel, "Magical and Neoplatonic Interpretation of the Kabbalah in the Renaissance."

ical studies. Human reason has attracted him to abide within its sphere; and he finds it difficult to accept as correct the teaching based on the literal interpretation of the Law, and especially that which he himself or others derived from those homonymous, metaphorical, or hybrid expressions. Hence he is lost in perplexity and anxiety.[76]

Neither does Arias Montano fit neatly within the tradition of Christian kabbalists such as Jacques Lefèvre d'Étaples, Guillaume Postel, Giovanni Pico della Mirandola, and Johannes Reuchlin.[77] In Arias Montano's printed and surviving manuscript works he never acknowledged kabbalist methodologies as valid exegetical tools. Neither did his methodology rely on the numerology or alphabet symbolism central to the exegeses of Reuchlin and Pico. As for word and letter manipulation, he does not use *temurah* (changing consonants), *gematria* (assigning numbers to words or phrases), or *notarikon* (permutation of consonants) as ways of wringing additional meanings from scripture, although his reliance on identifying Hebrew roots sometimes led him to equivocate.[78] On a few occasions he acknowledged that employing metathesis (transposing letters or words in biblical passages) might account for the changes in the meaning, but he attributed these to transcription problems.[79] Comparing Arias Montano's approach with that of Guillaume Postel shows that, although they agreed on a few points—such as granting Hebrew status as the primordial language that harbored secrets forgotten over time that other languages had not preserved—their methodologies for recovering these meanings were radically different. Postel developed a method he called *l'emithologie* that sought to recover the true (אֱמֶת, *ĕmet*) meaning of words by juggling their Hebrew and Greek etymologies and constructing anagrams

76. Maimonides, *Guide for the Perplexed*, 2.

77. Pico della Mirandola and Reuchlin were notoriously absent from the *Index* prepared in 1571 under Arias Montano's direction, even though, as Gómez Canseco points out, their kabbalistic works had been banned in Spain in 1559 and Rome in 1564: Gómez Canseco, "Los sentidos del lenguaje arcano," 84–85.

78. Sáenz-Badillos, "Benito Arias Montano, Hebraísta," 349, 353.

79. This is the way he explains how the Hebrew word for earth אֶרֶץ (*'ereṣ*) is similar to the Latin *terra*. (Arias Montano transliterated אֶרֶץ as ARETS.) He notes that when the final letter ץ (*tsade*) is removed and the letters are read backward the word corresponds to the Latin TERA. However, I consider this, instead of kabbalistic word manipulation, an example where Arias Montano's intention was to show how later languages, such as Latin, retained vestiges of the Adamic language: Arias Montano, *Libro de José*, 233, 467, [59]. Gómez Canseco, however, considers it an example of *temurah*: Gómez Canseco, "Sentidos del lenguaje arcano," 77.

based on their phonetics and representation in search of concordances that revealed the truth behind the word.[80] Arias Montano did nothing of the sort.

Gómez Canseco has identified certain aspects of Arias Montano's thought as drawn largely from the kabbalistic tradition. For example, he finds that his concept of creation and theory of language shared an objective with the theory of language of the Zohar, where the world is seen as having been created by the very letters of the Hebrew alphabet—that is, the world proceeds from the Hebrew letters and words spoken by God.[81] Yet it is difficult to ascribe this notion solely to the Kabbalah, since the notion that God's spoken word brought everything into existence is also central to Christian interpretation of Genesis and of John 1:1 ("In the beginning was the Word"), where John equates Christ with the Word of God. In Arias Montano the laden language of the Old Testament was studied from the perspective of humanism, with its emphasis on philological exactitude, and for the purpose of finding examples prefiguring Christ, not because particular letters had intrinsic, embedded meaning that the exegete had to decipher. But as Gómez Canseco points out, simply knowing about Kabbalah, as Arias Montano certainly did, raised to new levels his awareness of meanings embedded in certain Hebrew letters and numbers; these were aspects of the biblical text that an attentive Christian exegete could not simply gloss over without careful analysis. Gómez Canseco also situates Arias Montano's division of the human soul into male and female—an aspect developed at length in the *Magnum opus*—as Neoplatonic and kabbalistic.[82]

Having been educated in the Complutensian tradition of biblical studies, Arias Montano embraced the Hebrew language and likely some rabbinical approaches to biblical study. Yet he shared to some degree the perspective of other sixteenth-century Christian doctors who saw contemporary Jews as having a distinct approach to biblical exegesis; they considered Jewish commentaries mystical, secretive, and kabbalistic, a term often used as a pejorative (as Mariana had done). Only a few Jewish rabbis escaped this characterization, principal among them Maimonides and David Kimhi, whose Hebrew grammars and dictionary of roots proved invaluable to Christian Hebraists.[83] It was this tradition that found a home at the trilingual colleges of Louvain, Alcalá, and Paris. Indeed, Arias Montano went further than most in engaging

80. Secret, "Emithologie de Guillaume Postel," 404–5.

81. Gómez Canseco, "Sentidos del lenguaje arcano," 55–56, 75–76.

82. Ibid., 78–79.

83. Such as Kimhi's *Sefer Mikhlol* (a grammar) and *Sefer Shorashim* (book of roots): Fernández López, *Alfonso de Zamora y Benito Arias Montano*, 25. On Kimhi's rationalist approach to Jewish *peshat* or literal interpretation, see Talmage, "David Kimhi," 70–102.

with rabbinical literature, relying on the Talmud and Mishnah to elucidate the cultural aspects of Hebrew biblical actors in his biblical commentaries.[84] These were sources that allowed him to reconstruct the cultural and religious milieu of the Bible and thus to contextualize certain aspects of early Christian tradition and understand them historically.

Juan de Mariana's accusation that Arias Montano had borrowed much from the "kabbalah of the rabbis" in his discussion of the divine names in the *De arcano sermone* has also been taken as evidence that this was the model he followed in that discussion.[85] Although expositions on the meanings and uses of the divine names are an important component of Jewish and Christian Kabbalah, in the *De arcano sermone* Arias Montano limited his study to an etymological analysis of the names and the particular meanings they accrue in given biblical passages. The approach fits neatly into his exegetical methodology of wringing meaning from the root of Hebrew words and has more in common with the philological methods of the humanists than with the letter symbolism and allegorical readings typical of kabbalists. It is substantially different, for example, from the concept of atomization associated with Abraham ben Samuel Abulafia's method of creating elaborate compositions based on letter permutations and variations of the divine name and the meditations related to this practice.[86]

Finally and most fundamentally, Arias Montano's work does not reflect the kabbalist ethos that knowledge of nature can be attained only through illumination and transcendence. Neither the *De arcano sermone* nor the *Magnum opus* suggests paths of spiritual ascent to the godhead or a ritual of study to achieve this. He believed instead that careful observation of nature—informed by the knowledge of the arcane language of scripture—can help humanity attain the knowledge of nature necessary for living in concert with God and the world. Arias Montano did not acknowledge the Kabbalah as part of the received wisdom of the ancient Hebrews; thus it was not an element of the *prisca theologia* that had to be recaptured. As we will see in subsequent chapters, his cosmology was not kabbalistic either, nor does he ever refer in his printed works to anything akin to the ten spiritual emanations (Sefirot) or angel magic of the Kabbalah. In the votive elegy that opens his *History of Nature* he does suggest, however, that while deep in prayer he heard an otherworldly voice name the Sacred Scriptures as the source of knowledge of nature. This

84. Dunkelgrün, "Multiplicity of Scripture," 145–48.
85. Gómez Canseco, "Sentidos del lenguaje arcano," 79–86.
86. Idel, *Kabbalah in Italy*, 66–68.

is far from kabbalistic illumination and more in keeping with Christian mysticism.

*

In the Antwerp Polyglot, Arias Montano had unveiled a hyperliteral exegetical methodology that seemed to some of his contemporaries entirely divorced from church tradition and accepted methodologies (and therefore teetering on the edge of disobedience to the mandates of the Council of Trent, if not simply heretical). Others saw it, less ominously, as annoyingly unconventional because his citations, in addition to being so few, ignored the opinions of church fathers and later Scholastics. Arias Montano was—and still is—difficult to follow. Puzzled readers could be persuaded equally into believing that he was a Judaizing kabbalist, a sloppy scholar who did not cite his sources, or someone unwilling to tread carefully in the post-Tridentine landscape of biblical scholarship. Yet in his approach to the Pagnino edition and in the *De arcano sermone*, but most importantly in his response to the project's reception, Arias Montano's actions testify to an unflinching commitment to a particular theory of language and its resultant exegetical methodology—one he was willing to follow even if it meant dismissing centuries of biblical commentaries and, more to our purpose, even centuries of natural philosophical thought.

CHAPTER 6

Nothing New under the Sun

"He extremely loved solitude; his recreation was his garden and the flowers in it."[1]

The controversies surrounding the Antwerp Polyglot did not dissuade Arias Montano from an exegesis based on the literal interpretation of the Bible using Hebrew etymology. Even though he wanted to turn his full attention to his theological writings after concluding this project, he was, and would continue to be, a very busy man. Philip II summoned him back to Spain to continue organizing and building the collection of the glorious royal library at El Escorial; having Arias Montano so close at hand, the king repeatedly called on him for advice on a broad range of subjects. After his definitive return to Spain in 1576 and over the following decades, he served his king as envoy, librarian, and ad hoc adviser on cultural and political topics. He formed part of an intellectual elite of *letrados* that surrounded the king in the 1560s and 1570s. As Bustamante García has pointed out, these were all men born between 1508 and 1530 who had been educated in Spain and coalesced at the Habsburg court at a time when the House of Habsburg was consolidating its empire as a Catholic Spanish monarchy. The impressive series of projects that originated at court or found support there were part of a grand intellectual and unified enterprise (perhaps appearing disjointed only in that it was poorly articulated) that sprang from a group with common values rooted in their humanistic education and committed to the "construction and legitimation" of a modern state as defined by a patrimonial monarchy.[2]

---

1. "Amó por estremo la soledad, su recreación era su guerto i las flores": Francisco Pacheco, *Libro de descripción de verdaderos retratos*, no. 46.

2. This circle of bureaucrat–*letrados* also included Juan Páez de Castro, Francisco Hernández, Juan de Herrera, and Juan López de Velasco, to which I would add Antonio Gracián and Juan de Ovando: Bustamante García, "Los círculos intelectuales y las empresas culturales de Felipe II."

Arias Montano formed part of this extraordinary group despite his unrelenting desire to separate himself from life at court and devote himself to his biblical studies. He nonetheless found time to continue writing exegetical works and began to lay the foundation of his self-described *Magnum opus*. Once settled into a house near Seville—as well as maintaining his remote hilltop retreat at the Peña de Alájar—he assembled a large library and natural history collection that became a gathering spot for some of Seville's leading doctors, cosmographers, naturalists, and prominent members of the city's commercial elite. It was during these last eight years of his life that he finally found the time and quiet to write the *Magnum opus* detailing his philosophy of humanity and of nature.

Although Arias Montano's theological and exegetical works earned him a prominent place among European humanists, he repeatedly said he wanted to devote himself to developing a novel philosophy of nature. In the preface to the *Phaleg*, one of the treatises he included in the *Apparatus* of the Antwerp Polyglot, he noted that if he were free from other concerns, in only seven or eight years he could write volumes on all sorts of arts and disciplines using the Sacred Scriptures as his sole source; these works would equal or surpass anything written by the Greeks and Romans.[3] When he finally took on this project, his work reflected the empiricist approach characteristic of his Sevillian milieu as well as the disquiet of the broader community of European philosophers who questioned the ability of ancient natural philosophies to interpret nature. For Arias Montano, these two approaches—the purely empiricist and the philosophical—if left unreconciled, introduced an untenable tension into humanity's understanding of its place in a world that, from a sixteenth-century perspective, appeared to be rapidly changing. This chapter explores the many facets of his interest in reforming natural philosophy and the way he reconciled the learned empiricism of his humanist background, the evidence of empirical investigations, and his desire for a cohesive natural philosophy to explain the natural world.

INTERLUDES AT EL ESCORIAL

By 1575 the king wanted Arias Montano back in Spain as a member of a group commissioned to rebut the charges raised in the Lutheran history of the church, the *Magdeburg Centuries* (1559–74),[4] but the royal chaplain wanted to

3. Arias Montano, *Prefacios*, 158–61.
4. Pizarro Llorente, *Un gran patrón en la corte de Felipe II*, 265–67. The task would eventually be taken up in Rome and given to Cardinal Caesare Baronio (1538–1607).

remain in Rome or even Flanders to pursue his studies for another year or two. Comfortably installed in Cardinal Carlo Borromeo's palace of Saint Praxedes, he had access to all the Holy See had to offer a studious exegete.[5] Furthermore, Rome put some distance between him and his most virulent detractors in the circle of León de Castro during the years before the Mariana *censurae* was issued, and it kept him safe from his creditors in Antwerp, whom the Spanish monarchy, as was typical, had been very slow to pay.[6] After delaying the inevitable return—the king never gave him license to stay—he arrived at court in July 1576, and the king granted him a lengthy audience; from that point on the king intended to keep him close at hand. He became involved in the project to collect existing manuscripts of the work of Saint Isidore of Seville—later he would do the same for the works of Ramón Llull—but it was not until early 1577 that he was called on to take over duties as librarian of the then-embryonic royal library of El Escorial.[7]

Arias Montano's early involvement in the library of El Escorial is well documented; his correspondence with royal secretaries testifies to how assiduously he pursued and procured printed books and manuscripts for the royal collection during his sojourns in Antwerp and Rome.[8] Having returned to Spain, he took over the duties of organizing the growing collection from the royal secretary Antonio Gracián Dantisco after the talented young man died unexpectedly in April 1576. By March 1577 Arias Montano was at San Lorenzo of El Escorial to begin his duties as the future library's *librero mayor*. He would visit the monastery and palace periodically over the next fifteen years.[9] One

5. Morocho Gayo, "Trayectoria humanística, II," 260; Tellechea Idígoras, "Benito Arias Montano y San Carlos Borromeo."

6. Letter from Arias Montano to Philip II, Rome, 29 July 1575. AGS 583 doc. 107. For other letters along these lines, see DIE, 41:312–14, 316–26.

7. Morocho Gayo, "Trayectoria humanística, II," 267–69; Macías Rosendo, *Correspondencia de Arias Montano con Juan de Ovando*, 307–10.

8. Macías Rosendo, *Correspondencia de Arias Montano con Juan de Ovando*, 121–35; Bécares Botas, *Arias Montano y Plantino*, 194–222.

9. Gregorio de Andrés reconstructed the dates of most of Arias Montano's stays at El Escorial. What follows includes the most recent scholarship. The first extended stay began on 1 March 1577 and lasted until the end of the year, when he left on a mission to Portugal that kept him there until spring 1578, when he visited El Escorial for a few months. His next visit to the library was from 8 September 1579 to 9 March 1580. He was also there from February to August 1583, and again from January 1585 until April 1586, when he expurgated the books in the library in light of the new Spanish *Expurgatory Index* of Quiroga (1582). He returned for a few months in summer 1587. His last visit took place from January to April 1592 when he oversaw installing the books in the finished royal library: Andrés, *Proceso inquisitorial del padre Sigüenza*,

thing is certain, however—he intensely disliked spending time at El Escorial, particularly working as its librarian. When he was called back in 1579, not only did he not stay in the monastery, residing instead in the house of royal secretary Sebastián Cordero de Nevares y Santoyo,[10] but he complained bitterly to Gabriel de Zayas that the king had not received him. He looked down on his duties in the library, even once complaining to Zayas that his duties there amounted only to "Taunts, I meant rather, titles of principal librarian" (*Escarnios, quise decir, títulos de librero mayor*).[11]

Ultimately he was responsible for establishing the initial thematic arrangement of the books in the royal library into sixty-four disciplines corresponding to arts, philosophy, and theology. Additionally, he left a testament to his esthetic sensibilities on the granite monolith of El Escorial, principally in his decoration of the facade of the basilica facing the interior courtyard, which he suggested adorning with six Old Testament kings of David's line. Scholars have speculated about his involvement in designing the iconographic program of the royal library's frescoes to correspond with the arrangement he envisioned for the books, but we have little documentary evidence that he did so.[12] It was not before 1587, but definitely by the 1592 stay, that his duties included preaching and teaching Hebrew and Greek, as well as cosmography and mathematics, at the small arts college and seminary attached to the monastery.[13]

Especially during his last visit to the monastery during the first half of 1592, Arias Montano found a few willing Hieronymites within the conservative monastery community housed at El Escorial who wanted to learn his approach to biblical exegesis and his ideas about natural philosophy; chief among them was fray José de Sigüenza. His presence and profound influence on Sigüenza caused such a stir in the community that fellow Hieronymites leveled accusations against Sigüenza's orthodoxy. As soon as Arias Montano had left the

31–35. For a more nuanced chronology, see Morocho Gayo, "Trayectoria humanística, II." For Arias Montano's involvement in the library, see López Guillamón, "Benito Arias Montano y la Biblioteca escurialense"; J. Gil, "Montano y El Escorial"; and Flórez and Balsinde, *Escorial y Arias Montano*, 323–33.

10. Sebastián Cordero de Nevares y Santoyo (b. 1523) was *secretario de camara* of Philip II.

11. Arias Montano to Gabriel de Zayas, San Lorenzo de El Escorial, 10 October 1579, in DIE, 41:408.

12. Sigüenza, *Historia de la orden de San Jerónimo* (2000 ed.), 2:607–29; Hänsel, *Benito Arias Montano*, 176–81. On his alleged contribution to the decoration of the library, see Portuondo, "Study of Nature, Philosophy and the Royal Library of San Lorenzo of the Escorial."

13. Preliminary study by Luis Villalba Muñoz in Sigüenza, *Historia del Rey de los reyes y Señor de los señores*, 1:lxix.

monastery an inquisitorial investigation got under way. During the whole affair, which ended with Sigüenza being cleared of any suspicion, the friar stood as a proxy for Arias Montano's ideas concerning Scholastic theology, biblical exegesis, and personal revelation. I will discuss the particulars of Sigüenza's inquest in chapter 11.[14]

THE BIBLIOPHILE

As much as Arias Montano disliked working as *librero mayor*, he cherished books and amassed them as voraciously as his means and credit permitted. As early as his student days in Alcalá, Arias Montano owned 128 volumes, mostly on theology, but a good third on subjects such as logic, physics, geography, mathematics, astrology, and natural history. By 1553 he had added more books on astronomy as well as works by Georg von Peuerbach and Gemma Frisius.[15] In his 1582 will Arias Montano divided his impressive library into three bequests.[16] He left his collection of 33 Greek, Hebrew, and Arabic manuscripts to the royal library at El Escorial, to be deposited there after his death. To the conventual library of Santiago de la Espada in Seville he left all his printed books in folio and quarto. The remaining 150 smaller volumes he offered to sell to the two libraries then functioning at El Escorial, the royal library and the one attached to the seminary. These were largely books dating from his days as a student, some purchased in Seville from the estate of Sebastián Fox Morcillo.[17] The ones left unsold were bequeathed to his secretaries, Pedro de Valencia and Juan Ramírez. This 1582 will was only a waypoint in his book collecting; he continued collecting until the last years of his life, with agents in Antwerp having standing orders to send him 100 to 200 florins worth of books a year. These were to be the latest books on a number of topics: theology, moral philosophy, physics (and criticisms of such), medicine, law (if the books were praiseworthy), history, and ancient history, as well as poetry, cosmography, and geography.[18] Sadly, no inventory of his books at the time of his death has survived, or at least none has been found.

During his years abroad he acted as an unofficial book agent on behalf of

---

14. Andrés, *Proceso inquisitorial del padre Sigüenza*, 51–52; Flórez and Balsinde, *Escorial y Arias Montano*, 193–229.

15. For the book inventories, see Gil, *Arias Montano y su entorno*, 52–53, 163–81.

16. Ibid., 207–9.

17. Gonzalo Sánchez-Molero, "La biblioteca de Arias Montano en El Escorial."

18. Bécares Botas, *Arias Montano y Plantino*, 223.

friends.[19] Not only did he arrange for the publication of their works by the Plantin press, as we saw in the case of Francisco de Arceo, he also actively sought out and purchased books for them. Arias Montano had introduced the Sevillian medical doctor, botanist, and cosmographer Simón de Tovar (1528–96) to Plantin, who published his *De compositorum medicamentorum examine nova methodus* (1586). Over the years he would also coordinate substantive book purchases for the doctor, totaling over one hundred volumes in 1593 alone.[20] Among the purchases he made for friends interested in scientific topics, those for Tovar and Juan de Ovando stand out.[21]

Ovando's career had continued in the ascendant after the years in Seville where he made Arias Montano's acquaintance; he now served as an *oidor*, or judge, for the Supreme Council of the Inquisition and had recently completed a *visita*, or inspection, of the University of Alcalá. In 1568 he conducted a *visita* at the Council of the Indies and proposed a series of important reforms to that crucial body of imperial governance. In 1571 he was asked to preside over the council, which he did until his death in 1575. As part of his concerted effort to collect and maintain empirical information about Spanish possessions in the Americas and the Philippines, which he considered an essential part of good governance, Ovando implemented the well-known questionnaires that yielded the *Relaciones geográficas de Indias* and the first global attempt to simultaneously observe lunar eclipses as a way to determine geographical longitude. We have a rare glimpse of how Arias Montano perceived the functions and skills expected of someone holding Ovando's post, particularly regarding the scientific skill his friend needed to pursue the proposed reforms at the Council of the Indies. As Arias Montano relayed to Ovando in a letter, the Duke of Alba, Fernando Álvarez de Toledo y Pimentel, then governor of the Low Countries, had expressed (as part of the hours-long soliloquies he liked to bestow on Arias Montano) that it was necessary for a president of the Council of the Indies "to have a mind that can imagine lines and measures and angles and seaports and fields and animals and plants and natures that have not been seen by many, nor do they coincide with the ones here, and also their uses and purposes, because all of this concerns government and is subject to it, and those who are over there may fool those who govern from over here."[22] By

19. Ibid., 176–241.

20. Ibid., 225–33.

21. On Arias Montano's relationship with Ovando, see Paniagua Pérez, "Burócratas e intelectuales en la corte de Felipe II."

22. Letter of Arias Montano to Juan de Ovando, Antwerp, 6 October 1571, in Macías Rosendo, *Correspondencia de Arias Montano con Juan de Ovando*, 239.

repeating the duke's words to Ovando, Arias Montano was both delivering a compliment—he relayed that he had assured the duke that Ovando possessed this preparation and skill—and showing that he was well aware of the scope of the scientific projects Ovando was putting in place at the Council of the Indies.

Not long after arriving in Antwerp in 1568, Arias Montano wrote to Ovando offering to supply him with all sorts of goods (tapestries, chairs, table linens, and napkins), but most importantly with books, geographical descriptions, and astronomical instruments.[23] He could procure them at a good price from Plantin and in fact had already done so for his personal collection. To Ovando he offered terrestrial and celestial globes, large ones after designs by the renowned mathematician Gemma Frisius (1508–55) and others, more expensive and presumably more up-to-date, by Gerard Mercator (1512–94). Arias Montano anticipated Ovando's acceptance and said he had already ordered a pair of Mercator globes on his behalf. He also noted that he had purchased for himself a very large astrolabe, nearly a foot in diameter, made in Louvain. He had special tympan plates made for the latitudes of Andalucía and Extremadura, because it came with only a general plate[24] and some for northern latitudes. If Ovando was interested he could visit the maker in Louvain and have plates made for all of Spain at a third of the price it would cost to get such an instrument there. He was heading to Louvain to order for himself a metal sphere—likely, a celestial sphere—and an astronomical ring and offered to get Ovando these items as well. In the same letter he promised to inquire about astrology books and send a list separately. Ovando placed his order more than a year later, confident in his friend's knowledge of astronomical matters and trusting his judgment when selecting the proper instruments, as long as they were "well crafted" (*bien labrados*) and "with elegant bases and engravings" (*con peannas y encaxes muy galanes*). Ovando ordered an astrolabe, a pair of lacquered globes, a large astronomical ring (*anulo*), and one or two *radios astronómicos* or Jacob staffs. All of these were to come accompanied by the books explaining how to use each instrument. He also desired all the geographical descriptions and printed maps mounted on cloth that Arias Montano thought suitable. To this order he added a religious triptych suitable for an altar and

23. Letter of Arias Montano to Juan de Ovando, Antwerp, 14 June 1568. The letter is accompanied by a list of books by church fathers, "modern" Catholic authors, and philosophers available for sale by Plantin: Macías Rosendo, *Correspondencia de Arias Montano con Juan de Ovando*, 168–71.

24. Likely a reference to the "Rojas" orthogonal projection, after Juan de Rojas y Sarmiento, who popularized the projection in his *Commentariorum in astrolabium quod planisphaerium vocant* (Paris, 1551).

a half-dozen landscape paintings to decorate a study. Ovando was not just setting up a prestigious-looking study; he wanted useful reference works, so he asked Arias Montano to get him copies of some of the classics in astronomy such as Ptolemy's *Almagest* and "whatever you may find by Copernicus and Regiomontanus," as well as Gemma Frisius's book on the use of globes and Peter Apian's *Instrumentum primi mobilis* (1534).[25]

Ovando received most of the goods over the course of the next six months (the globes and descriptions took six months longer), and it took the intercession of six trusted agents and friends to get the goods safely to Spain. The care and effort Arias Montano expended to get them there reflect the difficulty of procuring these expensive and rare instruments in mid-sixteenth-century Europe, where travel routes were often closed owing to wars and thieves did not respect even shipments to royal ministers and inquisitors.[26] But it also reveals to us an Arias Montano perfectly in tune with the empirical projects taking place in the Philippine court of the 1570s and eager to help obtain the tools necessary for these projects. In addition to the projects associated with the Council of the Indies, a parallel project was under way to create a detailed geographical description, topography, and chorography of the Iberian peninsula.[27]

GEOGRAPHY

Arias Montano's interest in the science of meridians and parallels went back to his days as a young man under the tutelage of his father and his first tutor, Diego Vázquez Matamoros, who had traveled to the Holy Land and whose tales of the trip left a lasting impression on our biblist.[28] He put his interest in historical cartography on display in the *Apparatus* of the Antwerp Polyglot. Geography, and sacred geography in particular, provided textual and visual tools in the form of descriptions and maps that allowed readers of the Bible

25. "Lo que hallare de Copérnico y Ioanis de Monte Regio." List titled "Memorial de libros" in a letter from Juan de Ovando to Arias Montano, Madrid, 17 April 1571, in Macías Rosendo, *Correspondencia de Arias Montano con Juan de Ovando*, 219–27.

26. Macías Rosendo, *Correspondencia de Arias Montano con Juan de Ovando*, 204, 212.

27. On the genesis of the project that has come to be known as the *Relaciones topográficas*, see Alvar Ezquerra, García Guerra, and Vicioso Rodríguez, *Relaciones topográficas de Felipe II*, 1:29–36.

28. Preface of the treatise *Nehemias* of the *Apparatus sacer*, in Arias Montano, *Prefacios*, 229–33.

to imagine the Holy Land and immerse themselves figuratively in the places and cultural trappings of the chosen people of the Old Testament. The extent of his direct contribution to the several maps included in the *Apparatus* (the only ones that appeared among his published works) is difficult to ascertain, but it is clearly on display, as Shalev has noted, in the topographical plan of Jerusalem, *Antiqua Ierusalem*, in the treatise *Nehemias*.[29] Arias Montano attributed the map in the *Chanaan, sive De duodecim gentibus* (1572) to a very learned man from Mantua, an expert on the Hebrew language whom he had met while in Trent and who had the map made while in Syrian Palestine.[30] In his preface to the *Phaleg* he reiterated his argument about the Sacred Scriptures being a source for knowledge about the natural world and pointed out how ancient geographers such as Strabo, Mela, Stephanus, Solino, Ptolemy, and even Plato and Aristotle had written obscurely and enigmatically (*obscure, ac veluti per aenigmata*) about places they did not know.[31] There was not a single discipline, in Arias Montano's opinion, that could not be improved with knowledge of geography; philosophers seeking to know the nature and causes of things needed to know about the various types of lands and seas; medical doctors needed to take into account the variety of medicines found in other places; likewise for moral philosophers, merchants, navigators, and practitioners of the military arts.

It was in great part this interest in the whole range of geographical arts that was behind Arias Montano's lifelong friendship with Abraham Ortelius (1527–98). The oeuvre of the Flemish cartographer, in particular his *Theatrum orbis terrarum* (1570) dedicated to Philip II, and its subsequent *additamenta*, featured prominently in the book shipments Arias Montano brokered for his acquaintances. He made sure the cartographer became well known in the highest circles of the Spanish court and even asked royal secretary Zayas to petition the king for a special reward for Ortelius, remarking pragmatically, "Setting such examples and without incurring much expense . . . has a great effect [in Antwerp] and creates perpetual statues in glory of princes, one of whose greatest merits is having been a friend and patron of the learned and virtuous."[32] A golden chain and the title of royal geographer were granted to Ortelius in 1573, to which Arias Montano added a gold medal with the king's image on it.

29. Shalev, *Sacred Words and Worlds*, 43–63.

30. Preface to the *Chanaan* of the *Apparatus sacer*, in Arias Montano, *Prefacios*, 186–87.

31. Ibid., 171–73.

32. Arias Montano to Gabriel de Zayas, Antwerp, 31 December 1573, in DIE, 41:298.

Arias Montano was one of the first to sign the *Album amicorum* of Ortelius in 1574, gracing the book in folio 18 with a sketch of his personal motto: a young Archimedes pointing to a book and shouting Eureka![33]

Besides Ortelius's constant longing for Arias Montano's companionship, a common theme in their correspondence is the concerted effort by our biblist to provide Ortelius with any geographical information or maps the Flemish cartographer might find useful. While in Antwerp, he had written to Juan de Ovando about how Ortelius had added to his latest supplement to the *Theatrum* a few maps of Germany that had been sent to him.[34] Why not also add some maps of their native Extremadura? Arias Montano remarked that had he been in Spain he would "walk all of it to describe it." If Ovando knew of any such map, he asked, would he please send a copy with the name of the author so it could be included in the *Theatrum*. Likewise with any part of the ultramarine territories, but only if it could be published legally "for the common use of studies" (*ad communem usum studiorum*). He explained that he was driven by a desire to encourage the growth of this "good discipline," indicating that he was well aware that the maps of the New World were zealously guarded at the Council of the Indies.[35] In 1576, while in Rome, he made Ortelius aware of a map of China drawn by a Portuguese envoy to that kingdom and owned by Giambattista Raimondi (1536–1614), a professor of mathematics at the college of La Sapienza. Arias Montano "begged him to make me a copy, drawn simply and straightforwardly but with precision."[36] On another occasion he edited part of a description of Valencia by Fadrique Furió Ceriol (1527–92) concerning the region's Moorish population and advised him not to mention the Inquisition unless it was particularly germane to the geographical description.[37] In the years after Arias Montano's stay in Antwerp, Plantin sometimes acted as an intermediary between Ortelius and

33. Morales Lara, "Españoles en el 'Album Amicorum' de Abraham Ortels, 1036-37."

34. Arias Montano to Juan de Ovando, Antwerp, 20 January 1573, in Macías Rosendo, *Correspondencia de Arias Montano con Juan de Ovando*, 278–80.

35. It is difficult to ascertain the precise maps Arias Montano might have procured for Ortelius in this and other instances. For maps of the Americas, see Reinharts, "Americas Revealed in the *Theatrum*."

36. Arias Montano to Abraham Ortelius, Rome, 28 February 1576, in Morales Lara, "Cartas de Benito Arias Montano a Abraham Ortels," 166–67.

37. Arias Montano to Abraham Ortelius, Seville, 3 January 1590, in Morales Lara, "Cartas de Benito Arias Montano a Abraham Ortels," 179–83.

Arias Montano, writing in 1579 to inquire on Ortelius's behalf about maps of America and China, which the cartographer received in 1581.[38]

As far as we can ascertain, although Arias Montano might have provided some textual descriptions of Spanish lands, he never drew a map or provided Ortelius with geographical coordinates. Doing so was well within his capabilities and those of the instruments he had procured for himself while in Antwerp, yet it appears his collaboration with the royal geographer was limited to helping him obtain geographical information collected by third parties. As his treatises in the *Apparatus* of the Antwerp Polyglot amply demonstrate, his personal research interests in geography were directed toward contextualizing the historical events described in the Bible. The *geographia sacra* that resulted from this effort was purposely antiquarian rather than directed at creating an accurate description of the Holy Land as it existed in the present. In fact contemporary geographical descriptions, such as those by his childhood teacher Vázquez Matamoros and the unnamed Mantuan he met at the Council of Trent, were simply starting points for extended exercises in toponymic regressions that aimed to reconstruct in a map the Holy Land as it was when the biblical events took place.

NATURALIST FRIENDS

Arias Montano's interest in natural history has long been noted by his biographers; it is evident in his correspondence with the Flemish doctor and botanist Carolus Clusius (1526–1609) and his acquaintance with some of the most important naturalists who frequented Plantin's press in Antwerp.[39] Yet his circle of naturalist friends extended beyond the Low Countries as well. None other than the Italian naturalist Ulisse Aldrovandi (1522–1605) recalled the great honor he had received when Arias Montano visited his museum of natural objects in Bologna. The naturalist also kept the *Apparatus* close at hand and used it to inform his own brand of Mosaic philosophy. To what extent he consulted the first part of the *Magnum opus* is unknown.[40]

38. Dávila Pérez, *Correspondencia en el Museo Plantin–Moretus*, 2:443, 2:457, 2:473.

39. On Arias Montano's relationships in the Low Countries, see Dávila Pérez, "Arias Montano y Amberes," 199–212; Landtsheer, "Benito Arias Montano and the Friends from His Antwerp Sojourn"; Barona, "Clusius' Exchange of Botanical Information with Spanish Scholars"; and López Terrada, "Flora and the Hapsburg Crown."

40. Berns, *Bible and Natural Philosophy in Renaissance Italy*, 97–100. The reference to Arias Montano appears in Ulisse Aldrovandi, *Bibliologia*, Biblioteca Universitaria di Bologna,

Although Arias Montano's inventory of books from his student days shows a tangential interest in botany, the number of volumes on that topic pales beside his significant collection of mathematical and astronomical books. He owned many of the classics—Pliny the Elder's *Natural History* (a 1545 edition from Basel),[41] Aristotle's *History of Animals* and his *Parts of Animals*, Fernández de Oviedo's *La historia general de las Indias* (1547)—though curiously absent were editions of Dioscorides and Theophrastus.[42] Yet these early inventories do not reveal the extent of his botanical interests. During these years Philip II had installed a botanical garden in the Flemish style on the grounds of the Palace of Aranjuez. Its care, as well as care of the distilling house, was under the direction of Jehan and François Holbecq, brothers from Mechlin (Malines).[43] The king also sent agents to Seville in 1568 to procure plants from the New World. But it was Arias Montano's interactions with practicing botanists in Antwerp that catalyzed his interest in botany. In his *éloge* to learned men, the *Virorum doctorum de disciplinis benemerentium effigies XLIIII* (1572), he recognized the contributions of a number of botanists, beginning with Pietro Andrea Mattioli (1501–77), whom he acknowledges for his zeal in observing nature and his contributions to the botanical legacy of Greece.[44] His words in praise of Rembert Dodoens (ca. 1516–85) reveal close friendship with the important botanist and doctor who had a *hortus medicus* and library at his home in Mechlin. (Arias Montano's visit to Dodoens's home coincided with that of Clusius, who was writing a book on the flora of the Iberian Peninsula.) Just a few years before Arias Montano's arrival in Antwerp, Plantin had begun to publish Dodoens's latest botanical works. His earlier book, *Cruijdeboeck*

MS Aldrovandi 83, 1:426. The date of the visit is unclear, but it might have been in either 1572 or 1575. I thank Professor Berns for this reference.

41. Rekers indicated that Arias Montano's interest in natural history was in evidence in a heavily annotated copy of a 1569 Alcalá edition of Pliny's *Historia Natural* at the National Library in Madrid, BNE R/5381. Rekers was repeating an earlier citation by Miguel Colmeiro, *La botánica y los botánicos*, 3, item 13. For a study of the Pliny volume, see Pérez Custodio, "Plinio el Viejo y los progymnasmata," 4.2:974. Examining the BNE copy reveals the work of a humanist correlating this edition with a manuscript source of the *Naturalis historia*, but there is no evidence in the annotations that they are the work of Arias Montano.

42. Gil, *Arias Montano y su entorno*, 165–81. Otto Brunsfels's *Herbarum vivae eicones* (1530) was among the books he offered to El Escorial in 1583. Gonzalo Sánchez–Molero, "Biblioteca de Arias Montano en El Escorial," 101.

43. Rey Bueno, *Los señores del fuego*, 33–46; López Rodríguez, "Sevilla, el nacimiento de los museos, América y la botánica," 82–83.

44. Gómez Canseco, "Ciencia, religión y poesía en el humanismo," 132.

(1554), was the first to propose new classificatory schemes for plants based on medical properties, thus moving beyond the earlier systems of Leonard Fuch. From 1566 to 1585 Plantin published six works by Dodoens, including his masterpiece, the *Stirpium historiae pemptades sex* (1583).[45] In the *Stirpes*—which Arias Montano cherished, having received it in Spain the very year it was published[46]—Dodoens placed botany on a sound empirical foundation. He divided the plant kingdom into six groups largely defined by a plant's usefulness, a taxonomic criterion that would later inform Arias Montano's *History of Nature*. The history of science recognizes Dodoens's observational skills and meticulous descriptions and considers his emphasis on plant morphology to be his greatest contribution to early modern botany. His plant classification system allowed for organizing a greater number of plants in a manner that grouped them not just according to their utility but also into larger genera and families sharing similar characteristics.[47]

Clusius also understood the growing problem of plant identification and the need for a practical taxonomic scheme, given the marked increase of the number of specimens figuratively washing up on European soil. He would become a central node of the communications among European naturalists, gardening enthusiasts, and collectors during the latter half of the sixteenth century. Clusius used his botanical expertise to translate and annotate botanical works, incorporating these into his own *Rariorum plantarum historia* (1601) and subsequent works in hopes of creating a corpus of botanical information as a basis for plant classification. Clusius had visited Spain in 1564–65, and over the years he continued to cultivate Spanish contacts, translating their works, exchanging seeds, and keeping close tabs on new plants coming from the Indies. As early as 1569 he sent Arias Montano a list of specimens he wanted to procure, and the biblist replied that he was sending copies of the list to two or three friends in Spain who might help locate them.[48]

Contact with Clusius did not resume for several decades thereafter, though Christophe Plantin promoted Arias Montano's interest in growing exotic plants in his personal gardens during the intervening years. In a letter from 1583, Arias Montano lamented that a shipment of bulbs and plants Plantin had

45. Vande Walle, "Dodonaeus: A Bio–bibliographical Summary."

46. Arias Montano to Plantin, Peña de Aracena, 22 September 1583, in Dávila Pérez, *Correspondencia en el Museo Plantin-Moretus*, 2:489–502n19.

47. Visser, "Dodonaeus and the Herbal Tradition," 51–55. On the issues surrounding plant taxonomy during the early modern era, see Ogilvie, *Science of Describing*, 182–92, 215–28.

48. Arias Montano to Carolus Clusius, Antwerp, 22 April 1569; translated in Clusius, *Correspondencia de Carolus Clusius con los científicos españoles*, 105.

sent him had been rendered useless by delays in delivery. He asked him to send the seventy-six specimens again so he could plant them at the Peña and get to "know" them, explaining, "all my life I have had a temperament given to the contemplation and admiration of nature."[49]

Given Arias Montano's role as a knowledge broker, it should come as no surprise that Clusius drew him into the culture of exchange of American plants between naturalists in Seville and in the rest of Europe.[50] Arias Montano found himself a member of a group of collectors who trafficked in American *naturalia* from the city on the Guadalquivir River, with Clusius being a principal beneficiary of their access to New World specimens. We know of several botanical gardens in Seville that grew medicinal plants and specialized in cultivating American specimens, following in the tradition of Nicolás Monardes (1493–1588). Among them was the doctor Simón de Tovar, Portuguese by birth and deeply enmeshed in the commercial circles of Seville.[51] Arias Montano introduced him to Plantin, whose press published three of his books, two on pharmacology and one on navigation. Through Arias Montano's agency a list of American plants that Tovar grew in his Seville garden made its way into Clusius's famous catalog. Testament to the prominent role Arias Montano played in this circle is the inclusion in Clusius's *Rariorum plantarum historia* (1601) of an engraving of the *Narcissus indicus iacobeus*, the name Tovar had given the plant because it resembled the emblem of the order of the Knights of Saint James, which was, not coincidentally, Arias Montano's religious order[52] (fig. 6.1). A similar exchange took place between Tovar, the botanist Bernardus Paludanus (Berend ten Broeke), and Clusius, contacts Arias Montano brokered during the 1590s.[53] In fact, in friendship and gratitude Tovar bequeathed Arias Montano his garden in Seville, as well as "medicines, oils,

49. Arias Montano to Plantin, Peña de Aracena, 22 September 1583, in Dávila Pérez, *Correspondencia en el Museo Plantin–Moretus*, 2:489–502n22. Dávila Pérez has identified the specimens in their order in the catalogs of Arias Montano's acquaintances, botanists Mathias L'Obel and Rembert Dodoens.

50. I borrow the label "broker" from Lazure's work on the circle of friends surrounding Arias Montano in Seville, but also from a broader strand in the history of science on the role of the "go-between.": Lazure, "Building Bridges between Antwerp and Seville," 31–43; Schaffer et al., *Brokered World*.

51. Rey Bueno and López Pérez, "Simón de Tovar."

52. Clusius, *Correspondencia de Carolus Clusius con los científicos españoles*, 72–73; Egmond, *World of Carolus Clusius*, 40.

53. Arias Montano to Abraham Ortelius, 25 November 1594, in Morales Lara, "Otras tres cartas," 225–27.

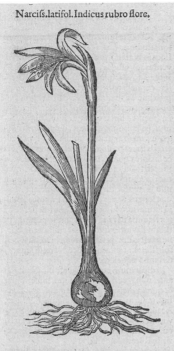

Narcifs.latifol.Indicus rubro flore.

Sᴇɴᴀ vel plura habet is Natciſſus folia, Narciſſi vulgaris foliorum inſtar lóga,ad quorum latus emergebat caulis lævis & enodis, intus concavus, ſummo faſtigio in nodum deſinens, membranaceum quoddam involucrum purpuraſcentis coloris ſuſtinentem, è quo unicus flos ſeſe exerebat ſex lógis & anguſtis folijs conſtans, qualia ferè in Narciſſi Autumnalis minoris (de quo cap. xvi. hujus libri) flore conſpiciuntur, non flavi tamen,ut illa,verùm rubri ſaturi & ſplendentis coloris, inſtar floris Arundinis Indicæ vulgo appellatæ, cui non valde abſimilis eſt, è quorum medio prodibant ſex ſtamina ejuſdem pænè coloris,oblonga (quibus inſidebant apices fuſci coloris quia forſitá ex attritu corrupti) & medius ſtilus, ſub quo rudimentum triangularis capitis,quod haud dubiè ſemen dediſſet, fortè etiam maturum, niſi ipſe flórem præcidiſſet, ut mihi conſpiciendum præberet : radicem, Paludano referente, habet bulbaceam, cæpis vulgaribus rubétibus prorſus ſimilem.
Vnicam autem ejus plantam habebat, ipſi miſſam ab eruditiſſimo viro D. Simone de Tovar Hiſpalenſi Medico,quæ flórem dabat Iunio ᴍ.ᴅ.xᴄɪɪɪɪ.alterum autem expeċtabat ſequente menſe,quoniam præcedente anno, eadem planta bis illi florem protuliſſet, menſibus Iunio & Quintili.
Aᴛ anno inſequente, ex Indice horti Tovarici, & epiſtolis quas ipſe Tovar ad me deinde ſcribebat,cognoſcebam Narciſſum Indicum Iacobæum ab ipſo nuncupatum.

Nᴀʀꜱɪꜱꜱᴠꜱ Indicus,Iacobæus mihi indigetatus, ut ſcribit in epiſtolâ Cal. Iunij ᴍ. ᴅ. xᴄᴠɪ. ad me datâ,ex Occidétali Indiâ (ubi *Azcal-Xochitl,* quod eſt,Bulbus flore rubro,vocatur)nobis delatus: nulli rei herbariæ ſcriptori haċtenus, quod ſciam, notus, cujus radix bulboſa,cepæ rotundæ ſimilima , verùm ſupernâ tunicâ pullâ , folia primùm emittit craſſa, obl longa,Narciſſi marini,tibi Hemerocallis Valentina diċti,æmula: ſecúdum folia yerò, atque
O adeò

Narciß.latifol. vɪɪɪ. qui & Narciß.Indicus.

Tempus.

Narciß.Iacus Iacobæus. Azcal-Xochitl.

Nataleſ

ꜰɪɢᴜʀᴇ 6.1. *Narcissus indicus iacobeus.* From Carolus Clusius, *Rariorum plantarum historia* (1601), 157. Special Collections, Sheridan Libraries, Johns Hopkins University.

balsams, roots, stones, woods, and other strange things" that Arias Montano and "other persons" had asked him to send to Clusius and to Pieter Ernst von Mansfeld, governor of Flanders and an avid collector of botanical specimens.[54] Arias Montano also mediated between Clusius and both Rodrigo Zamorano, cosmographer and *piloto mayor* at the Casa de la Contratación, and doctor Juan de Castañeda.[55]

Another member of his circle of naturalist friends was the humanist and medical doctor Francisco Hernández (1514–87). Arias Montano and Hernández likely met at court in the years before they both left to undertake two of the flagship projects of the early part of Philip II's reign. One project took Arias Montano to Antwerp, and the other, in 1571, sent Hernández on the first voyage of scientific exploration to the viceroyalties of New Spain and Peru.[56] During the seven years he spent in Mexico—he never made it to Peru—Hernández assembled a vast corpus of information on the country's natural history, which included not only materia medica—whose study was the official purpose of his expedition—but also Mexican antiquities, a history of the conquest of New Spain, and a commentary on Pliny's *Natural History* that incorporated observations about the flora and fauna of the New World. Sadly, on returning to Spain Hernández was not permitted to continue working on the material he had collected, a turn of events that shattered him. He was well aware that dealing with the exotic flora of Mexico had forced him to abandon the classification schemes traditionally used in Europe in favor of one informed by the etymology of the plants' names in Nahuatl. Who in Europe could make sense of his findings? Yet the task of producing an extract of the corpus focused on medicinal plants was given to the Neopolitan court physician Nardo Antonio Recchi (1509–90). The rest of the material—hundreds of drawings and maps and at least thirty-eight volumes of text—remained cherished but unpublished at the library of El Escorial until they perished when flames ravished the upper rooms of the library in 1671.[57]

We know little about Francisco Hernández's life after his return from Mexico. He repeatedly petitioned the king to let him continue his work, but permission was never granted. One of the few vestiges we have of these difficult years is a poem he dedicated to Arias Montano. It survives only as a draft among his papers that had been copied and deposited at the Colegio Imperial in Madrid. In the poem Hernández recalled their friendship, their parallel paths in service of the king, and the hardships he endured in Mexico. He complained that when he returned his work had been criticized by envious people knowing nothing of the subject. He lamented bitterly, "Oh, who can be judge, censor, and expert, who knows nothing of plants wherever they come from?"[58] The poem solicited Arias Montano's protection, but even beyond

56. The scholarship on the expedition is vast. An essential reading list includes Somolinos d'Ardois and Mirando, *Vida y obra de Francisco Hernández*; López Piñero and Pardo Tomás, *Influencia de Francisco Hernández*; Bustamante García, "Francisco Hernández, Plinio del Nuevo Mundo"; Hernández, *Mexican Treasury*; Varey, Chabrán, and Weiner, *Searching for the Secrets of Nature*; and Pardo Tomás, "Francisco Hernández."

57. Cabrera de Córdoba, *Historia de Felipe II*, 2:791–92.

58. Chabrán and Varey, "'Epistle to Arias Montano,'" 634.

this, asked him to please read Hernández's books and, "if they prove unworthy of honor, embrace the ideas just like those of a dear brother and thus favoring me, embrace me for eternity."[59] Hernández sought in Arias Montano someone who would validate his work and silence his critics by virtue of his reputation and his influence at court. Sadly, we have no traces of Arias Montano's opinion on the Hernández oeuvre, which our biblist would have been able to consult at the library of El Escorial, zealously guarded by Arias Montano's disciple José de Sigüenza.[60]

COLLECTING

Like scholars from time immemorial distracted from their studies by the minutiae of daily life, Arias Montano yearned for an idyllic retirement and longed for the solitude of his Peña de Alájar. In a famous letter to Zayas in 1578 offering the patronage of his humble chapel at the Peña to the king, Arias Montano described the idyllic setting of his retreat:

> Because I have never seen in all I have traveled in Spain, nor in other provinces, a place similar to this one of the Peña de Aracena [Alajar], in which so many natural things concur that when each is found by themselves are very valued, such as its elevation, the mildness of the sky, and the healthfulness of the site, the abundance of water, the expanse of the sky, and other things due to its comfortable isolation, I have been thinking for many days now that this is a place worthy of being owned by a king.[61]

59. Ibid.

60. Sigüenza's attitude is evident in his response to a petition by a certain Dr. León, who asked to borrow the Hernandian documents in order to prepare a book on medicinal plants. It reads in part: "Lo que me parece acerca de los libros de las plantas, arboles, peces, aves y serpientes y animales que estan en la libreria Real y de lo que pretente trabajar en ellos el Dr. León es lo siguiente. Lo primero se presupone que es bien estos libros sean de algún provecho y que no esten muertos. . . . Tras esto es forzoso darle los libros todos para que los tenga alla todo el tiempo que escriviese estara el riesgo de perderse y por lo menos maltratarse. . . . Segun esto mi parecer es que no se lleve a los libros para solo esto sin dexalos estar en la libreria que tanto fruto se saca dellos en que este aqui una cosa tan singular y tan rara, como estan otras muchas en las librerias que su fin no es mas que deleitar el ingenio y los sentidos del hombre y con esto puede tambien aprovechar de que con ellos sabremos lo que de nuevo nos trujeren de las indias.": BME MS Ç-III-3, fol. 391r. José de Sigüenza, "Borrador del P. Siguenza: Que los libros de aves, serpientes y animales–de Indias–que estan en la Libreria de S. Lorenzo no se entreguen para clasificarlos al Dr. Leon," n.d.

61. Arias Montano to Gabriel de Zayas, Peña de Aracena, 16 October 1578, in DIE, 41:369.

Retreating to the Peña meant severing himself from his more pressing duties as the king's chaplain and librarian, and other distractions that came with royal service at court. But most important, it meant physically retreating from the world. He achieved this retirement and a substantial pension only in 1585.[62] He then quickly set about improving the land around the small chapel (*ermita*), adding a number of buildings and making the site productive. Years after Arias Montano's death Rodrigo Caro visited the site and left the only eyewitness account of the retreat:

> The rooms at ground level of the houses were still habitable, although the upper stories were in disrepair. In the patio or area where the houses were located, toward the south stood a terrace surrounded by jasmines, with floors of white marble and with a table at its very center under which ran a trench of fresh and very cold water originating from a grotto near the entry to the building. Thus when eating at this table it was not necessary to fetch drinking water or any other service; the water then ran along the sides of the building and watered a garden that was adjacent to the main houses.[63]

In 1587 Arias Montano purchased a small estate with a *cortijo*, or farmhouse, northeast of the city wall of Seville and named it the Campo de Flores, and there he met with a select group of friends when he was in the city.[64] His retirement there did not last long, for in 1592 he was named prior of the monastery of Santiago de la Espada in Seville and served for the next three years, during which he wrote the second part of the *Magnum opus*, the *History of Nature*. We have ample testimony about a collection of *naturalia* and *artificialia* he assembled, which he would ultimately install in his home in the Peña de Alájar and at the Campo de Flores in Seville.[65] His interest in natural philosophy preceded him to Antwerp; as early as 14 February 1568, Plantin had

62. He was granted the *encomienda*, or commission, of Pelay Correa, which assured him a comfortable income: Morocho Gayo, "Trayectoria humanística, II," 283–84.

63. Caro, *Varones insignes en letras naturales*, 102.

64. Gil, *Arias Montano y su entorno*, 34–38, 365–87.

65. Caro explains, "Llevo a ellas mucha parte de sus libros y papeles, con gran numero de medallas y monedas antiguas y otras curiosidades de mucha estimación.": Caro, *Varones insignes en letras naturales*, 101; also Gil, *Arias Montano y su entorno*, 38–95. On Spanish collections, see Morán Turina and Checa Cremades, *Coleccionismo en España*, and Urquiza Herrera, *Coleccionismo y nobleza*. On New World curiosities, see López Rodríguez, "Sevilla, el nacimiento de los museos, América y la botánica."

offered to help him procure all sorts of mathematical artifacts, "astronomical rings, globes from Mercator mounted with its copper armillary rings, and all sorts of mathematical instruments, and of course books for any science," all of which, as we saw earlier, Arias Montano purchased.[66] But the first explicit reference to a collection per se is from a letter dated 1575 and addressed to Johannes Crato—the Protestant doctor of Emperor Maximilian II—where Arias Montano expressed his enthusiasm for natural wonders and other curiosities.[67] Arias Montano thanked the doctor for gifts he had received by way of Abraham Ortelius. These, he explained, had arrived just as he had been seized with a fervent urge to examine nature and the arts—he had been taken by an incredible desire to see all that had been created under the heavens.[68] To satisfy his "craving" for knowledge, he had installed the gifts on a pyramid he considered his small traveling museum (*museolum*). In addition to the great pleasure he derived from examining them, he also sought to use them to help him grasp the recondite (*non vulgaris*) secrets of the Sacred Scriptures requiring explanation. He entreated Crato to send him some items he listed on a separate sheet and asked for information about the country and region each item had come from, their names and uses, forces and efficacies, and how much was true and certain and what was considered common knowledge.[69] The official duties in Antwerp and on the Polyglot might have distracted him from his earlier interest in natural history, but now he was clearly eager to continue his studies in this area.

The exchange of exotic objects with other collectors came hand in hand with the economy of friendship in learned humanist circles throughout Europe. Exchanges with Ortelius are the best documented.[70] The cartographer sent Arias Montano printed maps, and the Spaniard reciprocated with manuscript maps of Spanish regions in the Americas and Asia, gems, and other "efficacious" stones.[71] Ultimately he sent him three bezoars, stones known to

66. Dávila Pérez, *Correspondencia en el Museo Plantin-Moretus*, 1:4–5.

67. Arias Montano to Johannes Crato, Antwerp, 21 January 1575. Cited in Almási, *The Uses of Humanism: Johannes Sambucus*, 86. The letter was published in its entirety in Clusius and de Ram, *Caroli Clusii . . . Cratonem epistolae*, 102–4.

68. Clusius and de Ram, *Caroli Clusii . . . ad Cratonem epistolae*, 103.

69. Ibid.

70. Macías Rosendo, "Correspondencia de Arias Montano con Abraham Ortelio," 551–72.

71. "Guardo una elegante piedra bezoar, elegida para ti, con algunas otras gemas o piedras de admirable eficacia.": Arias Montano to Abraham Ortelius, Seville, 30 March 1590, in Morales Lara, "Cartas de Benito Arias Montano a Abraham Ortels," 187–89.

be an excellent cure against black bile—one mounted in gold to wear as an amulet, and two others to be given to friends.[72] He also sent a medicinal stone to the bishop of Antwerp, Levinius Torrentio; later, another gift of a "ring made by Indians, but not without elegance," was lost in transit.[73] Ortelius's gifts were accompanied by American silver nuggets that could be assayed or converted into very pure metal. The last recorded exchange between the two took place in 1596. On that occasion, as a token of gratitude for having received the fifth *additamentum* of the *Theatrum*, Arias Montano sent Ortelius a small parcel from what he described as his "theater of nature and art" (*nuestro teatro de naturaleza y arte*).[74] The small silver coffer included five rings with exotic gems, a golden ring with an emerald (for Ortelius's sister), and a fragment of a *chalcanti* stone from the West Indies[75] with admirable efficacy against oral ulcers. This shipment included another bezoar, this one powerful against poison; a golden figurine of an Amerindian deity found in a chieftain's grave; a similar silver idol; a silver effigy of an animal made in the Indian style; another made of hematite jasper; a very sharp double-edged Mexican knife; and even a handful of fleece from the (unfortunate) vicuña that had yielded the bezoars.[76]

Shells and mollusks were among Arias Montano's other favorite *naturalia* to collect. He marveled at the "architectural" ability of the tiny animals and eventually dedicated many words to them in the *History of Nature*, noting that his attention had been drawn to them initially while studying the Sacred Scriptures, in what he considered a divine gift (*divinum beneficium*).[77] The shells were the key to his interpretation of the verse, "They that go down to the sea in ships, that do business in great waters; These see the works of the LORD,

72. "Una de color dorado difuso, incluida para ti para que la lleves como amuleto, cuya eficacia se afirma que es excelente contra las molestias de bilis negra": Letter of Arias Montano to Abraham Ortelius, Seville, 10 April 1590, in Morales Lara, "Cartas de Benito Arias Montano a Abraham Ortels," 193–97.

73. L. Torrentius to Arias Montano, Antwerp, 17 May 1585, in Charlo Brea, *Levino Torrencio: Correspondencia con Benito Arias Montano*, 17–19; Arias Montano to Abraham Ortelius, Seville, 3 January 1590, in Morales Lara, "Cartas de Benito Arias Montano a Abraham Ortels," 179–83.

74. Arias Montano to Abraham Ortelius, near Seville, 26 April 1596, in Morales Lara, "Cartas de Benito Arias Montano a Abraham Ortels," 193–97.

75. Perhaps a reference to chalcanthite, a copper mineral that, when dissolved in water, turns it deep blue.

76. For a detailed study of these exchanges and the resulting collection, see Hänsel, *Benito Arias Montano*, 203–24.

77. *Corpus*, 286–94; *HistNat*, 396–404.

and his wonders in the deep" (Ps. 107:23–24 KJV).[78] The shells' intricacy was for him a clear sign that the handiwork of God extended even to the bottom of the sea. Not having seen shells and mollusks explained elsewhere, he spent years observing many examples of the creatures, which were to him an amazing testament to the ingenious spirit with which God had endowed them.

In February 1578 Arias Montano made a trip to Portugal to visit friends and acquaintances and used this opportunity to collect unusual mollusks. The Spanish ambassador, Juan de Silva, had been alerted to the interests of his illustrious guest and after the visit reported to Zayas that Arias Montano had returned home "loaded with mollusk shells" (*cargado de conchas de caracoles*).[79] Arias Montano had apparently circulated a list before his trip describing marine organisms he hoped to collect. His acquaintances catered to his interest in these watery invertebrates. In 1577 Francisco de Aldana (1537–78) dedicated a poem to him, the *Carta para Arias Montano sobre la contemplación de Dios y los requisitos della*, in which he devoted thirty-three verses to shells.[80] In the poem Aldana proposed to take Arias Montano on a sea voyage where, among other things, he would see "a thousand twisted shells" of amazing colors, and others having "the light that God stores in the painted celestial arches, of various operations, of various enterprises, shooting sparks in a rich mixture of gold and turquoises."[81] Arias Montano returned to the image from the poem years later in one of his sermons at El Escorial. When establishing a simile comparing the grandeur of God and the sun, Arias Montano mentioned that mother-of-pearl—which grows inside shells in the depths of the ocean—is a testament to the enlivening and penetrating power of the sun.[82]

What other objects made their way to Arias Montano's *museolum*? Luckily

---

78. Arias Montano explained that the Hebrew version of the psalm uses the word MAGASSE (possibly a reference to מַעֲשֵׂי, *maʿăśê*, deed or work), which to him meant a specific thing made with technical skill, of precise and defined form, such as the heavens: *Corpus*, 287; *HistNat*, 397.

79. The historiography has assigned political overtones to this trip, arguing that Arias Montano was sent to Portugal to gauge local sentiment toward the Spanish, but evidence of this purpose is scant. Arias Montano definitely intervened on behalf of Castilian merchants who were being forced to pay a tax levied on the *converso* population. For contrasting views, see Alvar Ezquerra, "Benito Arias Montano en Portugal," and Lara Ródenas, "Arias Montano en Portugal." Lara Ródenas has explored this facet of Arias Montano's collecting in great detail.

80. Lara Ródenas, "Arias Montano en Portugal," 358.

81. "Los unos del color de los corales, los otros de la luz que el sol represa / en los pintados arcos celestiales, / de varia operación, de varia empresa, / despidiendo de si como centellas, / en rica mezcla de oro y turquesas.": Aldana, "Carta para Arias Montano," 456.

82. Arias Montano, *Sermones castellanos*, 147–48.

we have an inventory he wrote in 1597 when he planned to donate the collection to his assistants Pedro de Valencia and Juan Ramírez.[83] He wanted the collection not to be dispersed, a wish he shared with many collectors of his time, who saw the whole of their collections as a personal manifestation of their learning. It was an impressive collection indeed, not because of the sheer number of objects but because of their exquisite selection. The inventory highlights the monetary value and artistic merits of Arias Montano's mathematical instruments, both those he had collected over the years and those he acquired from the estate of Simón de Tovar. He had celestial and terrestrial globes by Gerard Mercator and Gemma Frisius, the most outstanding cosmographers of their time, as well as three astrolabes engraved in Latin, Hebrew, and Arabic. Alongside these he displayed other *artificialia*, such as polished precious and semiprecious stones, and more than 250 items described as being either "old" or "new."[84] Items that he considered products of nature unaltered by human hands were displayed in the "natural" studio, among them "soils, stones, metals, minerals, and mineral samples [*medios minerals*] of different types, resin woods, liquors and roots, fruits, large animals, parts of animals, and a number of forms and types of natures, and likewise maritime and marine things that I have in my studio called 'the sea.'"[85] The donation also included oil and tempera paintings by Pedro de Villegas, Pieter van der Borscht, and Francisco Aledo as well as a great number of prints. In addition to the artifacts, the collection had several antiquities, an essential component for any humanist collector with a desire to historicize the material legacy of the past. Sadly, the inventory does not mention the provenance of these artifacts, but if we consider the many references in Arias Montano's correspondence to procuring New World products, we can safely assume that many of them came from the Americas. Take as a case in point Arias Montano's fifteen-year search for the *ungüento de Bálsamo* mentioned in the Bible (Gen. 43:11, Jer. 8:22). He was convinced he would find it in the Indies. When he believed he had finally located the "true balsam," it came through some friends in Antwerp, although we do not know if in fact it came from the Indies.[86]

83. The document at the Archivo Notarial de Zafra was first identified in Salazar, "Arias Montano y Pedro de Valencia, 487-93." Also Gil, *Arias Montano y su entorno*, 287-92.

84. Salazar, "Arias Montano y Pedro de Valencia," 490.

85. Ibid., 491.

86. From Arias Montano's *De optimo imperio, sive In librum Iosuae commentarium* (Antwerp, 1583), 174, as cited in Gil, *Arias Montano y su entorno*, 44. On efforts during the early sixteenth century to determine whether a balsam produced in Santo Domingo was the "classical" balsam, see Barrera-Osorio, *Experiencing Nature*, 15-23.

If we examine the collection in its totality it is clear that, unlike so many others of the time, this was not simply an assemblage of curiosities brought together to satisfy prevailing notions of good taste or to elicit visitors' admiration. For Arias Montano, as historian Juan Gil has pointed out, the *museolum* formed an integral part of a reference library designed to serve one principal function—his vocation of biblical exegesis.[87] His personal library was well stocked with geography, cosmography, and astronomy books by ancient and modern authors. Along with his *museolum*, these were the textual and material resources he mined in his effort to bring the American reality, among other things, into concert with the biblical narrative.

ARIAS MONTANO, "SEVILLIAN EMPIRICISM," AND THE NEW WORLD

Arias Montano's life after his return from Antwerp revolved around the city of Seville. As significant as his sojourns at El Escorial might have been, it was among the community of merchants, cosmographers, doctors, and learned humanists in Seville that he truly found his intellectual home in Spain. During the sixteenth century the city was an unequaled vantage point for witnessing natural marvels of the world arriving on European soil. It was the site where the task of rationalizing and organizing the material vestiges of the New World began so they could ultimately serve and inform the empire. Travelers would make their way to the home of physician Nicolás Monardes bringing animals or plants from the Americas. At his home and *hortus medicus* Monardes studied the medicinal properties of the plants and tried to situate them within taxonomies inherited from Theophrastus and Dioscorides. It was in Seville that Francisco Hernández finally settled after seven years immersed in the natural and cultural world of Mexico and tried to conclude his natural history. It was also where Arias Montano set himself the challenge of devising a new natural philosophy that could encompass the old and the new, what was known and what would be learned in the future, and thus could lead humanity to God and eternal salvation.

To understand the role of Seville in the history of early modern science, it helps to think of the city as a site where American reality, in all its conveyable aspects, came into direct contact with the conceptual framework of European natural philosophy. Of course, contact between the two also took place in the American space, often with fascinating and intriguing results. But for Seville,

87. Gil, Arias Montano y su entorno, 40–44.

the "conveyable aspect" is an important qualifier; conveyable aspects of a particular thing can much more easily be turned into a commodity, whether material or conceptual, and once commoditized they can circulate. Those realities that remain imprinted in the psyche or memory of the witness—unarticulated and unconveyed—do not circulate, or their circulation is limited or delayed altogether until they are finally and fully articulated and conveyed. The result was that materials brought from the New World arrived in Seville largely decontextualized—signs without referents—or, at best, situated by their intermediaries within highly mediated contexts. Perhaps the bearer of an American curiosity knew something about its use and shared this information when handing it to someone like Monardes. This very well might have been the origin of a phrase Monardes uses so often in his work when giving the history of something: "they say that" (*dizen que*).[88] The phrase manifests a desire to identify, describe, and begin to sketch a natural history of the item. If the bearer's description did not satisfy, then it was necessary to elaborate new methods of observation that would re-create firsthand experience. Throughout the early modern era we find these efforts in historiography associated with the phrase "to make an experience" (*hacer experiencia*). Monardes, in fact, described this process of re-creating the experience others might have had with the product using the phrase "seen by experience" (*visto de experiencia*). By reenacting the experience, Monardes could begin to fashion a referential narrative situating the product or artifact within the intellectual framework of European natural philosophy.

In most cases these processes of experiential re-creation coexisted comfortably with the paradigms of Aristotelian natural philosophy and rarely required venturing into philosophical speculation. Aristotle sufficed for Sevillian empiricists, whose concerns were largely pragmatic. Although it is true that they used empirical observation as an epistemological tool in compiling useful information about nature, for the most part this did not entail formulating new natural philosophical postulates. Neither do we find in Seville a concerted effort to create a new natural philosophy based solely on empiricism. This point bears emphasizing, since it has not been made sufficiently clear in the historiography discussing empirical practices in sixteenth-century Seville. While for some historians the empiricism characteristic of Seville was a sort of proto–Scientific Revolution, the philosophical implications of this empirical posture have not received as much attention. The gradual but definitive divorce of empiricism from natural philosophy did not go unnoticed by Arias

88. Monardes, *Historia medicinal*, 18, 42, 48.

Montano. Although he never clearly articulated the feeling, it was essential to him that the whole of the natural world—encompassing both the Old World and the New World—be fully reconciled by one natural philosophy derived from an incontrovertible set of first principles. As we will see in the next three chapters, this was precisely the objective of the *Magnum opus*, and the *History of Nature* in particular. His disquiet is also evident in his dealings with the New World and its apparently inexhaustible supply of novelties.

At first glance the New World and its natural novelties seem to occupy a marginal place in Arias Montano's natural philosophical work. In the *History of Nature* he makes no more than three or four references to American natural products.[89] For example, when describing different types of tubers, he mentions a certain "foreign potato" (*batata extranjera*) that is sweet like "the ones we know and is brought from the islands of the Ocean."[90] In a section where he discusses a particular class of trees, we find a reference to trees that flower rapidly, "like foreign ones and those sought and brought in ships from other places, which we hear are called cinnamon, canes, and tamarinds" (*canelos, cañas y tamarindos*).[91] He exhorted his naturalist friends to continue their labors compiling descriptions of such specimens, but he did not engage in this work.[92]

His attitude toward New World novelties was informed by a deep-seated conviction that the New World must have been prefigured in the Bible; his predecessors had simply failed to notice. He believed that the Bible made repeated references to the New World and that it was known to the ancient Israelites, who seemed to have sailed there frequently.[93] Consider the treatise *Phaleg* from the *Apparatus* of the Antwerp Polyglot; there Arias Montano joined others before him who—following Christopher Columbus's lead—associated America with the biblical land of Ophir. Arias Montano interpreted the relevant biblical passages mentioning Ophir in genealogical and geographical terms. The sons of Joktan, Ophir and Jobab (descendants of Noah through the line of Shem), settled the lands bordering the Pacific Ocean. Jobab settled even beyond Ophir, and his descendants lived along a mountain range known as Sephar (Gen. 10:29–30). Arias Montano employed two types of evidence

89. Arias Montano's references to American artifacts and natural products have been studied in Navarro Antolín, Gómez Canseco, and Macías Rosendo, "Fronteras del humanismo: Arias Montano y el Nuevo Mundo."

90. *Corpus*, 242, 247; *HistNat*, 350, 355.

91. *Corpus*, 255; *HistNat*, 363.

92. *Corpus*, 241; *HistNat*, 349.

93. Arias Montano, *Prefacios*, 161.

to reinforce his interpretation; the first was empirical, the second philological. The articles brought from the New World testified to the abundance of gold, pearls, and precious woods in those lands. He went on to explain that the word Parvaim (פַּרְוָיִם, *parwāyim*) had become confused through a commutation of letters with the word Ophir (אוֹפִיר, *'ôpîr*).[94] Furthermore, in Hebrew Parvaim is spelled using an ending that signifies duality (ים); thus, Arias Montano surmised, the word originally indicated *two* regions. These two regions had to be those that were the source of most of the New World gold and now went by the names Peru and New Spain. After the time of Solomon, however, Ophir, through common usage, went on to designate only one region. This is why the word in the Bible sometimes appears in its original plural form. Furthermore, he placed the lands of Jobab in the New World, and in particular the region of Paria (Venezuela)—a region renowned for its abundant gold and pearls. Another of Jobab's sons settled near the biblical Sephar Mountains, which Arias Montano took to be a reference to the Andes.[95]

His philological approach to biblical interpretation also yielded new insights into other scientific disciplines. The second part of his *Magnum opus*, the *History of Nature*, fully displays his approach. For example, he believed there was an error of usage, as well as of translation, in the way the Bible designates certain trees with precious woods. According to Arias Montano, over the years the words ALMUGIM and ALGUMIM had become confused and amalgamated. As he explained, the ALGUMIM (אַלְגּוּמִּים, *'algûmmîm*) were trees that grew in the forests of Lebanon, while ALMUGIM (אַלְמֻגִּים, *'almuggîm*) were the precious woods that Hiram's fleet brought back from Ophir (1 Kings 10:11). (Furthermore, the combinations of five letters used in these words were in themselves very rare in the biblical language, indicating a foreign origin.) Just as VPHIR had become PIRV over the course of many years, the names of the trees had become confused.[96]

How did Arias Montano hope to fold these New World novelties into his biblical natural philosophy? As we will see in the next chapter, his interpretation was based on a historically informed hermeneutics of nature that in practice worked to nullify the temporal dimension of any discovery. Given that Adam had seen, named, and been given domain over *all* of nature, the apparent novelty of the New World and its products was simply a historical accident. A wholly reformed natural philosophy in concert with the Word

94. Arias Montano, *Phaleg*, in AntPoly, 8: Praefatio, fol. a2v and p. 12.

95. Arias Montano, *Phaleg*, in AntPoly, 16.

96. *Corpus*, 270; *HistNat*, 380.

would situate and explain all "novelties," since the form, function, and place of every plant, every animal, and every natural phenomenon had been predicated and predicted in the Bible.

EPISTEMOLOGY AND EMPIRICISM

Arias Montano was a lifelong admirer and curious student of nature. His admiration for the natural world went beyond the reflections of someone who approached nature solely from the contemplative perspective of natural theology. His engagement with nature was also an active enterprise, as evinced by his interest in collecting, the friendships he maintained with naturalists and geographers, and his own contributions that he would unveil in the *History of Nature*. Yet these activities belie how deeply he had reflected on the relation between natural philosophy and empiricism for most of his life. His skepticism toward received natural philosophical systems would become the launching point for his *Magnum opus*, but his reflections about empiricism must be pieced together from episodes in his life when he felt compelled to discuss the attitude toward knowledge he had gained from experience. One such episode was Arias Montano's involvement in the Lead Books of Granada affair. Its study yields one of his clearest expressions concerning his approach to ascertaining matters of fact and the function he ascribed to empiricism.[97] His attitude toward empiricism was buttressed, as we would expect, by his interpretation of key biblical passages, particularly those he believed had been written by Solomon, wisest of the biblical kings.

During the years he devoted to writing the *Magnum opus*, Arias Montano became involved in the thorny issue of the mysterious parchments and lead books found in the Sacromonte neighborhood of Granada. These writings supposedly demonstrated that the earliest Arab settlers of Granada had practiced an early version of Christianity.[98] The discoveries took place during the fraught years between the Morisco rebellion of the Alpujarras and the eventual expulsions from Spain of this population of Muslim converts to Catholicism beginning in 1609, at a time when the Morisco population in Spain was un-

97. Arias Montano's involvement in the affair was initially studied by Cabanelas and by Domenichini. Recently, Montanian scholars have also pointed out the importance of the correspondence that resulted from it and what it says about his attitude toward knowledge. See Dávila Pérez, "Correspondencia inédita de Benito Arias Montano," 1:70.

98. On the Lead Books of the Sacromonte of Granada, see Harris, *From Muslim to Christian Granada*, 1-7, 28-46, and García–Arenal and Rodríguez Mediano, *Orient in Spain*, 22-29, 172-73.

der increasing pressure to adhere more closely to Catholic orthodoxy and to assimilate culturally. Thus the historical revision the material implied carried significant political consequences, since it ultimately seemed to suggest that the Moorish population of Granada descended from early Christians.

The first cryptic manuscript from Granada—a parchment containing a prophecy purportedly by Saint John the Evangelist—was found along with some relics in a lead box under the Turpiana Tower in 1588. This first find was followed in 1595 by the discovery of twenty-two books inscribed on lead sheets or plates and a number of other relics. Arias Montano apparently saw a copy of the first manuscript, but no record survives of his initial assessment. Five years later, though, at the insistence of Pedro de Castro, the new archbishop of Granada, Arias Montano was asked again to weigh in on the matter. His refusal to travel to Granada was not seen as a reason to excuse his participation, and a canon from the cathedral was dispatched to Seville with the parchment and instructed to wait there until Arias Montano examined the original document and issued an opinion on its authenticity.[99] His report was addressed to Luis de Pedraza, president of the cathedral chapter of Granada, on 4 May 1593. The passage quoted below is part of the preamble to Arias Montano's detailed analysis of the original parchment, which concluded that the document was a forgery based mostly on philological evidence as well as on material evidence such as the age of the parchment, the nature of the ink, and the scribe's hand. After explaining that he had studied the parchment carefully, read its contents and marginal annotations, and taken into consideration an opinion written on the subject that had been sent along, Arias Montano stated:

> It might be due to my limited understanding and ability or because of the way I have always approached my studies by not admitting uncertain things as certain and surrendering my opinion to such things that I remain undecided about how to resolve two types of issues because I do not dare—as I have never dared—to lean toward what I most desire to lean toward. Because I, in matters of faith, have always taken the Sacred Scriptures and the church's pronouncements concerning scripture as foundations. Besides these two [issues]—which constitute a sole firm one—other matters do not even raise strong opinions in me. Concerning natural things, I have always sought to know what scripture teaches, and when I am unable to ascertain this, I try to ensure that the intellect does not contradict experience. In histories I ordinarily follow authors of the same time period or close to the same, devoid of affectations. And as to other

99. Cabanelas, "Arias Montano y los libros plúmbeos de Granada," 13.

things I am unable to investigate following these avenues, I take them as uncertain or as opinions and point this out.[100]

In the preamble to his opinion Arias Montano wanted to willfully detach both emotionally and intellectually from the implications of his findings. He did this by assuming a posture that was simultaneously dogmatic and skeptical in his approach to his assessment—dogmatic in that he professes to follow only scripture and church doctrine in matters of faith while embracing a skeptical attitude toward everything else. To the extent that he can draw from scriptural knowledge about the natural world he has done so, but otherwise "I try to ensure that the intellect does not contradict experience." In the intellectual climate of the Spanish disquiet, claiming that one did not let intellect (*la razón*) overrule experience meant a deliberate effort to set aside ingrained patterns of thought in favor of learning by means of an attentive evaluation of what was gleaned from the world by sensory experience.

After concluding, based on the material evidence, that the parchment was a forgery, Arias Montano discussed what he considered to be the "cloaked" (*disimulado*) style of the document. Its style had a lot in common with other types of genres meant to appear legitimate but that were really hoaxes: papers usually manipulated to arouse admiration or terror or to elicit comments and interpretations. The style reminded him of the enciphered writings of those who prognosticate using astrology and other such "curiosities." He considered these simple ciphers more akin to *jerigonza* (gibberish, or pig latin), but

100. "[Yo] habiendo mirado con atención el pergamino original con todas sus partes y menudencias, y leído y considerado así el texto como la interpretación o advertencia de la margen, y, después de esto, pasado con atención lo que el Sr. canónigo Lorca escribe con muy mucha diligencia en su libro, o sea cortedad de mi entendimiento y poca habilidad mía, o sea el modo que yo he tenido siempre en mis estudios de no admitir las cosas inciertas por ciertas y rendir mi sentido a las tales, me he quedado y estoy irresuelto en dos géneros de resolución que no me atrevo, como jamás me atreví, a inclinarme a lo que más desearía poderme inclinar. Porque yo en las cosas de fe siempre he tomado por fundamento la divina Escritura y la declaración de la Iglesia concerniente a la Escritura. Fuera de los dos fundamentos, que se reducen a uno firme, lo demás no me hace ni aun opinión fuerte. En las cosas naturales también he procurado saber lo que la Escritura enseña, y, cuando no alcanzase esto, procuro la razón que no contradiga a experiencia. En los historiales ordinariamente he seguido los autores del mismo tiempo o cercanos a él, desnudos de afecto. Y a lo que por estas vías no averiguo téngolo por incierto o por opinión, y así lo refiero. Testigo es vuestra Merced que en el concilio de Trento seguí este uso, y en los de Salamanca y Toledo, como lo saben los que se hallaron en ellos; y en mis escritos y coloquios lo hago así.": Cabanelas, "Arias Montano y los libros plúmbeos de Granada," 17.

the one used in the parchment did not fit this mold. The style of the parchment reminded him instead of the recipes of alchemists or of Paracelsian empirics who like to perplex followers who want to learn their mysteries.[101]

When the lead books surfaced at the Sacromonte in 1595, Arias Montano was again asked to issue an opinion. Again in a letter couched in modesty and humility, he argued against the authenticity of the find. This time he took on the matter, as he explained, because he recognized how important it was to know what was true and to define all its conditions, since this was what must be done for all types of things "lest we take uncertainties as certainties and assent to them blindly."[102] He resolved several specific questions asked by the archbishop and even inspected some rather poor imprints of the lead plates, but he never claimed to have been able to read them, let alone understand them. Ultimately Philip II, wishing to resolve the growing polemic and various interpretations circulating around the authenticity of the finds and their message, issued a royal order or *cédula* on 9 August 1596 convening a group of experts, including our biblist, and ordering them to travel to Granada to examine the finds.[103] Arias Montano managed to not comply with the order by alleging various difficulties: illness, both sudden and chronic, frailty due to age, and limited expertise on certain subjects. The conferences went on without him, and after the examiners came to the opinion that the lead books were likely authentic, the archbishop once again consulted Arias Montano.

In response to the archbishop, Arias Montano reiterated many of the problems he had noted earlier concerning the Turpiana parchment, especially the style of Arabic used and the problems it presented to any translator. Although one of the principal objectives of Arias Montano's response was to avoid becoming embroiled in the matter, this time he tried to explain to the bishop his epistemic commitments and how they limited what he could contribute

---

101. "Aun mas disimulado estilo llevan y usan los que por astrología u otras tales curiosidades pronostican y escriben la manera que llaman aquí cifra. No lo es sino entre las invenciones de jerigonza; es de las mas simples y de menos arte y menos cuidado para se entender. Parece o semeja, aunque con menos arte, a las recetas de los alquimistas y a [las] de algunos empíricos paracelsistas que, con poca ciencia, desatinan a los que los siguen hasta entender sus misterios.": Cabanelas, "Arias Montano y los libros plúmbeos de Granada," 19.

102. "Por lo que importa saber la verdad y definirla por todas sus condiciones, pues en todo género de cosas es común regla . . . *ne incerta pro certa habeamus iisque temere assentiamur.*" Letter of Arias Montano to Pedro de Castro, archbishop of Granada, 3 May 1595, Seville, in Cabanelas, "Arias Montano y los libros plúmbeos de Granada," 17. The phrase is borrowed from Cicero, *De officiis*, 1:18. Arias Montano swapped *incerta/certa* for *incognita/cognita*.

103. Domenichini, "Quattro inediti di Benito Arias Montano," 53–54.

to the evaluation of the lead books. It was very important to him to establish before his superior a clear distinction between what drove his studies and the "curiosity" that drove others. His strategy was to present the approach he had followed in his studies as rigorous and rational and to explain that it was not driven by curiosity about obscure things. He began, "I, Sire, profess to be no more than a poor student eager to be taught and not an enthusiast of the opinions of any of the discipline I have studied."[104] He expressed a preference for what was straightforward, clear, and pure, therefore showing that he was not a curious person, especially if one considered that the ancients defined curious persons as those who sought to know things that did not pertain to them. The implication was that a curious person studied recondite, obscure, and impure things. Instead, his mission had been to question and investigate (*preguntar e inquirir*) the principles and bases of different subject matters and to see whether they conformed to the Sacred Scriptures, with the natural sense (*sentido natural*), or with both. What had driven him to this approach was the conviction that there was no other way to determine anything with any certainty.

In studying the natural sense of things he had guarded against being driven by curiosity and had been careful to see and experience (*ver y experimentar*) natural and artificial things. If the subject of inquiry did not pertain to natural and artificial things, he considered it a curiosity and set it aside for those who had a taste for such things. In this statement Arias Montano was deliberate in placing curiosities that could be explained in their natural sense into an ontological category distinct from natural and artificial things. This leads me to believe he was referring to supernatural curiosities beyond those described

---

104. "Yo, Señor mío, profeso ser no mas que un pobre estudiante deseoso de ser enseñado, y no afeccionado a opiniones en disciplina alguna de las que he estudiado sino a lo llano, claro, y puro de ellas, y conforme a esto jamás he sido curioso; que los antiguos llaman curiosos a los que procuran saber lo que no les toca. Lo que yo he trabajado ha sido preguntar e inquirir los principios y fundamentos de las materias y procurar de ver si conforman con la divina Escriptura, o con el sentido natural, o con ambas partes, por no hallar certeza en otra manera, y para esto del sentido he sido no curioso sino cuidadoso de ver y experimentar cosas naturales y artificiales, y en lo que a esto no tocaba, dejar la curiosidad para quien de ella gustasse. Para saber de veras conozco ser grande la ayuda de las lenguas, y alabo mucho y doi gracias a Nuestro Señor por las que por merced suya he aprendido, que yo reconozco quando me han aprobechado; empero también conozco que no esta el fundamento de el saver en ellas sino en la naturaleza propia de las cosas que se quieren aprender; y sin éstas lo demás es adherente, o accidente sin su acomodada substancia.": Letter of Arias Montano to Pedro de Castro, archbishop of Granada, Campo de Flores near Seville, 10 November 1596, in Domenichini, "Quattro inediti di Benito Arias Montano," 61.

in the Sacred Scriptures, perhaps the range of magical arts and practices he associated with astrologers and prognosticators in the earlier letter. The issue of language as a means of deciphering this type of knowledge was clearly in Arias Montano's mind as he addressed the archbishop, because he immediately explained his understanding of the role languages played in seeking knowledge. Yes, knowing many languages had been of great help to him in his studies, "but I also know that the basis of knowledge is not in them [languages] but in the very nature of things that we want to learn about, and without these the rest is an adherent or accident without its associated substance [*adherente, o accidente sin su acomodada substancia*]."[105]

The message to the archbishop was clear: if these lead books addressed "curious" subjects or were written in the style associated with them, they were beyond his area of expertise, and even interest. The language they were written in—whatever it might be—was not the way to reach a clear understanding of what the books contained. The path to any knowledge lay in examining the very nature of things; the language used to describe them was just an ancillary adornment, or simply meaningless—as he implied by referring to them as an accident without substance, a theoretical impossibility in Aristotelian-Thomistic natural philosophy.

By 1597 Arias Montano was describing the issue of the lead books as a serious matter (*negocio gravíssimo*) because the junta of examiners Philip II assembled had deemed them to contain revealed doctrine dictated by the Holy Spirit. He explained to the archbishop that he had not seen or learned about the books' contents, nor did he know anything definitive about the style or phraseology, which to him were essential in order to understand how things were referred to.[106] In other words, although it went unsaid, he had not studied the text and was therefore not able to subject it to the kind of philological analysis that had yielded most of his exegetical works.

This philological excuse was not lost on the archbishop, who wrote later that year inquiring about the meaning of some specific Arabic words on some of the plates. Arias Montano translated the phrase as "No Dios sino Dios Jesús espíritu de Dios" (Not God but God Jesus spirit of God). The word for spirit (RUHU) had proved particularly difficult for the translators, and the archbishop wanted an explanation. Arias Montano wrote that it derived from the Hebrew word RUAHH (רוּחַ, *rûaḥ*). The problem for him lay in a translator's interpret-

---

105. Domenichini, "Quattro inediti di Benito Arias Montano," 61.

106. Letter of Arias Montano to Pedro de Castro, archbishop of Granada, 30 April 1597, n.p., in Domenichini, "Quattro inediti di Benito Arias Montano," 62.

ing the word to mean much more than its simplest definition: spirit (or in some rarer cases soul or *anima*). Instead the translator alleged it could mean "alma, vida, ser, hijo charíssimo, natural, muy amado, de la misma substancia y ser que el Padre." Arias Montano was mystified as to where anyone would have found such varied meanings for such a simple word. He explained, not without a little irritation, that it could be done only by way of a gloss or deductive logic. Anyone employing this means of analysis—even himself, with his modest intelligence—could find so much to say about the word that he could fill two folios and their margins with gallant interpretations. He, however, was not permitted to do so by God's law, the Gospel, Saint James, and the religion that had taught him how to deal with sacred things.[107] With this not too veiled criticism Arias Montano charged against a Scholastic interpretation of the Sacred Scriptures that, through logical gymnastics, produced an exegesis far too removed from the intended literal meaning of the holy interpreters. He preferred to stay close to the simplest meaning of a word as defined by what he understood to be its root. For Arabic this meant the root of the Hebrew word. To extend a definition of something as simple as "spirit" into the realms that evoked Thomistic Aristotelianism, such as notions of substantiality and ontology, seemed to him a travesty.

His caveats and narrow arguments were nonetheless widely discussed, and a few months later he was forced to respond to the archbishop defending his position (or purported position) in the polemic and begging again that he not be dragged into the fray. He insisted that he had not offered a categorical opinion in private or in public about the authenticity or contents of the lead books, and that he did not intend to do so. In the last letter of the exchange—Arias Montano died five months later—the archbishop sent copies of some characters nobody seemed to know how to interpret, to which a clearly exasperated Arias Montano responded, "Some of the figures are like Samaritan ones, others Ethiopian, and even others, Greek; there are also Aramaic ones. Some seem simple while others seem to be forming syllables or are even engraved as in words. When you do find someone who can read or interpret them, I beg of Your Honor to have word sent to me with some explanation about them."[108]

Arias Montano's attitude toward ascertaining matters of fact and his investigations into nature must be considered in concert with his attitude toward the

107. Letter of Arias Montano to Pedro de Castro, archbishop of Granada, Seville, 3 December 1597, in Domenichini, "Quattro inediti di Benito Arias Montano," 64.

108. Letter of Arias Montano to Pedro de Castro, archbishop of Granada, Campo de Flores near Seville, 9 February 1598, in Domenichini, "Quattro inediti di Benito Arias Montano," 66.

purpose of knowledge and of the studies that led to knowledge of nature. They were two sides of the same coin, impossible to dissociate from our understanding of the motivation for our biblist's activities described in this chapter, and ultimately for his writing a *History of Nature*. For Arias Montano the purpose of attaining knowledge of nature was in line with the purpose of natural theology: it was a means of approximating an understanding of the nature of the Divinity by studying his creation. But perhaps the clearest expression of his attitude toward knowledge of any kind appears in a commentary on Ecclesiastes that he left unpublished.[109]

As was widely thought at the time, Arias Montano believed that the author of the sapiential book of Ecclesiastes was Solomon, and his exegesis interpreted the text as words spoken by the wisest of kings. In this book Solomon acknowledged the futility and vanity of amassing wealth, pleasures, and knowledge; in doing so humanity neglects its duties to God. Commenting on Ecclesiastes 8:17, Arias Montano portrayed a Solomon who knew humanity could attain only limited knowledge of nature, and he warned against the arrogance of those who think they can attain such knowledge.[110] He cautioned, we find it difficult to explain things that are right before our eyes, so how can we explain the works of God? Those who say they know and understand many things are only fooling themselves. Even natural things, referred to by Solomon as those things "under the sun," still defy explanation despite being constant and permanent and experienced for so long:

> Even with this we still know nothing, because who has figured out the trajectory of the heavens, the birth of its stars, etc.? The nature of fire and of air and of water and earth and of the animals and things that are born of these ele-

109. The manuscript has been preserved at the library of El Escorial in three copies attributed to Arias Montano. In the introductory study to the first edition of this text Núñez Rivera clarifies the relation between the three versions and explains that MS g.IV.32 (fols. 8–83v) can be considered the archetype by Arias Montano and the other two are copies expanded on by his disciples at the monastery (Mesa 22-1-9, pp. 1–39, and MS I.III.24 (fols. 1–142v). In addition to lexical changes, the later texts include a few longer digressions commenting on how the biblical narrative relates to the situation in Spain. These additions, some of which might be by José de Sigüenza, do not comment on the natural philosophical aspects of the text: Arias Montano, *Discursos sobre el Eclesiastés de Salomón*, 11–16.

110. "Then I beheld all the work of God, that a man cannot find out the work that is done under the sun: because though a man labour to seek it out, yet he shall not find it; yea further; though a wise man think to know it, yet shall he not be able to find it" (Eccles. 8:17 KJV).

ments? And who can say he knows these things, wise as he may be? . . . what will he make of those secret ones about His judgment and providence . . . about which men talk so freely?[111]

And though at first glance it might seem Arias Montano's interpretation of these passages from Ecclesiastes suggests that he had a defeatist attitude toward knowledge of nature, it is important to keep in mind the overarching point he was trying to make. Humanity cannot forget its ultimate purpose—to know and honor God—and must not allow earthly diversions, even seemingly virtuous ones like the study of nature, to distract it from this purpose. In Arias Montano's opinion, humanity could be tempted away from God by human-built philosophical systems, the lure of sensual perception, and curiosity. But rather than considering the status quo inevitable, he saw it as his mission to supply a positive plan for knowledge. It had to provide a sound philosophical foundation—one that was biblically based—for interpreting what was learned about nature by means of the empirical methodologies he embraced.

*

In the midst of the natural philosophical turbulence of the late sixteenth century, Arias Montano saw an opportunity to continue what he thought was the ecumenical, theological, and philosophical mission launched with the Antwerp Polyglot. From the distance of his hilltop retreat at the Peña de Alájar, he thought about the bustle of novelties in Seville and realized that the task of making sense of this expanding world could not be tackled solely with the empirical tools of observation, description, and classification. In the *Magnum opus* he attempted to institute new metaphysical principles and a natural philosophy that liberated humanity from the misguided ideas of the ancient philosophers and also from the distraction of novelties. His hermeneutics of nature was intended to serve as a guide for empirical observations, but placing this activity within the context of a Mosaic philosophy constantly reminded the naturalist what the true purpose of this enterprise should be. It also implied rationalizing the New World through a hermeneutics that normalized any appearance of novelty; Solomon himself had done as much in Ecclesiastes 1:9–13 when he said there was nothing new under the sun. For recent discoveries and novelties this meant, as Arias Montano explained,

111. Arias Montano, *Discursos sobre el Eclesiastés de Salomón*, 240.

New kingdoms are discovered, or something new happens suddenly or an extraordinary invention is found or something also along these lines. "So he fools himself—he replies—because we have already seen things similar to this during the first centuries or its equivalent, as we can tell by experience from how things change." . . . It looked to us like a new world and thus we called it so, what was discovered of the Indies; it is true others had discovered and found them first, but memory of it was lost.[112]

Eliminating the "new" from the New World was one of the consequences of seeing the world through Arias Montano's hermeneutical lens.

112. "Hanse descubierto nuevos reinos o sucedido esta extrañeza repentina, o hallóse una invención extraordinaria, o otra cosa semejante. Y engañanse–responde–porque en los siglos primeros se ha ya visto otra cosa semejante a esa y esa misma o su equivalente, como lo vemos por experiencia en las mudanzas de las cosas. . . . Pareciónos nuevo mundo, y ansí le llamamos, lo que de las Indias se descubrió y es cierto que otros las descubrieron y hallaron primero y perdióse la memoria.": Arias Montano, *Discursos sobre el Eclesiastés de Salomón*, 101.

CHAPTER 7

Premises of the *Magnum opus*

The thing that hath been, it is that which shall be; and that which is done is that which shall be done: and there is no new thing under the sun.

Ecclesiastes 1:9 (KJV)

Once Arias Montano was granted his long-sought retirement from service at court in the latter part of 1592, he could finally turn his full attention to the project he had nurtured for over thirty years. A lifetime of study and observation had prepared him to tackle one of the greatest problems of his era, the reform of natural philosophy. He would undertake it in a trilogy he referred to as his *Magnum opus*. Other projects—and what he described as constant unwelcome interruptions—also occupied the last six years of his life. He wrote a book of poetry, *Hymni et secula* (1593), and a theological commentary that he had begun while in Antwerp in 1574 and finally completed in 1594, *Commentaria in Isaiae prophetae sermones* (1599). His assistant, Pedro de Valencia, oversaw the posthumous publication of Arias Montano's *In XXXI Davidis Psalmos Priores Commentarium* (1605), in which the biblist dedicated commentaries on the first thirty-one Psalms to friends and associates. Arias Montano's main concern, however, was the *Magnum opus*; unfortunately, his death in 1598 kept him from completing the task.

This chapter begins the detailed study of the two surviving parts of the *Magnum opus*, the *Anima* and the *Corpus*. After discussing the publication history of the books, their style, and their structure, the chapter revisits Arias Montano's exegetical methodology and examines his notions about what constituted "nature" and the true knowledge of it. The remaining sections highlight the fundamental premises—some historical, other epistemological—that informed his biblical natural philosophy.

THE *MAGNUM OPUS*: STYLE AND SCOPE

Arias Montano dedicated the *Magnum opus* to the Most Holy Roman Church as a work of piety, but also as the reflection of a deep desire to instruct others in what he considered a synthesis of the arguments of all his writings. The *Magnum opus* was conceived as a work in three parts, which he referred to as the *Anima*, the *Corpus*, and the *Vestis*. The first part, titled *Liber generationis et regenerationis Adam, sive De historia generis humani: Operis magni pars prima, id est Anima*, hereafter called *Anima*, consisted of eight books and was published by Christophe Plantin's successors at the press in Antwerp in 1593. Arias Montano began the *Anima* before settling at the Peña; it was in the hands of Jan Moretus, Plantin's successor, by late 1591.[1] As its title suggests, it is the history of humanity from its generation at the Creation through to its regeneration with the coming of Christ.

The second part of the trilogy, the *Naturae historia: Prima in magni operis corpore pars* (hereafter *Corpus* or *History of Nature*) would soon follow. He signed the dedication in January 1594, and Moretus received the manuscript in July 1594.[2] Only this first of the three projected parts of the *Corpus* made it into print. It tells the history of the natural world beginning with Creation and ending with the physical creation of humanity and a description of its attributes. Arias Montano acknowledged in the dedication of the *Corpus* that he had hurried to bring this first part to print, aware that he was presenting only the "head" (*caput*) of the project but confident that some would find it instructive. It seems that once well into the project he realized how ambitious he had been and yearned for collaborators who would help him develop the various disciplines associated with the natural philosophy he was setting forth. The first part of the *Corpus* received approval from the censor Henricus Sertus Dunghaeus in March 1595. After Arias Montano's death, Pedro de Valencia explained to Moretus that Arias Montano had written several sections for the planned second and third parts of the *Corpus*. He had collected these from among his mentor's papers soon after his death and, finding them significant works of erudition and doctrine, was intent on publishing those as well.[3] These were never published, however, and now appear to have been lost. Publication of the first part of the *Corpus* was delayed until 1601, in part

---

1. Dávila Pérez, *Correspondencia en el Museo Plantin–Moretus*, 2:776–77.

2. Ibid., 2:819–23.

3. Pedro de Valencia to Jan Moretus, 18 October 1598, Zafra, in Dávila Pérez, "Correspondencia Latina inédita de Pedro de Valencia," 233–40, 236.

because Moretus was reluctant to publish Arias Montano's work without first receiving adequate financial incentives.[4]

The third part of the *Magnum opus*, the *Vestis*, remains a mystery. It most likely was never written, since Pedro de Valencia does not refer to it in his correspondence with Moretus about the posthumous publication of his mentor's works. No documentation from Arias Montano survives that fully explains the scope of the third part of the *Magnum opus*. References in his other books suggest that the *Vestis* would have been a history of the "adornment of the human condition" and would surely have discussed language and perhaps religion. Some references to the *Vestis* in the *Corpus* suggest that it also dealt with categories and nomenclature, since the sections of the *Corpus* that discuss taxonomy had many references to the *Vestis*.[5] In a letter to Moretus in 1594, Arias Montano even hinted that he had considered extending the *Magnum opus* to include treatises on all the scholarly disciplines.[6] It seems it was also to include a series of treatises on dialectic and rhetoric: one of them, *Abigail*, was sent to Moretus yet never published and now seems to be lost.[7]

Arias Montano's objective in the *Magnum opus* was to institute a reform of knowledge disassociated from any philosophical system and in complete concert with the Sacred Scriptures. He began the *Anima* and the *Corpus* at a point that could be seen as constituting a philosophical clean slate: the very beginning of the world. Furthermore, since he chose to reconstruct natural philosophy using as his guide the historical narrative contained in the Bible, he had to start with the first event described there—the Creation—and proceed chronologically. So it is likely that, like the *Anima* and the *Corpus*, the *Vestis* would have also been organized as a historical narrative, but in this case about how humanity developed varied ways of "dressing," "embellishing," or "shrouding" itself, the natural world, and perhaps the divine since its creation.

The *Corpus* and *Anima* are written in similar styles, with perhaps their most noticeable feature being the profusion of biblical quotations used to support practically every declarative statement. This profusion makes the *Magnum opus* stand apart from other natural philosophical treatises of its time and renders it stylistically more akin to a biblical commentary. Here, as in a commen-

4. On Moretus's reluctance to publish Arias Montano's works, see ibid., 222–26.

5. Gómez Canseco, "Estudio preliminar," *HistNat*, 21–22.

6. Dávila Pérez, *Correspondencia en el Museo Plantin–Moretus*, 2:823n2.

7. On the lost treatise titled *Abigail, sive De ratione dicendi ex sacrorum eloquiorum observatione*, see Dávila Pérez, *Correspondencia en el Museo Plantin–Moretus*, 855–59, and Dávila Pérez, "Correspondencia Latina inédita de Pedro de Valencia," 223.

tary, biblical citations serve as both wellspring and ultimate authority, but Arias Montano rarely called on extraneous authorities to adjudicate the merits of a particular interpretation—a significant deviation from a biblical commentary *sensu stricto*. While the *Anima* seems to follow more faithfully the Bible's order of narration, Arias Montano does not comment on every verse as a traditional commentator would have done, but instead pays close attention to the sequence of events, electing to elucidate mostly events emphasizing the relationship between humanity and God as creator. The *Corpus*, more than the *Anima*, is arranged thematically. He follows the text of the Bible only for the Creation narrative; the rest of *Corpus* is a systematic exploration of the composition, arrangement, and inhabitants of the world—"world" understood in the early modern sense of the Latin *mundus* to include the terrestrial and celestial realms.

The style and thematic arrangement of the *Magnum opus* is a consequence of Arias Montano's objective: to search for a natural philosophy in the interstices of the Sacred Scriptures. He marshals biblical citations, especially those from the Old Testament, as testimony to the prior existence of the natural philosophy he was trying to recapture. But more important, he approaches them as textual evidence; biblical words and formulations contain vestiges of the lost philosophy he was intent on recovering. He subjects key quotations to careful philological and historical analysis, in some instances weaving together up to half a dozen passages from the Old and New Testaments to illustrate the many meanings a certain word could have by showing the various contexts in which it was used or how his etymological analysis was validated in different contexts. For Arias Montano the elegant humanist, biblical quotations are also called on to offer rhetorical emphasis, as in the case of the divine dictum *et factum est ita* (and it was so [KJV]), which Arias Montano deployed frequently to buttress arguments about the immutability of nature.

The *Anima* and the *Corpus* are *historias* in the humanistic sense, works where philological erudition, observation, careful description, and chronology coexisted easily in a single project. What makes the *Magnum opus* distinct from other works of this humanistic genre, however, is Arias Montano's refusal to consult—or acknowledge in print—any authority other than the Bible. And yet the work is deeply informed by an empiricist understanding of reality. On this aspect, authority derives not from the works of ancient philosophers, doctors of the Roman Catholic Church, or contemporary scholars, but from his personal observations of natural particulars. Within the *Magnum opus* project, Arias Montano acknowledged the Bible and experience of the natural world as his sole guides. In contrast to the historical genre of his era, the objective of the *Magnum opus* was not to dwell on *exempla*, but instead to teach a *praeceptum* or rule that explained the nature of humanity and of things.[8] Arias Montano deliberately chose the historical genre for his *Magnum opus*. Like other sixteenth-century biblists, he understood the relationship between God and humanity as evolving and changing through time. The sequence of historical events in the Bible was as important as the events themselves. In fact, *when* something happened was essential to the exegetical exercise. Therefore preserving the temporality of the relationship between God and humans meant telling the history of humanity's creation and its regeneration after the Fall. This is the *historia* told in the *Anima*. Similarly, the natural world also had a history. It had a beginning at the Creation, it unfolded during the six days of Creation, and it continues unfolding to this day. The nature of things, however, did not suffer temporal vagaries—only free will imposed these of essential mutations on human history. Instead, in Arias Montano's conception of nature, it unfolded according to a predetermined plan. This is the *historia* told in the *Corpus*.

Thus the *Magnum opus* trilogy was conceived as a series of *historiae* telling three distinct stories but drawing on the same sources: the Bible and experience. It was an uncontested presupposition of Christian exegetes since Saint Augustine that the Sacred Scriptures were written to tell the story of God's relationship to humanity. Whatever it said about the natural world—except for the benevolent act of the Creation itself—had been relegated to a secondary plane by the biblical authors. Nonetheless, Arias Montano believed the plan of nature was partially described in the Bible, albeit in a manner accommodated to humanity's ability to understand it and articulated in passing. He explained that Moses's principal objective—as it was for other prophets—was to transmit to humanity divine doctrine and law. But since the best doctrine is the one that explains the reason for things from the beginning (*rerum rationes a principio tradit*), Moses began the Pentateuch from the beginning, with the Creation, and proceeded in chronological order.[9] Moses's reason for following this order was not just that he was divinely inspired to do so, but also that he had been educated among the Egyptians, who claimed to be the oldest people on earth, so it was important that this audience also learned how the world began. Thus, even before mentioning humans, Moses described what makes up the world and how it is maintained—its principles, its parts

8. Pomata and Siraisi, *Historia*, 20; Maravall, *Estudios de historia del pensamiento español*, 2:191–212.

9. *Corpus*, 149; *HistNat*, 251.

and their arrangement, its different eras—including those aspects that can be perceived at present and those that cannot but that can be comprehended by means of their description and by reason.[10] Arias Montano understood that the entire act of Creation was encapsulated in the first sentence of the book of Genesis, "In principio creavit Elohim caelum et terram." In what constituted for him an example of the hermeneutics of accommodation that informed the Hexameron, Moses then separated the acts of the Creation into separate stages or "days," each taking place in time, so that the nature of created things could be understood more easily.

The emphasis in the *Anima*, as its title suggests, is on the spiritual aspects of humanity; humans as biological beings are discussed in the *Corpus*. Understanding the relation between spirit and body had important consequences for establishing humanity's relation to the knowledge of nature. In the *Anima*, Arias Montano examined the history of humanity and its relationship with God for clues about what the relationship between spirit and body should be. He identified its archetype in the relationship between God and Adam before the Fall, after which human history was an ongoing—and largely failed—attempt to regain this state of harmony. Arias Montano framed the story of humanity against an essential tension that resulted from the Fall—the recurring conflict between the interior and exterior man.[11] The interior man—gendered masculine—resided in the spirit and strove for those things that led to knowledge of God, while the exterior man—gendered feminine—belonged to the realm of bodily materiality and was driven by lower, sensual appetites. The interior man and exterior man were caught in a veritable life-and-death struggle, with the interior man wanting to lead its host to knowledge of God while the exterior man constantly tempted him away from this knowledge. Yet it was this impulse of the exterior man that seemed to drive humanity's interest in knowledge of nature. How could this curiosity with all its pitfalls be turned into a virtuous pursuit?

Arias Montano found in the Bible a historical explanation that allowed him

10. "Totius opificii, quo mundus constat, principia, seriem atque tempora, et certa atque stabilita iura, ante omnem de hominis natura sermonem, disertè descripsit: omniaque, praeter Deum, quaecumque sunt, ac videri humano sensu possunt, quaeque non possunt etiam; sunt tamen, et ratione atque intelligentia comprehenduntur.": *Anima*, 25; *LibGen*, 125.

11. Saint Paul speaks of the inner man (*interior homo*) who communes with Christ (Rom. 7:22–23, Eph. 3:16–17). For Saint Augustine the inner man is in constant conflict with the exterior, earthly body (*City of God*, 13.24). The association of the interior/male and exterior/female can also be found in his *On the Trinity* 12.1.1–4 and in *The Literal Meaning of Genesis*, 3.22. Saint Thomas would then characterize femininity as passive and thus inferior to masculinity (*Summa Theologica* 1.92.1).

to shift curiosity about the natural world away from the domain of the exterior man. To establish the historical relation between God, humanity, nature, and the role that knowledge of nature has for humanity's quest for knowledge of God, the *Anima* begins by recapitulating the events of the Creation, or the making of the world (*constitutio mundi*). This section included an exploration into the nature of God through his divine names and his angels, how God fashioned the world, assigned the order of things, and gave humanity its unique nature. The rest of the *Anima* follows the biblical story, with Arias Montano emphasizing periods when humanity seems to be approaching God, inevitably followed by periods when humanity is tempted and wanders away from the Creator. It is only the coming of Christ—presented as the final redemptive gift of a patient God to his constantly wavering principal creation—that humanity received definitive instructions about how to attain salvation.

Among the many temptations that had led humans away from God, Arias Montano considered the quest for knowledge of nature one of the most treacherous paths humanity has had to navigate, and in his estimation it had done so very poorly. For in the quest for knowledge of nature, humanity has neglected the proper study of the Sacred Scriptures. Instead, history abounded with stories of false prophets—teachers who lured (and still lure) with philosophies of their own invention that are not meant to lead people to God. In Arias Montano's view Moses's legacy as prophet and author of the Pentateuch had important consequences for the pursuit of knowledge about the natural world; the Mosaic books taught what it was essential to know about human nature and the natural world. Thus, recapturing the philosophy of nature embedded in this relationship promised to redirect humanity toward the divine.

Having set down in the *Anima* the premises that informed his theory of nature, in the *Corpus* Arias Montano turned to the description of natural particulars and cosmology. It begins by recapitulating the principal points of the preface and the first few chapters of the *Anima*, although this time he supports his claims with copious biblical citations.[12] He must have been responding to criticism about the profusion of classical authors, including a number of Stoics—Ovid, Juvenal, Horace, Persius, Pindar—whom he employed in the preface of the *Anima* to emphasize his arguments. Here we find again a restatement of the purpose of nature and of mankind: how the soul makes humanity unique in the world and how knowledge of God and nature has been corrupted through time, its true purpose having been forgotten and the message having fallen prey to false prophets.

---

12. "In praefationem historiae generis humani annotationes," in *Corpus*, 1–5; *HistNat*, 101–5.

The first sections of the *Corpus* are devoted to exploring the nature of spiritual entities—God and angels—and establishing the validity of scripture as a testimony of their divine natures. Arias Montano also explained how prophetic dreams and visions are valid means through which the Divinity communicates with humanity. The exploration of the world begins with a general account of Creation, *De natura disputatio*, where he derives a series of metaphysical first principles that will later inform his study of nature. The description of the parts of the world and its inhabitants (except humanity) takes place in an extensive section titled *De rerum natura—Observationes*, where he explains the cosmos and the nature of things according to their order of creation: plants, fish, birds, and beasts. For each category—for example, stars, jellyfish, or mountains—the first task is to examine key words in relevant biblical passages using the Hebrew lexicon and etymology. The exercises consist largely of choosing the appropriate meaning for a given context that, when properly deciphered, reveals the "arcane and occult . . . properties, nature, and essence of things" (*arcanae et latentes . . . rerum proprietates, natura ac vis*).[13] The resulting descriptors are then corroborated using experiential knowledge and observation.

A MATURE EXEGETICAL METHODOLOGY

Throughout the *Magnum opus*, and particularly in the *Corpus*, Arias Montano would finally get to fully develop the exegetical approach he had rehearsed in the *De arcano sermone*. Still committed both to literal interpretation and to resorting to the Hebrew philology and lexical interpretations as a way of unlocking a new level of meaning from the Sacred Scriptures, he gave little justification for choosing this approach, perhaps convinced that after a lifetime of employing this methodology he had little left to explain. In the *Anima* he made scant use of his philological exegesis; it features only in his treatise on the divine attributes implied by the different names of God used in the Bible.[14] Given that his purpose in the *Anima* was interpreting historical events that tell the story of humanity, he did not need his philological approach to develop or sustain novel arguments; in his view, human and divine actions spoke for themselves.

It is in the *Corpus*, however, that Arias Montano's exegetical methodology lays the foundation for his arguments in support of his novel natural philos-

13. Arias Montano, *Libro de José*, 91, 403, [1].
14. *Anima*, 13–17; *LibGen*, 113–17.

ophy. He took key phrases, especially from the first two chapters of the book of Genesis and, after deconstructing key Hebrew words, searched in their etymology for alternative meanings that the Latin Translator—Arias Montano's sobriquet for Saint Jerome—had missed. This type of analysis yielded a lexical tree that presented a much larger canvas for interpretation. Sometimes he systematically explored all the meanings and applied each of them to the thing or activity in question; on other occasions he selected one among the possible meanings and constructed a new interpretation of the biblical text based on it. These selections are often the result of his either confirming the meaning of the words by comparing them with other instances where they are used in the Bible, or more interesting, by applying to the terms an empirical "test" based on observation of the natural world. We will explore this aspect of his methodology in the next chapter.

His literal exegetical methodology was a strategic move as well. It allowed Arias Montano to wander far from the text of the Vulgate in search of new interpretations yet still remain faithful to the Sacred Scriptures. He often accused the Latin language of being *parca*—of not having enough words—and used this to excuse "the Translator" for misinterpreting some sections. On a few occasions Arias Montano suggested that Saint Jerome had perhaps not been aware of other possible meanings—richer meanings—of the Hebrew words. In sum, using this philological methodology, Arias Montano could always return every natural philosophical interpretation, no matter how farfetched, to the biblical source.

As we have seen in earlier chapters, the cornerstone of Arias Montano's philological exegetical methodology was his belief that biblical Hebrew was—if not the divine language itself—the oldest language that retained vestiges of the very words God had spoken to Moses and other prophets. He must have assumed his readers shared this belief, because he gave no justification for his assumption other than a reference to another of his works.[15] The ability of Hebrew words to encapsulate the essence of their referents—as illustrated by Adam's act of naming—was a derivative of the tremendous capacity of that language to reveal the true nature of things. Arias Montano acknowledged that the Bible did not mention or name everything in the world, thus making some essential natures difficult to recapture. This placed some serious limitations on his project, yet he never fully addressed these concerns in the *Magnum opus*.

15. As the editor of the Spanish edition suggests, in this instance Arias Montano might also be referring readers to his unpublished treatise *Adam, sive Humani sensu interprete lingua*: BNE MS 149, fols. 1–15, published now as Arias Montano, *Tratado sobre la fe*.

As we will see later on, he seemed to vacillate between two opinions: that the Bible reveals all that humanity needs to know about the natural world versus the belief that the Bible teaches how the study of nature should be undertaken. For example, he finds in the Bible only the names of the sun, the moon, and Venus. Since the names of the other four planets that had been known since antiquity were not mentioned in the Holy Book, Arias Montano explained that their properties could not be known and their natures should not be a source of speculation.[16] Yet by the same token he encouraged his readers to be like Solomon and observe nature and try to understand God's design.

To illustrate the power encapsulated in the arcane language of the Bible, he began the *Anima* with the most comprehensive example of this: an exploration of the Hebrew names of God and what they reveal about the nature of the Divinity. He noted that human language is simply unable to give a name to the singular majesty of divine nature; even Adam failed to find a name for God. Yet God had chosen to reveal his own name in Hebrew and spoke it directly to the prophets. This direct linkage between name as descriptor and the entity it seeks to describe suggested to Arias Montano that studying the names of God was one way of learning whatever humanity could ever possibly come to comprehend about the nature of God (*natura Dei*). Its study could reveal aspects of the divine nature that prophets and later exegetes had left unsaid or that had been poorly translated. The two names of God that appear in Genesis were particularly meaningful for Arias Montano. In Montanian exegesis ELOHIM stood for the creation, providence, and governance of the natural world, while the secret meaning of the ineffable name (YAHVEH), if revealed by God, inspired the recipient to a life of piety, faith, and desire for salvation.[17] Arias Montano's Spanish audience was familiar with this type of exegetical exercise. When expounding on the names of God, he joined an important tradition of Spanish exegetes who had written on this topic, among them fray Luis de León in his controversial *De los nombres de Christo* (1583–85).[18]

Montanian "Nature"

Throughout this book I have used the word nature, according to its modern usage, as shorthand for "natural world" or to indicate the character of some-

16. *Corpus*, 190; *HistNat*, 296.

17. *Anima*, 138–39; *LibGen*, 241.

18. Fernández Marcos, "De los nombres de Cristo de fray Luis de León y *De arcano sermone* de Arias Montano."

thing, as in "human nature." Arias Montano used it in these two senses, but also in ways that might be unfamiliar to a modern reader. In the *Magnum opus* "nature" also signifies a verity that was divinely ordained and intrinsic to all things. Simply put, the "nature" of something implies the "how," the "what," and the "why" of something, while the "who" is implied in the thing's name. A thing's nature stood for its "realness," its essence, and the characteristics—both sensible and insensible—that differentiated it from other things. Identifying a thing's nature meant finding its true being. The Montanian correspondence between nature and truth left little room for relativist interpretations of physical reality (a premise that would surely cause consternation in present-day metaphysicians). The relation between truth and nature rested on the notion that only one explanation (*ratio*) of nature existed and that it could be discovered with God's guidance. "Nature" designated what was in the world, but also things that lay beyond it. Therefore it was licit to refer to the divine nature of God despite understanding that it was unique, singular (and yet in three parts), and distinct from everything else in the world.[19]

Arias Montano also implied something slightly different from this when he used the word nature in the term *rerum natura* (usually translated into English as "the nature of things"). It did *not* mean, as our modern definition suggests, "the characteristics of nature," where it implies a set of identifiers that can be found and described. For an early modern audience the *natura* of something suggested a complex mix of visible and invisible properties intrinsic to the thing. These could be Platonic forms, inherent energies and forces, an ingrained teleology, or even a thing's capacity to change (e.g., transmute, reproduce, die), because it was in the nature of the thing to do so. To fully comprehend nature entailed comprehending the *rerum natura* of particulars in an exercise that consisted of moving the veil of understanding back to the point where the underlying metaphysical principles of everything (earthly and divine) were clearly identified. Given the use of inferential logic in expositions that attempted to explain *rerum natura*, it is no surprise that Arias Montano placed his metaphysical treatise toward the beginning of both the *Anima* and the *Corpus*. The metaphysical principles he defined in these sections serve as the foundation for the descriptions of the *rerum natura*—including the *natura Dei*—that follow it. These were phrases that would have been familiar to Arias Montano's audience, the former having survived since antiquity in the European lexicon, primarily through Lucretius's *De rerum natura* and

19. *Anima*, 5; *LibGen*, 106.

the many attacks of the atheism implied by its atomistic materialism, while the latter recalled Cicero's *De natura Deorum*.

Arias Montano's Disquiet about Wisdom and Knowledge

For Arias Montano the terms wisdom and knowledge were interrelated but clearly differentiated. Wisdom implied an exercise of judgment based on discernment and could be either an attribute of the Divinity (as in God's wisdom, *sapientia Dei*) or exercised by humans as a divine gift. Knowledge (*cognitio*) was largely a human endeavor, the result of an arduous intellectual exercise undertaken since Adam's Fall. He began to explore the question of true wisdom in the *Anima*, where he explained that knowledge derived from observing nature is very different from the knowledge God transmits when he instructs via oracles and utterances.[20] *Cognitio*, on the other hand, leads to knowledge of things and is achieved only with much study and sacrifice (unless it had been bestowed in the form of a divine gift, as with Solomon).[21] The other type of knowledge—more properly called wisdom—carries with it discernment that allows a greater understanding of God. To obtain it requires obedience, devotion, and a simple heart. In the Montanian version of the history of humanity—as in the Spanish natural theologies discussed in an earlier chapter—the paths to wisdom and knowledge were under constant threat of ambush. Of the two, the more treacherous was the path taken by those who studied nature without seeking discernment or desiring a God-bestowed dose of wisdom.

From the compositional structure of the *History of Nature* it appears that this issue increasingly disquieted Arias Montano as he wrote the book. Halfway through, he abruptly stopped the ongoing discussion about the material composition of the earth and inserted a section titled *De sapientia et scientia brevis admonitio* (Brief consideration of wisdom and knowledge).[22] The section appears right after a discussion of the meteorological relation between heaven and earth but before he began one on the "fruits of the earth." While the section might simply signal the division between the "general" description of the world and the "specific" discussion of living things—plants, animals, and humans—its placement midway through the text is curious. If Arias Mon-

---

20. *Anima*, 124; *LibGen*, 226.

21. "Neque quisquam eruditus et sapiens natus homo fuit, neque repentè doctus evasit, nisi rarissimo aliquando et singulari Dei beneficio ac privilegio mirabiliter auctus fuerit, cuius unicum nobis hactenus exemplum in Salomone exstitit.": *Anima*, 125; *LibGen*, 227.

22. *Corpus*, 235–37; *HistNat*, 343–45.

tano did in fact purposely put it in the very middle of his section of observations, it suggests this was intended as a timely admonition, a reminder of the true purpose of this history of nature to readers about to become engrossed in investigating natural particulars. What pitfalls lay ahead for someone intent on the study of nature he advocated but who might forget the true purpose of this study?

In the *Admonitio* Arias Montano first defined wisdom as the knowledge of things human and divine. Divine wisdom encompasses all eternal things, God and the spirits, but also all corporeal things—all that is in heaven and earth—which share in the divine nature by virtue of being created and sustained by God. Acquaintance with things human and divine should be sought from two sources: the knowledge of something and its utility. Knowledge resides in science, while utility is the foundation of prudence or discernment.[23] One who has attained both science of nature and prudence can be called truly wise, Arias Montano counseled. Yet scripture frequently reminds us that attaining both is beyond what humanity is capable of achieving alone and can come only as a gift from God. This is the kind of wisdom Job yearned for when he wrote, "But whence can wisdom be obtained, and where is the place of understanding? Man knows nothing equal it, nor is it to be had in the land of the living" (Job 28:12–13 NAB). Although inferior to the wisdom Job asked for, when accompanied by piety the study of nature allowed humanity to attain a "pious wisdom," endowed with its own kind of knowledge and discernment. Solomon was the best example of someone who had attained this type of wisdom. In addition to the wisdom to govern wisely, he had been given knowledge about the composition of the world and what may be observed about the world's adornment and the nature of things (*mundi ornatus* and *rerum naturae*), that is, "what wise men call 'philosophy.'"[24] Solomon had prayed to God for wisdom of this type: "Therefore I prayed, and prudence was given me; I pleaded and the spirit of Wisdom came to me" (Wisd. of Sol. 7:7 NAB). What God had granted Solomon was extraordinary, as Arias Montano pointed out by citing the entire biblical passage: knowledge of existing things, the organization of the universe and the force of its elements, the beginning and the end and the midpoint of times, the changes in the sun's course and the variations of

23. *Corpus*, 235; *HistNat*, 343.

24. "Atqui non ea tantùm quae ad populos gubernandos conferrent, verùm etiam ea quae mundi ornatum ac rerum spectarent naturas, id est, cognitionem omnem, quam philosophiae nomine humani sapientes definiunt; viro huic divino munere constitisse, innumerae disputationes, et varia ab eodem relicta scripta declararunt.": *Anima*, 328; *LibGen*, 431–32.

the seasons, cycles of years, positions of the stars, natures of animals, tempers of beasts, powers of the winds and thoughts of men, uses of plants and virtues of roots (Wisd. of Sol. 7:7, 7:17–21 NAB).

Other biblical authors also testified to the vast knowledge Solomon had been granted:

> Moreover, God gave Solomon wisdom and exceptional understanding and knowledge, as vast as the sand on the seashore. Solomon surpassed all the Cedemites and all the Egyptians in wisdom. . . . He discussed plants, from the cedar of Lebanon to the hyssop growing out of the wall, and he spoke about beasts, birds, reptiles, and fishes. Men came to hear Solomon's wisdom from all nations, sent by all the kings of the earth who had heard of his wisdom. (1 Kings 5:9–14 NAB)

Yet few could aspire to the same favor before God that Solomon enjoyed, and Arias Montano felt he had to caution those wishing to start their study of nature. The pursuit of pious wisdom—either granted as a divine gift like Solomon's or sought through study—should never lead away from or contradict the Sacred Scriptures, nor should it posit anything that cannot be reconciled with or deduced from them. The pious sage also had another responsibility: to share knowledge only about things that are true and certain (*vera et certa*) and to withhold knowledge of those that are superfluous. Arias Montano's pious sage had to be teacher and censor, a shepherd and an inquisitor concerning a body of knowledge about nature that was both vital for humanity's salvation and potentially dangerous. Discernment and prudence were essential. Furthermore, knowledge of nature should never be pursued for its own sake. The only purpose for seeking this knowledge was to gain greater knowledge of God, for only God could guide humanity to the "tranquil and secure port of certain and investigated knowledge" and show the true purpose of understanding nature.[25] In the well-known Tertullian reference to Thales of Miletus Arias Montano found a suitable admonition to philosophers who lose sight of the true purpose of knowledge of nature:

> For it might happen (and has happened to those of our kind) that while someone pursues distant and peregrine things he deviates too much from those he

---

25. *Anima*, Praefatio, fol. †4v; *LibGen*, 87.

should know well, avidly observing with fixed gaze celestial things that are so far that he barely notices the ground beneath his feet and falls into a well.[26]

Perhaps Arias Montano's most eloquent example of the proper purposefulness he saw as essential to the study of nature was his interpretation of the story of Moses and the burning bush, which he included in the *Corpus*.[27] Moses, he explained, was the first person God selected to learn the true meaning of his arcane name, YAHWEH. He had come to this knowledge in part by following his curiosity about the natural world—a curiosity God encouraged and approved. Moses's curiosity had been nurtured in Egypt, where he had been educated by the mathematicians and magi and learned the nature of things and their causes. Alone with his flock one day, while busy thinking of such things he noticed a green bush on fire. Aware that this was outside the common course of nature, his curiosity drew him to investigate why the bush burned but was not consumed. He wanted to understand not only the reality of the thing itself but what this might mean (*res ipsa et ratio*). Since he understood better than most the "reasons for fire and thorns" (that is, their natures), he knew that there was no better fuel for a fire than thorny bushes. Realizing just how odd it was that the bush remained green and unburned—yet not wanting to believe this was something beyond nature—he wanted to investigate whether it was due to the soil, the air, or the bush itself. Approaching the bush, he heard God call his name: "Moses! Moses!" He answered, "Here I am." God said, "Come no nearer! Remove the sandals from your feet, for the place where you stand is holy ground" (Exod. 3:4–5 NAB). Arias Montano interpreted God's response to Moses as an admonition. God's command "Come no nearer!" meant that inquiring into natural phenomena only from curiosity was against God's will. God desired that phenomena be approached for the right reason, as if one were treading on holy ground. The motivation had to be a desire to be nearer to God, not simply to satisfy curiosity.

Once the pious sage was committed to the prudent study of nature and the grander objective of such a study had been explained, Arias Montano turned

26. "Accidere enim potest (quod iam maximo nostri generis detrimento accidit) ut, dum quis aliena ac peregrina curiosè discenda et cognoscenda sectatur, quàm maximè dimoveatur, et quàm longissimè discedat ab iis quae maximè cognita explorataque habere debuit: et caelestia quae procul distant, constanti immotoque vultu conspiciendi avidus, ea, quae ante pedes iacent, minimè videns, in foveam puteúmve sese dedat praecipitem.": *Anima*, Praefatio, fol. †2v; *LibGen*, 84.

27. *Corpus*, 38–40; *HistNat*, 139–40.

to why one should study natural particulars. In his view, all natural events witnessed in the world were the result of divine providence and design. The pious sage needed to find the natural causes behind the *rerum natura* in order to recognize the divine presence in them, but also to know when to understand them as manifestations of God's benevolent will or, alternatively, manifestations of his ire against humanity. In biblical history, Arias Montano identified three persistent attitudes of humanity toward God. There were those who honored God in a righteous and true manner, those who completely disregarded God and followed their own opinions, and finally those who attempted to honor God in rites and ceremonies yet gave preference to their desire for well-being and thus learned only what was needed to attain it.[28] Nature provides only comfort to those of the first group, while God uses nature to punish those of the second group. Those in the third group are often confused by nature; it will sometimes appear useful and good and sometimes harmful and evil. When the situation warrants, God will also use natural effects to punish this last group. Thus nature cannot but appear perplexing to the religiously insincere and to those who fashion their own impious explanations about nature and its purpose. Because of his attitude toward knowledge of nature, Arias Montano dismissed any attempt to fashion a natural philosophy based on first principles conjured by pagans (or even Christian philosophers); he was profoundly skeptical of the fruits any such philosophical system yielded. He also dismissed the study of nature that relied on an epistemology whose end was not knowledge of humanity's relationship with God or that was empiricism for its own sake. Yet, as we have seen, he had a very positive view of humanity's capacity for attaining knowledge of nature.

HISTORICAL AND EPISTEMOLOGICAL PREMISES

We can identify a few core premises that serve as the point of departure for Arias Montano's exploration of the history of humanity and its quest for knowledge of nature. Although he went to great lengths to explain these premises, they essentially go unquestioned and are taken either as verities, such as the relationship between God, humanity, and nature, or as the metaphysical principles he found at work during the Creation. Other premises he considered were the consequences of historical events, such as the fate of prelapsarian knowledge, the determinism implied by Creation, and the utility of nature. These premises provided the basis for the Montanian reform of natural phi-

---

28. *Corpus*, 228–33; *HistNat*, 336–41.

losophy and significantly dictated the tenets of this reform. Let us examine each of these premises in turn.

Purpose of the Study of Nature

The point of departure for Arias Montano when establishing these premises was clearly defining the relationship between God, humanity, and nature, largely defined by each entity's purpose. In Arias Montano's conception God does not need to have an intrinsic purpose; he simply emanates love, and this act creates humanity and the world. Thus the creative act is the material manifestation of God's goodness. As God's greatest creation, human beings are uniquely endowed with the capacity to attain knowledge of God. Thus humans' final purpose is, by design, attaining this knowledge so they may earn salvation. This capacity places humans at the apex of creation, since the world was created principally to shelter them and only secondarily to testify to God's existence.

The preoccupation with purpose is the organizing premise of the *Magnum opus*. In the *Anima* this takes the form of asking about humanity's purpose on earth. Likewise, the guiding theme of the *Corpus* is the purpose of nature. Arias Montano's answers are unambiguous and not unexpected. Since the purpose of humanity is to earn salvation, which can be achieved only by knowing God, his first premise is that man (and presumably woman) cannot attain knowledge of God without first understanding his own human nature.[29] Yet human nature is in itself complex—as Saint Paul proposed in Romans 7:14–23—composed of a superior/internal/masculine nature that inclines toward the divine and is the seat of the soul, and an inferior/external/feminine nature that is earthy and subject to appetites and is the seat of the senses. After the Fall these natures were destined to be in constant conflict.[30] Given this complexity, Arias Montano acknowledged that understanding oneself and the whole of human nature was immensely difficult, but humanity was helped by an innate curiosity that inclines the soul toward this introspection; we need only follow a good teacher, preferably one who abides by the Sacred Scriptures. Arias Montano proposed that the best way to embark on this quest to understand human nature was to begin at the very beginning, a time when for once the true first principles and causes of the nature of humanity were understood; the rest should follow more easily.[31]

29. *Anima*, Praefatio, fols. †2v-3r; *LibGen*, 84.
30. *Anima*, 43; *LibGen*, 142–43.
31. *Anima*, 24; *LibGen*, 124.

The Corruption of Knowledge of Nature

Within the Judeo-Christian tradition the historical narrative of the Bible is understood as relating a pattern of divine revelation to humanity and the subsequent forgetting of the divine message. For Arias Montano this had been so not solely for matters of faith, but also for seeking knowledge of nature. His second historical premise was his conviction that postlapsarian knowledge of nature was ephemeral, easily forgotten from one generation to the next, and—unless cultivated for the right reasons—easily supplanted by other explanations presented by fabricators and charlatans that had little to do with revealed wisdom. Biblical history taught that knowledge of nature could be attained only through revelation or through investigation.[32] Following the Augustinian tradition of two roads to faith, he considered revelation and investigation two parallel paths. The path through divine revelation was exemplified by Solomon, while the other path employed God-given human intelligence and consisted of an exertion of the mind (*agitatio mentis*). It required investigating by observing and examining the things around us, but it should always remain within the bounds of pious wisdom.[33]

Arias Montano found the way of revelation illustrated in the instances when God spoke directly to a man, such as God's conversations with Adam, Moses's ascent of Mount Sinai (Exod. 19:16–25), or the response to Solomon's prayers cited earlier. For Arias Montano every instance of a divine utterance was taken as literally true (for God can only speak truthfully). With these utterances God had transmitted true knowledge to his elect, the kind of knowledge needed to know God *and* the nature of things.[34] Yet the way of revelation was not without pitfalls. In the section of the *Corpus* on dreams Arias Montano once again turned to the story of Solomon.[35] The knowledge of nature Solomon had prayed for came to him in a dream. Yet dreams can also be "false," because they may result from an agitated mind, certain bodily conditions, or the imposition of an external spirit. Likewise, dreams can also be the work of false professors or "conjurers" (*coniectores*), some of whom hide their art behind the benign label of "curiosities" (*curiosa*). Even if all these pitfalls could be avoided, Arias Montano pointed out that gaining great knowledge of nature is not in itself redemptive. Even Solomon struggled mightily to overcome the appetites of the

32. *Anima*, 1; *LibGen*, 101.
33. *Corpus*, 236; *HistNat*, 344.
34. *Anima*, 18; *LibGen*, 118.
35. *Corpus*, 121–22; *HistNat*, 222–23.

exterior man. No measure of knowledge of nature, once divorced from divine grace, could keep Solomon among the righteous.[36]

For all the wisdom the Bible attributed to Solomon, it was prelapsarian Adam, more than any man to come, who had wholly understood the nature of things. God had granted him knowledge of "the forces and efficacies" of natural productions so that humanity could sustain itself in nature.[37] But with the Fall this knowledge had been forgotten and dominion over nature lost. For Arias Montano the key to recovering this knowledge lay in understanding the history of this type of revelation. Thus he explained that after his disappointment with Adam, a patient God had again communicated the nature of things to the Israelites through Moses, again in a revelation to Japheth (son of Noah and progenitor of the Greeks), and yet again to Solomon. They had laid down vestiges of this knowledge in the Bible, albeit accommodated to a particular audience and in what seems to be arcane language.

For Arias Montano the Bible testified to the ephemerality of knowledge of nature. The first step toward forgetting was not realizing that the purpose of knowledge of nature was to find the one true God; the second step was giving free rein to the appetites of the exterior man. After Adam had eaten the fruit of the tree of paradise offered by Eve, Arias Montano explained, the floodgates of the imagination opened, setting forth many schools of thought, each professing its own wisdom, but now without a guide and having lost its true purpose. All the schools of human wisdom had emerged, with each one claiming, deliberating on, and worshiping its own senses as teacher, its own ingenuity as master, and its own appetites as god. The result was a multitude of definitions and divisions, all contradicting each other and often themselves.[38] The confusion and conflict caused by each man's following his own opinion first came to a head with Cain. He was an example of the trap waiting for those who pursue expertise for a purpose other than approaching God. Cain was so intent on working the land and inventing instruments for cultivating it that he neglected to cultivate his interior man and thus drew away from his family

---

36. *Anima*, 327; *LibGen*, 430.

37. "Nullum enim animalium genus illius obedientia exceptum voluit Deus: quare etiam nullum prorsus illi imperaturo cognoscere negatum. Praeterea herbarum, arborum, plantarumque, naturas formasque omnes, et in suas, et in animantium ceterarum escas et alimentorum copias, qui suppeditare voluit, idem quoque vim et efficacitatem, et succorum rationem, opportunitates, maturitates, historiamque omnem, sponte sua pernoscenda praebuit: ut quemadmodum animantibus ceteris suus cuique cibus ac pastus, nullo praeter Verbum Dei docente, notus constat.": *Anima*, 33; *LibGen*, 133.

38. *Anima*, 73–74; *LibGen*, 174–75.

and from God. His sin was therefore a direct consequence of having forgotten his purpose.[39] His descendants followed suit. They established cities and created the technical arts to construct and protect them. Among Cain's descendants was Jubal, the inventor of music, to whom Arias Montano attributed the first awareness of the importance of number, weight, and measures in nature. Jubal in turn taught his half-brother Tubal-cain, whose ability in surveying and determining proportions and mastery of metalworking led him to invent the compass and the ruler.[40] Ultimately Cain's descendants turned to religions based on false gods and came to believe that celestial bodies held power over their lives.

Arias Montano recognized in the Deluge an important moment of renovation not only of humanity's relationship with God, but also of knowledge of nature. His interpretation of the divine gifts received by Japheth—one of the three sons of Noah—was central to his reconstruction of the chronology of the history of prelapsarian knowledge. Japheth was given not just eloquence—as traditional interpretations of the Vulgate maintained—but more precisely, the gift of speaking with eloquence about knowledge of nature. Arias Montano's explanations of these events show his exegetical methodology hard at work. He first examined Noah's blessing of Japheth, "ιερητηε ελοηιμ ιαρητη" (יַפְתְּ אֱלֹהִים לְיֶפֶת, *yapt 'ĕlōhîm lĕyepet*, "May ελοηιμ expand Japheth"; Gen. 9:27 NAB). After unpacking the Hebrew etymology of יַפְתְּ (*yapt*) into the root פָּתָה (*pātâ*) and finding that it meant "to expand" or "stand out" and in the Hiphil tense also "to dilate," he noted that the word shares its root with the name Japheth. He then concluded that Moses had played with the phonetics of the words ιερητηε and ιαρητη to encode in the phrase a much more expansive message, rendering Saint Jerome's sparse "Dilatet Deus Japheth" as "Decoret, amplificet Deus creator et gubernator decorum et amplum" (May God creator and governor adorn and amplify the beautiful and spacious).[41] Arias Montano's interpretation built on the etymology of the name Japheth to mean all that was described by that name would be expanded and amplified. He also interpreted any verbal action applied to humans but referred to by the word ελοηιμ to mean "wisdom." The addition of the verb "expand" meant this wisdom must be able to be transmitted to others, which Arias Montano takes to be the gift of eloquence. This meant that Japheth would be able to speak eloquently about all things in ελοηιμ's domain, which included both heaven and earth.

39. *Anima*, 75; *LibGen*, 176.
40. *Anima*, 85; *LibGen*, 185.
41. *Anima*, 127; *LibGen*, 228.

Arias Montano's interpretation of the role of the descendants of Cain and Noah was not entirely novel. It borrowed aspects from Abraham Ibn Ezra's (1089–1164) interpretation of God's blessing of Japheth and from rabbinical literature that makes Japheth the eldest of Noah's sons.[42] Arias Montano's interpretation also reinforced the tradition, punctuated by Saint Isidore of Seville, that considered the Greeks and Romans Japheth's descendants, a topic he had mentioned in the treatise *Phaleg* of the Antwerp Polyglot.[43] What makes his approach distinctive is that he relied solely on the biblical text to support his argument, referring to no authority other than his etymological definitions and deductive logic.

The history of humanity's relationship with knowledge of the natural world continued after the Deluge. Humanity still had access to some vestiges of prelapsarian knowledge of nature through the line of Japheth, while the other descendants of Noah repopulated the earth without this knowledge. For Arias Montano this was a period characterized by utter confusion, an era where there were almost as many beliefs as there were people, symbolized in the Bible by the Tower of Babel. Again, knowledge of nature had drifted away from its true purpose, and as a consequence distortions arose that resulted in many sects led by men who called themselves wise and sophists, and later philosophers. To bring the biblical story into line with history, Arias Montano pointed out that these "philosophers" were the people who came to be known as Greeks, but their ideas had arisen much earlier among people who called themselves wise: the Chaldeans and Egyptians.[44]

Arias Montano derived several important lessons about humanity's relationship with knowledge of nature from this genealogy of prelapsarian knowledge. The principal one concerned the never-ending struggle between the interior man and the exterior man. These dual natures drive humanity to question the way things are and to pursue novelties without realizing that unless God is the guide, humanity will be unable to distinguish good from evil.[45] Inevitably, history shows that humans will make bad choices. Even Solomon, a man who had sought to be closer to God and had received knowledge of

42. Arias Montano picked up the interpretation of "to dilate" from Ibn Ezra, although the analysis of the name Japheth seems to be original: Ibn Ezra, *Ibn Ezra's Commentary on the Pentateuch*, 1:130.

43. In the *Etymologies* (9.2.26–38), Saint Isidore of Seville traced the origins of the Greeks and later Europeans to the line of Japheth: Isidore of Seville, *Etymologies of Isidore of Seville*, 193–94; Arias Montano, *Phaleg*, 15.

44. *Anima*, 193–95; *LibGen*, 293–96.

45. *Anima*, 326–44; *LibGen*, 429–36.

nature through revelation, had failed to preserve his kingdom and perpetuate his knowledge. For those who have no knowledge of God, Arias Montano found that human curiosity invariably led different groups to fabricate their own beliefs and philosophical systems that he considered completely untrue and divorced from reality. Ancient philosophers had based all their ideas on incorrect principles, and others built on these or twisted them so much that most philosophy became shrouded in myths and enigmas. In a not too veiled reference to Scholasticism, Arias Montano concluded that subsequent generations of philosophers further degenerated from the original models of learning. They added disputations, and their disciples caused further confusion with their inexplicable, prolix words and interpretations. Little of what these schools taught, he maintained, was worth teaching. In Arias Montano the Spanish disquiet was more like full-blown anxiety.

The ephemeral nature of knowledge, furthermore, made it easy prey for those who might pursue it for reasons other than a genuine desire to know. Arias Montano's historical perspective led him to disdain the notion that the works of even the most ancient sages, Zoroaster and Hermes, were worth studying. In the Montanian scheme, transmitted knowledge—the very wisdom of the ancients so valued by humanists—had been subjected to the temporal vagaries of humanity and was therefore corrupted. Here Arias Montano found Horace an eloquent witness to the futility the whole enterprise had experienced:

> Parcus Deorum cultor et infrequens,
> insanientis dum sapientiae
> consultus erro, nunc retrorsum
> vela dare atque iterare cursus
> cogor relictos.
>
> [I was a stingy and infrequent worshipper of the gods
> all the time that I went astray,
> expert that I was in a mad philosophy.
> Now I am forced to sail back
> and repeat my course
> in the reverse direction.][46]

Only three things met Arias Montano's criteria for valid sources of "certain knowledge": divine revelation, the Word as revealed in the Bible, and nature.

---

46. *Anima*, Praefatio, fol. †4v; *LibGen*, 87; Horace, Carminum 34, lines 1–5, in Horace, *Odes and Epodes*, 84–85.

Therefore Arias Montano in the *Magnum opus* was not advocating a program to recover a *prisca sapientia* from remnants of prelapsarian knowledge that might have survived in some of the works of the ancient Chaldeans and Egyptians. He likewise disdained the knowledge of nature received from various philosophical schools, stating clearly that his objective was not to establish a new theory or school, nor was he concerned with arguing against "Epicureans, Stoics, or Peripatetics."[47] Neither, we may add, was it a Ficinian program; Arias Montano makes no attempt to rediscover, let alone Christianize, an ancient philosophy. Neither does he subscribe to the objective inherent in the philosophical syncretism of Giovanni Pico della Mirandola's *Heptaplus* (1489), Agostino Steuco's *De perenni philosophia* (1540), or the work of his Spanish contemporary Sebastián Fox Morcillo. He was also critical of exercises like the one undertaken by Francisco Valles de Covarrubias in his *Sacra philosophia*. As Arias Montano explained in his commentary on Isaiah, he thought contemporary rationalizations about natural phenomena described in the Bible, such as the Passion eclipse, were misguided even if explained with careful experiments.[48]

It merits pointing out that Arias Montano had a positive attitude toward humanity's capacity for recapturing some measure of the prelapsarian knowledge of nature. Adam's folly had ensured that this would be far from an easy task, but humans retained enough capacity and ability so that in the search for knowledge of nature they could still tell "the thorns and weeds from the wheat."[49] Were we to situate Arias Montano within Peter Harrison's explanatory scheme—which draws a distinction between an Augustinian skepticism about the postlapsarian human's ability to grasp true knowledge of nature versus the Thomistic confidence that it was possible—we would have to situate Arias Montano's human firmly in the middle.[50] In the Montanian scheme, each human dragged along the heavy ball and chain of concupiscence that could at any moment sidetrack even the loftiest endeavors toward the temptations of the flesh. Only if this danger could be avoided *and* if a suitable natural philosophy was available—which he proposed to provide in the *Magnum opus*—could humans learn what they needed to know about the nature of things, or at least enough for one who had this knowledge to secure salvation.

47. *Corpus*, 405; *HistNat*, 527.
48. Arias Montano, *Commentaria in Isaiae Prophetae sermones*, 831–40.
49. *Anima*, 70; *LibGen*, 171.
50. Harrison, *Fall of Man and the Foundation of Science*, 31–46.

A Deterministic Creation

A third fundamental historical premise of the Montanian project entailed a conception of nature as the result of a divine action that made nature perfect in design, wholly determined and complete yet still unfolding according to a divine plan: "For God has made nothing in vain, nothing by chance, nothing useless in the nature of things."[51] Arias Montano devoted chapter 10 of the *Anima, De lege naturae* (On the law of nature), to developing these notions. In the Montanian interpretation of the Hexameron, God's actions, besides creating everything in the world, also gave each thing, living or not, a natural law that defined its nature and efficacy. This was what Moses meant by the often-repeated phrase of the first book of Genesis, "and it was so" (*et factum est ita*). This phrase in the sacred language meant "truth" or "perpetual law." God did not want anything in the "heavens or below the heavens" to be free to act and change in ways that did not replicate the archetype he initially set during Creation.[52] Thus the nature of everything remained as when it was first created, unaffected and uncorrupted by the passage of time. So, whereas all natural things existed solely because of God's will, they maintained themselves by keeping the law they had been assigned at the Creation.

The perpetuity of this natural law, according to Arias Montano, was explained in the Bible by the use of the word GHOLAM (עוֹלָם, *ʿōlām* or *saeculum*), which suggested to him that everything was created with a predetermined purpose, duration, and life cycle:

> Therefore we shall call the *saeculum* or GHOLAM the exercise and ministration resolved then by divine plan and prescribed unto either every nature or thing: as well as its duration. And so the *saeculum* as prescribed to each thing shall be its continual course until its consummation, that is, until the fulfillment to its full and proper office, from which derives that the *seculum* of God has no end.[53]

51. *Anima*, 73; *LibGen*, 174.

52. "Singulis generibus à se creatis et perfectis, certas naturae et efficacitatis leges Verbi virtus et efficientia dedit; nihil enim vel in caelo, vel sub caelo actionis, aut lationis expers, nihil otiosum esse voluit, nihil quod pro ratione ac modo virtutis atque sortis suae, authoris ipsius semper viventis, semperque agentis, non imitaretur exemplum.": *Anima*, 36; *LibGen*, 139–40.

53. "Itaque praescriptam unicuique vel naturae, vel rei, ac divino consilio decretam tum exercitationem et ministrationem: tum etiam durationem, seculum sive GHOLAM dicemus, eritque seculum praescripta unicuique rerum decursio ad consummationem usque, hoc est, ad perfectum ac praestitutum proprium munus, unde fiet, ut Dei seculum nullum habeat finem.": *Corpus*, 164; *HistNat*, 270.

For Arias Montano the word GHOLAM encapsulated the life cycle of every-thing, or the "order" everything must follow in perpetuity within the "sphere" (or cycle) that defines its existence. In his discourses on the book of Ecclesias-tes, he noted that the root of the word (that is, עָלַם, *ʿālam*) meant "to hide and conceal, and to lock up, hide from view, and cover up; from its meaning it is then clear that it indicates the passage and progress, the entries and exits, the beginnings and endings of natural things."[54] The GHOLAM of something thus determines when it is going to emerge into view and when it is to disappear. This notion was also what the apostle James meant by *rota nativitatis*, which Arias Montano translated as *naturalis sphaera* (presumably—he does not ex-plain why—because the heavens move in a constant and cyclical manner), and was the reason the Greeks called man a microcosm.[55]

GHOLAM became a central tenet of Arias Montano's hermeneutics of nature (discussed in the following chapter), where it served as the basis for his ideas about biological change and about generation and corruption, although it is only fair to point out that he studiously avoided describing actions implied by GHOLAM in the Aristotelian language of change, such as "being," "becoming," or accidental or substantial change. For example, the fertility of the earth was ensured by everything's fulfilling its GHOLAM; plants would always conserve their abundance and seed, animals would carry out their charge, and the stars would perpetually circle the heavens. And as long as a natural thing remained unrestrained and was not impeded by an external force, its GHOLAM ensured that it operated according to its own laws and would continue to fulfill its purpose.[56] This natural law applied also to the elements themselves: air, wind, water, and fire. These not only contributed to the design of the world, each one according to its species and form, but were endowed with the ability to weaken or strengthen the others.[57]

The necessary consequence of the teleology implied in a divinely designed

54. "Absconder y encubrir, y encerrar y trasponer y atapar, de cuya significación se echa de ver luego el discurso y el progreso, las entradas y salidas, los principios y los fines de las cosas naturales.": Arias Montano, *Discursos sobre el Eclesiastés de Salomón*, 92.

55. James 3:6 VUL. It is translated as "course of nature" in the King James Version and as "course of our lives" in the New American Bible. *Corpus*, 172; *HistNat*, 278.

56. Arias Montano also develops the notions of DOR (lifetime) and THOLEDAH (generations or history) in support of this notion of life cycle: *Corpus*, 165–66; *HistNat*, 271–72.

57. "Denique nec corpora ipsa, vel quae motu ac sono exercentur, ut aër, venti et aqua, atque ignis; vel quae muta, fixa, atque solidiora sunt; segnes otiosásve in mundi integritate partes obtinent; imo singula non specie modò atque forma orbis ornatum adiuvant, sed vel afficiendi, vel aliorum effecta referendi ac sustinendi facultate praedita sunt": *Anima*, 37; *LibGen*, 136–37.

world was the immutability in the order and genus of all things in nature. These must remain unchanged, since every living thing has been given a nature with a specific purpose and would otherwise not fulfill its purpose. After they live out their GHOLAM, a new generation replaces those whose cycle has been fulfilled. Even the simplest intellects among us, Arias Montano remarked, can notice that animals have been made with parts suited for their purposes: legs so horses can run and wings so birds can fly. Everything—with the notable exception of humanity—blindly obeys the laws and the order assigned at the Creation.[58]

This teleology designed into nature was the reason that when Adam gave names to the spectacle of nature the names reflected the use and singular nature of each thing.[59] Part of the knowledge of nature God had bestowed was knowing the purpose and life cycle of everything. Arias Montano's conception of a static and self-replicating world allowed him to assume that the original relation Adam had observed between thing and word persisted through time. Therefore insights gained from the Bible about the nature of things could— and should—be informed and supplemented by the careful observation of this world, which was not unlike the one Adam saw. This understanding of the natural world as having an immutable nature and behaving according to divinely ordained laws that were intrinsic to things' natures allowed Arias Montano to begin building a natural history that relied on the Bible as evidence—as the testimony of an eyewitness, if you will—but that can be augmented with observation of the world as it is now. He also found ample justification for this view in the exegesis of the Solomonic declaration "there is no new thing under the sun" (Eccles. 1:9 KJV).[60]

How then were observations about the natural world to be incorporated into knowledge of nature? Or, as Arias Montano articulated it, how were we to account for the visible properties of things? Here again he turned to the words of Solomon and the book of Wisdom. The key to interpreting sense perception lay in the Solomonic statement that "omnia in mensura, et numero et pondere disposuisti," which Arias Montano interpreted as "Everything has

---

58. "Quidquid de unoquoque animantum genere pronunciavit ac declaravit Adamus, id verè ac reapse pondus et ingenium perpetuum, munusque statum et immutabile censendum est, suis temporibus opportunè, ubi res postulaverit certa actionum, vel passionum opera exhibendum et praestandum. Nam quam diu unumquodque singulorum generum caput vivet, propriam re, et certa obedientia tuebitur vocationem, perinde atque sol, luna, stellarumque multitudo suam etiam à Deo nominatam vocationem tuentur.": *Corpus*, 368; *HistNat*, 486.

59. *Anima*, 36; *LibGen*, 136.

60. Arias Montano, *Discursos sobre el Eclesiastés de Salomón*, 99–100.

a measure, number, and weight"[61] (Wisd. of Sol. 11:21 VUL). This phrase revealed how all things in nature had been conceived, the means by which they had been organized by order and genus, and thus how they should be understood and explained. The word "measure" in the phrase suggested the limits and description (*fines et descriptiones*) of each form and thing. By saying these also have "number," the prophet was declaring that everything in nature has a place in a particular order. Although the abundance of natural things might make nature seem disorderly, as if made by chance, they were created according to distinct "order and species." Within this order, each species was given forces, proper faculties, and efficacies designed to appear at opportune times according to its life cycle (*maturationes*) and answering to its own causes. Divine wisdom referred to these characteristics as "weight."[62] Once the world was understood as unchanging and operating in response to certain divine laws, observations of nature could further explain God's creation, given that it was understood that everything had a measure, number, and weight.

Utilitarian View of Nature

The final historical premise at the core of the *Magnum opus* lay in Arias Montano's belief that the nature of things must be investigated from an entirely anthropocentric perspective and that the relationship between humanity and things was determined by utility.[63] The purpose of nature was clear: nature had been created to serve humanity—to be useful—and only secondarily to bear witness to God. Among all the species of beings there was not one, he believed, that did not have a singular nature, approved and designed by God to be useful to humanity. The very sequence of the Creation testified to this: God created man on the second half of the fifth day of the Creation *after* creating the plants

61. *Corpus*, 175; *HistNat*, 281.

62. "Ita omnia quae a Deo condita sunt, ad eos certos generum numeros redacta constitere: neque vanum quidquam, otiosum aut inutile à conditore extitit, sed suae cuique generi vires, facultates propriae, atque efficacitates tributae, suaque temporis momenta, suae maturationes, ad causae propriae rationem adscripta et indicata, has facultates et efficacitates certo virium examine trutinatas divina sapientia Pondus vocat.": *Corpus*, 175; *HistNat*, 281.

63. "Caelum sursum, et terra deorsum. Cùm primis verò illud statuendum, omnia quae à Dei sapientia ac virtute facta, conditaque fuerunt, certam naturam obtinuisse certo rerum genere comprehensam ac definiendam, singulaque actionis sive muneris et officii cuiuspiam faciendi causa fuisse edita: ita ut nihil sit otiosum, nihil ludicrum, nihil quod non ad mundi concinnitatem et usum faciat; ac tam generis quàm facultatum et virtutum ratione.": *Corpus*, 175; *HistNat*, 280.

and animals (Gen. 1:29–30) so that they would recognize man as their master and as more powerful than they were. The natural world's subservience to humanity was clear: nature would not exist otherwise. Humanity was created to be nature's guardian, dweller, cultivator, and governor.[64]

Since nothing in nature had been made in vain, part of the essential nature of things was their usefulness to humanity. This view embedded a utilitarian teleology in the nature of things that was often the starting point for an empiricist investigation. Thus a proper understanding of nature consisted in identifying the utilitarian aspects of natural things. The most obvious were works of the Creation used for food or shelter for either man or beast, but also, as we will see in the next chapter, the existence of the heavens or the relation between the sea and land could all be justified from a utilitarian perspective.

Sense Perception of Fallen Humanity

The four historical premises discussed thus far were the basis for biblical interpretation of the philosophical precepts Arias Montano sought to extract from the Bible, but he also made observing nature an essential part of his project. In fact, one objective of the *Magnum opus* was to provide a methodology that reconciled empirical observations with a natural philosophy derived from the Sacred Scriptures. This made his project distinct from the pragmatic tasks of the cosmographers who worked at the Casa de la Contratación (House of Trade) or from the approach of his many naturalist friends. Despite being immersed in the culture of Sevillian empiricism, he was also driven by the disquiet about received natural philosophical systems. Scholars of Arias Montano's ilk were fully aware that empirical observations *had to be* interpreted through a conceptual natural philosophical framework and had to be in complete concert with this framework. In his case, however, the natural philosophical system had to be created anew from vestiges left in the Bible.

Arias Montano recognized the importance of sense perception as the means of observing, but he also noted that this human capacity presented problems to any natural philosophical program. In the *Corpus* sense perception is always treated as a diminished human capacity. Whereas prelapsarian humans had enjoyed infallible sense perception, after the Fall they operated guideless, that is, without the divine wisdom that had allowed Adam to grasp the essential principles of nature with no threat of sensory deception. Arias Montano considered that prelapsarian humans had enjoyed an ability to understand reality that

---

64. *Corpus*, 348, 374–75; *HistNat*, 464–65, 493–95.

was "universal" and were free from having to "beg" the senses or to learn by "observing" experiments.[65] In fact, prelapsarian Adam did not learn through his eyes and ears. Each perception appeared to Adam as something new, and once understood it could be accepted or rejected. "The rule and the mastery" that permitted this discernment lay within him, and from there it descended by degrees to exterior potencies that acted accordingly.[66] Postlapsarian humanity had forgotten this "rule and mastery" and thus had to learn anew how to evaluate knowledge gained through the senses.

Once again Arias Montano gave a historical explanation. He found the historical key in the description in Ecclesiastes of Adam's thinking as "straightforward" or IASAR (יָשָׁר, *yāšar* or *rectum*) (Eccles. 7:29). This meant that his thinking at first had been free of any vacillation of mind or need to question or inquire into the nature placed at his disposal. This rectitude of thought resided in the superior/internal/masculine nature of man, the part that brought him into close communion with God. (Arias Montano had also read Ecclesiastes 7:28, where the author mentions not having found a single woman with this capacity; therefore his exegesis hinges on the distinction between the interior masculine and external feminine natures.) Adam's proximity to the divine served as a guide that subordinated sense perception to the soul and thus functioned without the fear or even possibility of falling into falsehood.[67] Furthermore, Adam knew the nature of things before he experienced them because God had imbued him with the knowledge of first principles of all things.[68] Adam's initial nature did not "condemn" or "obligate" him to the process of reasoning; he had no need to judge or choose among things that appeared before his senses because he had been given a "great" and "open" faculty of knowing. He did not need his senses or clever exercises to know the form of things and discover their causes.[69] Neither did Adam have to discern between good and evil, since all the works of creation were good, useful to him, and under his dominion.[70] To Arias Montano, although Adam had eyes and could see, they were as if closed to all sensorial acts of cognition.[71]

65. *Corpus*, 375; *HistNat*, 494.

66. Arias Montano, *Discursos sobre el Eclesiastés de Salomón*, 106.

67. *Corpus*, 405; *HistNat*, 527.

68. *Corpus*, 375; *HistNat*, 494–95.

69. "Ita ut animus hominis magna et aperta cognoscendi facultate praeditus, nec sensus, nec agitationis solertiaeve ad rerum capiendas formas, aut inveniendas causas egeret.": *Corpus*, 405; *HistNat*, 527.

70. *Corpus*, 376; *HistNat*, 495.

71. *Corpus*, 426; *HistNat*, 550.

This happy condition—at least in Arias Montano's view—was not to last. The devil Hilel, embodied in a serpent, knew how to exploit the division between the superior/internal/masculine part of Adam and his inferior/external/feminine part. This inferior nature, once an integral part of the Adamic human, became embodied when the first woman was created from Adam's rib. (Arias Montano does not use the name Eve.) She, although also possessing a superior/interior nature akin to man's, was nonetheless driven by her inferior/external nature.[72] This external nature was ruled by the senses, which served her well in her capacity as caregiver and in her subservience to man, but for Arias Montano this essential feminine nature was inherently weaker than the interior masculine nature. The serpent sensed this weakness and took advantage of it once it was able to find the woman alone, without man at her side to guide her inferior nature to loftier aims. The trap was set.

Arias Montano presented the conversation between the serpent and the Adamic woman as a long paraphrase of the serpent's arguments that led her to disobey God. The serpent tells the woman, "Until now, you truly have been blind; definitely you do not see all, or know all the things you may see and know."[73] To fully appreciate all the world, the serpent explained, it was also necessary to know how to tell good from evil. Yet discernment comes not only from knowledge, but also from sensory experience and reason. Too bad, the serpent said, that God, by making humans so perfect and claiming to have endowed them with all knowledge, had in fact shortchanged them by depriving them of the need to use their senses and reason to experience all there was in the world: the good and the evil alike.[74] As a result, humans were isolated and would never know all there was to know. God had hidden this for only one reason, the serpent intimated, for if they knew the difference between good and evil they would equal God in knowledge. Having aroused the woman's arrogance, all the serpent had to do was let her senses guide her to the beautiful tree: "The woman saw that the tree was good for food, pleasing to the eyes, and desirable for gaining wisdom. So she took some of its fruit and ate it; and she also gave some to her husband, who was with her, and he ate it" (Gen. 3:6 NAB).

The paraphrase frames the serpent's duplicitous act as one that drew the woman away from her interior nature and then let her senses lure her the rest of

---

72. *Corpus*, 372–73; *HistNat*, 491.

73. "Vos verò hactenus caeci estis, certè non omnia videntes, non omnia scientes quae scire poteratis, et videre.": *Corpus*, 414; *HistNat*, 536.

74. *Corpus*, 414; *HistNat*, 536.

the way. The Montanian serpent, however, deftly points out the limitations of the senses and lets the promise of enhanced reason and sense perception—not just the knowledge of good and evil—become the temptation. As Arias Montano explained, the concupiscence of her inferior nature, once unbridled by the serpent's promise, was also enough to overcome Adam's superior nature and lead him to perdition. In his retelling, the temptation that led to the Fall is a cautionary tale of what happens when sense perception and the empirical knowledge that results from it are left unchecked or become misguided. The actions that immediatcly followed eating the forbidden fruit were for Arias Montano more evidence of what happens once humanity has to exercise reason based only on empirical knowledge, without the interior discernment God had granted Adam. "The eyes of both of them were opened, and they realized that they were naked; so they sewed fig leaves together and made loincloths for themselves" (Gen. 3:7 NAB). The very first exercise of reason led them to see and then to judge what they saw as shameful, and as a consequence they had to discover ways to cover their shame.[75] They had "experienced" and judged the results, opening the way for humanity's tribulations about deciding what was best, to endless trials before choosing anything, and to none-too-certain choices. Fallen man now had to rely on his inferior/external/feminine nature to learn "by the sweat of his brow" what he needed to know to survive in the world. This meant that the type of cognition that relies on the senses was the only means left for understanding the world. It was, however, a capability so diminished and darkened that humanity could no longer perceive the magnificent and the common (*amplum et commune*) except through the external/feminine part. From the open and simple comprehension of all things that allowed Adam to know the universal truths of everything under the heavens, humanity now had to rely on the senses and to submit this evidence to reason and reflection, subject as these were to all types of opinions and errors.[76]

Arguing from precepts based on sense perception, or even arguing based on reason alone, was to Arias Montano a heavy cross humanity had to bear. To illustrate the tragic consequence of being left with only sense perception as a guide, he described the birth of a child who, on entering the world, first opens his hands and eyes, signifying an awareness of his senses, and then, as if realizing the terrible toil that awaits him, immediately begins to cry.[77] The type of wisdom available to fallen humanity was "wisdom of the flesh," which began

75. *Corpus*, 429–30; *HistNat*, 553–54.
76. *Corpus*, 450; *HistNat*, 576–77.
77. *Corpus*, 427; *HistNat*, 551.

with the aid of the senses and then called on the intellect.[78] Thus postlapsarian humans could not understand or desire anything unless they became aware of it through the senses. This, Arias Montano noted, was what the "philosophers" meant by "nothing is in the intellect that was not first in the senses" (*nihil est in intellectu, quin primus fuerit in sensu*).[79]

A Good Dose of Skepticism

Arias Montano thus concluded that depending on sense perception led humanity to rely too heavily on reason for guidance. Given that, in his view, the capacity for reason came as a result of the Fall, we can expect he found little good would come from it. In fact, he made the whole philosophical apparatus of Western knowledge the fanciful fabrication of the Greeks. Far from considering reason one of the defining characteristics of humanity, as "philosophers" maintained, he saw "reason" as a construction of Greek philosophers who did not understand the essential nature of mankind.[80] For them reason was the act of weighing one known thing against another and judging how they coincide or differ.[81] For Arias Montano this entailed a contradiction between the philosophers' claim that reason is humanity's defining characteristic and their claim that humans enter the world with their minds a blank slate (*tabula rasa*), first learning everything through sense perception and later organizing it in the intellect. How, he wondered, can the intellect and the ability to reason be the principal attributes of humanity and yet be subservient to the senses? Furthermore, philosophers had judged humanity's reasoning capabilities too

78. "Atque ita sapientiam et prudentiam carnis dici eam, quam homo sensuum opera inceptam etiam intellectus munere et officio comparat, eandemque mundanam, interdum terrenam dici.": *Corpus*, 423; *HistNat*, 546.

79. "Id est, nihil intelligere, nihil diligere possit, antequam ab exteriore et inferiore portione cognoscendarum, expetendarum vel fugiendarum rerum copias imaginesque acceperit. Id quod naturae, non illius quidem prioris ac rectae, sed posterioris ac depravatae istius contemplatores, quos philosophos dicimus, diligenter observatum suis formulis docendo tradiderunt. Nihil, inquiunt, est in intellectu, quin primus fuerit in sensu. Namque sensum illi idem ipsum quod nos exteriorem hominem sive inferiorem animae appellamus portionem." Arias Montano attributed the Scholastic paraphrase of the Peripatetics to a general category of 'philosophers': *Corpus*, 441; *HistNat*, 566.

80. *Corpus*, 404; *HistNat*, 525.

81. "Rationem autem hoc loco, ex horum etiam explicatione vocamus compositionem rei unius notae cum alia etiam nota, ex quarum vel convenientia, vel discrepantia aliquid praeterea aliud, quod nondum notum erat, vel convenire, vel discrepare cognoscatur.": *Corpus*, 404; *HistNat*, 525.

auspiciously. Given that after the Fall the ability to know was inherently flawed, how could philosophers of any school claim to have arrived at the true philosophy, let alone claim to have a true understanding of human nature, since they had at their disposal only vestiges of a corrupt and fallen humanity?[82] This attitude made Arias Montano very skeptical of the claims of all philosophical schools, whether of natural or moral philosophy. He found in the ancient—and ongoing—polemics between different skeptical schools of philosophy all the evidence he needed to argue for the futility of any type of natural philosophy that relied solely on sense perception and human reason.

The skepticism debates of the ancients, which had gained new traction with the publication of Sextus Empiricus's *Hypotyposes* in 1562, did not escape our biblist. That text allowed Arias Montano's contemporaries to contrast various stances deriving from Academic and Pyrrhonian skepticism. Pedro de Valencia, Arias Montano's secretary while he was writing the *Magnum opus*, wrote an exposition of the debates in his *Academica, sive De iudicio erga verum* (1596).[83] This was far from coincidental. Valencia systematically explored ancient skeptical postures and the way they treated the subject of truth (Academic, Pyrrhonian, Stoic, that of Carneades and the New Academy, and even Epicurean).[84] Valencia ultimately found these debates fruitless, if not disquieting, for he concluded the book with a statement in line with Montanian natural philosophy: "For God hides true wisdom from the lovers of false wisdom and certainly reveals it to [those who are like] children."[85]

Despite Arias Montano's belief in the fallibility of speculative philosophy and sense perception, his skeptical posture did not imply a negative and dogmatic attitude about the possibility of asserting anything as true. Quite the contrary, the Montanian program never abandoned hope of finding intractable truths about the natural world. Arias Montano, however, had also learned from the Sacred Scriptures that some things could be known better than others; complete knowledge of the *rerum natura* was not always within humanity's reach. Some of nature lay physically beyond human beings' imperfect sense perception, imposing limits on what could be said with certainty about some

---

82. *Corpus*, 443; *HistNat*, 569.

83. In the book—believed to have been written as early as 1590—Pedro de Valencia compared various ancient postures regarding the criteria for truth. He leaned heavily on Cicero's *Academica* but also discussed other skeptical philosophies.

84. Thorsrud, *Ancient Scepticism*, 36–58, 123–30. On the influence of these ancient philosophies in early modern Europe, see Popkin, *History of Scepticism*, 38, 64–79.

85. "Abscondit enim Deus veram sapientiam a falsae sapientiae amatoribus, revelat vero parvulis.": Valencia, *Obras Completas, Academica*, 3:444.

things. This, coupled with the human tendency to overreason—and the Bible's silence on some topics—placed some things beyond comprehension.

In the annotations to the preface of the *Anima* that serves as the preface to the *History of Nature*, Arias Montano pointed out that some works of nature cannot be known: "Some things are able to be known to some extent; they permit nothing beyond this to man, such as the movement and efficacy of the heavens and stars whatever they might be."[86] He supported his view with Job's remarks about the futility of trying to attain knowledge of the heavens. "Have you fitted a curb to the Pleiades, or loosened the bonds of Orion? . . . Do you know the order of the heavens; can you put into effect their plan on the earth? . . . Can you send forth the lightnings on their way, or will they say to you, 'Here we are'?" (Job 38:31–35 NAB). In Arias Montano's opinion, certain realms were created with the intent that they remain beyond humanity's purview. Why? The answer was clear: God never intended humanity to exercise domain over them. Thus God had not intended humanity to acquire *scientia* of these parts of nature or, as Arias Montano understood it, to gain causal knowledge about them. For example, the heavens could be observed and the stars and their movements studied and quantified, but the type of knowledge this exercise yielded was intended for only utilitarian purposes: reckoning time, forecasting meteorological events, and understanding the seasons. This attitude about the limitations of causal knowledge was in concert with prevailing views on what it was possible for astronomers and natural philosophers to discover about the heavens.

Yet Arias Montano was not a defeatist, for he believed there was still a vast realm of things humans could and should learn. The first, as I discussed earlier, was attaining an honest knowledge of oneself. Yet a close second was learning about things that shared in some aspect of human nature, whether because they were necessary for humanity's sustenance, shared in humanity's earthly/corporeal nature, or influenced humanity's dwelling place: the terrestrial realm. In the *admonitio* on wisdom and science discussed earlier, he made it clear that this lower order of knowledge, in the form of the "inferior pious wisdom" of Solomon, was attainable by postlapsarian humanity.[87] Rediscovering it was, in fact, the principal objective of the *History of Nature*. Despite how

86. "Quaedam tantùm cognosci possunt, nihil praeterea hominibus ex se permittunt: ut caelorum stellarumque cursus et efficacitates quaecunque illae fuerint. Quaedam verò et sciri et à cognoscente agi, gubernari, et corrigi, ut quae hominum curae et industriae patent. [Q]ualia agrorum cultus, animalium tractatio et agitatio, atque praecipuè omnium, sui ipsius institutio, et animi morumque formatio.": *Corpus*, 3; *NatHist*, 103.

87. *Corpus*, 235–37; *HistNat*, 343–45.

difficult this pursuit seemed, the quest for the *rerum natura* was a worthy and divinely ordained pursuit. Even the prophet Isaiah had admonished those who abandoned the contemplation of the heavens, the earth, and the composition of the world in favor of pursuing pleasure.[88] As Isaiah proclaimed, "Woe to those who demand strong drink as soon as they rise in the morning, And linger into the night while wine inflames them! . . . But what the LORD does, they regard not, the work of his hands they see not" (Isa. 5:11–12 NAB).

*

I have identified five core historical premises or suppositions that inform the *Magnum opus* and that Arias Montano derived directly from the historical narrative of the biblical text. They were all well within church tradition, albeit influenced by Renaissance humanism and its reacquaintance with Neoplatonic, skeptic, and Stoic notions. Yet for Arias Montano their stature as premises derived from what he considered to be their origin as unassailable historical verities resulting from divine actions and words. All of them had direct consequences for how the Montanian program thought the study of nature should be pursued. The fundamental premise concerned the purpose of humanity—to know God—which also entailed knowledge of human nature and of the natural world. His second premise was the belief that prelapsarian Adam had enjoyed complete knowledge of nature, largely lost after the Fall. Yet Hebrew—the language of Adam—had retained traces of it. Moses, as prophet and scribe, although not writing for this purpose, had recorded vestiges of the *rerum natura* in the Sacred Scriptures. This knowledge could be recovered by deciphering the "arcane language" of the Bible. Arias Montano's third premise concerned the immutability of the order and genus of all things in nature as God established them at the Creation. This assured him that observations made now corresponded to the same reality described in the Bible. The fourth premise derived from God's having placed all of nature under humanity's dominion. This implied that the purpose of everything in the world was to be useful to mankind. Therefore finding a thing's essential nature also meant identifying its utility. The fifth premise—this one an epistemological issue informed by historical circumstances—concerned the role of sense perception in the Montanian program. Sense perception was a diminished capacity in humans, helpful for living in the world, but since it was intrinsically flawed it could not serve as the sole basis for natural philosophy.

88. *Corpus*, 176; *HistNat*, 281.

These premises gave Arias Montano the necessary scaffolding for building a biblical natural philosophy that would finally offer a viable and pious solution to what I have described as the disquiet that had overtaken natural philosophy. The history of human knowledge about the natural world and the resulting state of human reason and sense perception raised epistemic issues that simply could not be resolved save by bringing to bear a greater authority. Given the fallibility of any human philosophy, only God and the Bible had the necessary attributes to be this final arbiter, though the vestiges of the *rerum natura* revealed to the prophets and laid down in the arcane language of the Sacred Scriptures could serve as a proxy. These premises informed every exegetical exercise and every observation or experience that Arias Montano applied to this project. His next step was to identify the first principles of nature that would guide his literal exegesis of the *rerum natura*.

CHAPTER 8

Montanian Hermeneutics of Nature and Cosmology

Arias Montano found a mandate to reform natural philosophy in the pre-ordained and deterministic world he imagined, one God had shown time and time again that he wanted humanity to understand. Driving his work was the realization that by deciphering the arcane language of the Bible he had devised a way to achieve this understanding that liberated humanity from the misguided paths and lies of the philosophers. What remained was to put forth a methodology that led to the true nature of things and was in concert with the world as described in the Bible. He conceptualized this part of his project as a new natural philosophy though he never described it as such, perhaps not surprisingly given his aversion to the trappings of philosophy. But as this chapter will show, it was a natural philosophical program informed by a particular hermeneutics of nature—in other words, a theory and methodology for interpreting the natural world that allowed him to systematically fill all the conceptual categories of early modern natural philosophy. To do so he borrowed ideas from existing philosophical systems, bent existing notions of natural philosophy, and fabricated some of his own—all while never acknowledging what he did and always turning to the Bible for validation. The result was what we might consider a distinct Montanian natural philosophy that addressed the central concerns of the Spanish disquiet.

The *Corpus* or *History of Nature* was his attempt to put this hermeneutics of nature to work in a systematic manner. Yet the book was never meant to be a comprehensive exploration of all aspects of natural philosophy; it simply sought to illustrate how first principles identified in the Bible could be used to explain nature. Once the metaphysical underpinnings were in place, Arias

Montano moved on to illustrate how his system explained some weighty aspects of cosmology, astronomy, astrology, meteorology, matter theory, physics, botany, zoology, and biological anthropology. This chapter focuses on the first three topics—cosmology, astronomy, and astrology; other fields are considered in the next two chapters. But we begin with an exploration of the metaphysical principles Arias Montano derived from the book of Genesis and used as the basis for studying the nature of things.

Along with the premises explored in chapter 7, these principles are the foundation for Arias Montano's hermeneutics of nature. But readers beware: this consideration of his metaphysics skips over his discussions of the spiritual and moral dimensions of the human condition—an important subject to be sure, and one he considered to go hand in hand with the natural philosophy he had identified in the Bible. Yet Montanian moral philosophy, it bears noting, merits a far lengthier treatment than this book permits. The omission is mainly a response to modern sensibilities, which are comfortable considering physiological matters separately from moral aspects in anthropology. Yet we should be aware that this division was inconceivable to a devout Christian humanist like Arias Montano.

MONTANIAN BIBLICAL METAPHYSICS

For Arias Montano, although the essential nature of things had been determined at the Creation, the continuing unfolding of the natural world was very much within the realm of what humans should try to understand. The key to understanding this unfolding and the natural world that resulted from it was to base all interpretations of what was happening in nature on a proper set of first principles. Despite his well-known disdain for Scholastic Aristotelianism, this aspect of his project was fully in concert with that approach: all knowledge should be based on a metaphysics or set of first principles that were known to be certain. Arias Montano assumed these first principles had been known to Moses, so it was reasonable to search for them in Genesis.

Arias Montano undertook the quest for these principles in a paraphrasis and commentary on the first six days of Creation, hints of which may be also found in the *De arcano sermone*. He repeated the exercise twice in the *Magnum opus*, first as a short section in the beginning of the *Anima* titled *De proposito Dei* and again as a much longer exposition in the *History of Nature*. The section in the *Anima* paraphrased the historical sequence of events of the six days, and in the *History of Nature* he examined the Hebrew lexicon of Genesis in search of first principles. In the *Anima* this paraphrase suffices for him to begin exploring the history of humanity; he makes no attempt to explain the underlying metaphysical principles involved in his interpretation of the Creation. Readers of the *Anima* may have been puzzled by his repeated references to unexplained Hebrew terms in this section, but they would have to wait until the *History of Nature* for an explanation. There, in the section *De natura disputatio*, subtitled in Greek "Generalities," Arias Montano explained the exegetical exercises that supported his interpretation of the Hexameron and how he derived the metaphysical principles that inform the *Magnum opus*.

Causa, Iehi, Elohim, and Maim

Arias Montano identified four metaphysical principles in the first three verses of Genesis (fig. 8.1). The first principle or first cause of everything was the goodness of God (*bonitates Dei*).[1] He attached this meaning to the Latin word CAUSA or *finis*, that is, the reason something is or will be. Thus CAUSA simply stands for the goodness of God implied in the act of a purposeful creation and was indicated by the Hebrew word RASHITH (רֵאשִׁית, *rēʾšît*), "beginning." Only God in his goodness existed before the second principle was made manifest in an act of volition by the Deity, the first action that set in motion the creation of the world. This principle is evident in the verb IEHI (יְהִי, yĕhî), "to be" (*erit* or *fiat*). Arias Montano explained that it indicated the creative force in nature, a word that, when used by or in relation to God, gives perpetuity to form and constancy to things. IEHI has a continuing role in the world by virtue of the perpetuity implied by being made according to God's will.[2] Arias Montano also pointed out that the true nature of this word was unknown to other philosophers (*philosophi exteri*), implying that this is why they did not understand how the world came to be and continued to exist just as it was created. He will later go so far as to state that the force and virtue that renews everything in the world and that philosophers call "nature" is all contained in the command IEHI, "be!"[3] To him it was a command that could not be rescinded except by God. The first result of IEHI was light.

1. The section that follows draws from *Corpus*, 149–58; *HistNat*, 252–61. He also discussed similar subjects in *Anima*, 24–29, and *LibGen*, 124–29.

2. "Divina igitur vox haec IEHI constat constabitque semper singulis rerum generibus pro cuiusque ratione, veritate, ac propria duratione creandis, producendis, constituendis, et conservandis praesens atque efficax": *Corpus*, 149–50; *HistNat*, 252.

3. Arias Montano uses the Latin *esto*, the future imperative of "to be.": *Corpus*, 188; *HistNat*, 294.

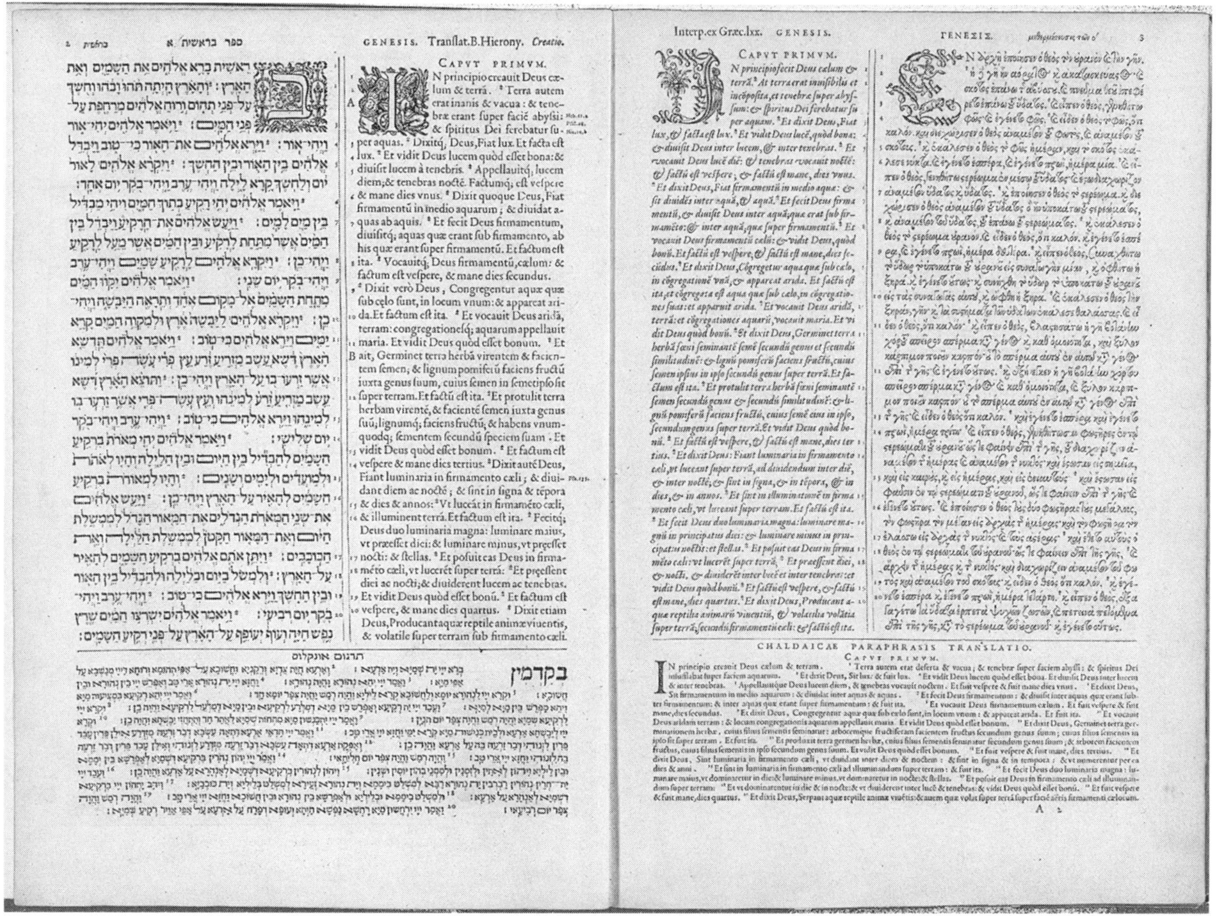

FIGURE 8.1.
Genesis 1:1–20.

Antwerp Polyglot, 1:2–3.
Special Collections,
Sheridan Libraries,
Johns Hopkins
University.

The third principle was the spirit ELOHIM (אֱלֹהִים, *'ĕlôhîm*), "God," a divine force that is diligent, is present, and extends over everything, permeating and encompassing everything, animating and giving shape to everything according to its use.[4] Whereas IEHI commands the creation, the designer and overseer of this creation is ELOHIM. The spirit ELOHIM prepares all forms, distinguishes among them, establishes them, and directs them. It preserves the types (*genus*) of all things with the aid of its ministers, who can be either corporeal or incorporeal spirits.[5] When the word ELOHIM appears in the Bible along with the word MERAHHEPHETH (מְרַחֶפֶת, *mĕraḥepet*), "moved," "hovered," it suggests setting things in motion, as in agitating the surface of the abyss or the heavens. To clarify the concept Arias Montano resorted to an image, one of the few in the *History of Nature*, depicting ELOHIM agitating the waters of the abyss (fig. 8.2). He explained that this agitation imparted to lesser spirits the power to determine the favorable moments for things to appear, increase, or diminish. Thus action from ELOHIM resulted in all the actions needed for things to generate, decay, and regenerate. The fourth principle was MAIM (מַיִם, mayim), "water," a form of matter capable of creating anything when subjected to the action of ELOHIM. Arias Montano noted that the sacred text used the Latin *aqua* as a synonym for MAIM, but in his opinion this was not entirely accurate. The word appears in the Hebrew Bible modified with an ending (ם), indicating duality or plurality, which to him suggested something with a double nature. So in Arias Montano's exegesis MAIM became a kind of primordial liquid with a double nature. One of these dual liquids is fertile, sweet, and somewhat flexible; it can flow anywhere and not condense, and it resembles milk or oil; it is greasy (*pingue*), can catch fire, and is akin to sulfur. The other liquid is humid and salty (*salsum*): it penetrates and fills voids; it is subject to cold and heat and can condense; it is anathema to fire and akin to mercury. These two liquids, the unctuous and the aqueous, mixed in proper proportions, account for all corporeal things; everything in the world, both in heaven and on earth, is made of MAIM. Since by its nature MAIN does not suffer decay, all things made from it may persist into eternity or otherwise exist to the limits imposed by what they are, their utility, and how long they are meant to last. The liquids can change forms, dissipate, or conserve themselves following the designs of the

4. "Erat verò in initio Spiritus ELOHIM, id est, vis quaedam divina, agilis, ac praesens, per omnia pertinens, omnia complens: et quocumque Deus intenderit, subitura: remque omnem pro veritatis cujusque usu animatura, atque informatura.": *Corpus*, 150; *HistNat*, 252.

5. *Corpus*, 188; *HistNat*, 294. The idea that ELOHIM is aided by spirits might have drawn from Ibn Ezra, who understood ELOHIM's actions as carried out by angels: Ibn Ezra, *Ibn Ezra's Commentary on the Pentateuch*, 1:25.

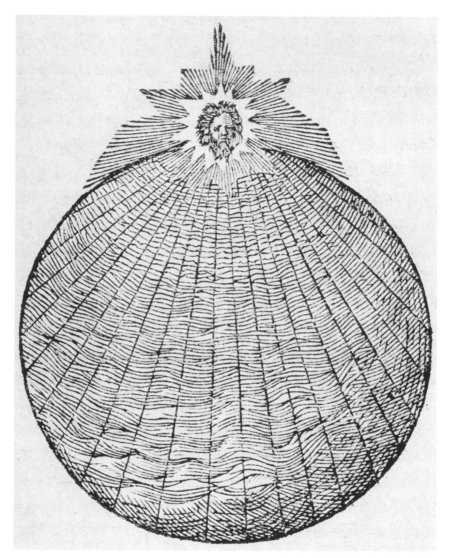

FIGURE 8.2. Spirit ELOHIM agitating the waters. Arias Montano, *Naturae historia*, 187. Biblioteca Histórica de la Universidad Complutense de Madrid (BH FLL 14614).

supreme Creator and the spirit ELOHIM. To show that his conclusions about the nature of MAIM were correct, Arias Montano brought to bear a small dose of empiricism. These dual fluids (*liquores gemini*), he explained, can easily be perceived by our senses in compound things. This is evident when a body is presented to fire, since fire always separates the salty from the greasy. The greasy liquid feeds the fire and makes it grow, while the fire rejects the salty part and seems to spit on it, because salty liquids can sometimes extinguish fire.[6]

In the narrative of Creation everything began with the word IEHI, but the arranger of all creation (*concinnator*) and its form giver was the spirit ELOHIM.[7] IEHI created the conditions necessary for preparing earth to house humanity— heat, cold, warmth, cooling, contraction, and expansion—but ELOHIM was, and continues to be, the source of motion responsible for bringing these at opportune times. In the Montanian version of Genesis these principles un- folded on a temporal dimension and in a particular order, as did their effects on earth. Thus the first thing created was light—"an incorporeal reality that conserves the image of the Divinity"—so that time could be measured and so the sequence of events recounted at the Creation would take place in a given order.[8] The next action was initiated by ELOHIM: the spirit agitated the sur- face of the abyss so that the MAIM could separate or coagulate and otherwise transform itself into the different bodies that make up the world. The offices of ELOHIM continue to unfold in the temporal realm to the present, but acting for the most part through other products of the Creation. For example, it was the sun and the moon—the two ministers (*ministri duo*)—that were directly responsible during the Creation for determining the favorable moments for other things to appear on the earth. Through their cycles of day and night they properly "seasoned" the earth, a complex process analogous to weather that suggested to Arias Montano that many celestial "helpers" in addition to the sun and moon took part. These agents brought about the "increase" and "diminution" necessary so plants could germinate and grow.[9]

Montanian Syncretism?

As much as Arias Montano despised the consequences of ancient philosophies for humanity's prospects for salvation, we cannot expect his natural philo- sophical proposal to have been wholly divorced from the conceptual frame- works that informed fundamental epistemic questions of his time. The very fact that he "found" four metaphysical principles apparent in the first moments of the Creation tells us not only that he expected to find them but, even more tellingly, that he expected Moses to have described the Creation with them in

6. *Corpus*, 150–51; *HistNat*, 253.

7. *Corpus*, 153; *HistNat*, 256.

8. "Rebus in ordinem seriemque referendis initia sumpta sunt.": *Corpus*, 151; *HistNat*, 253.

9. *Corpus*, 153; *HistNat*, 256.

mind! So ingrained were the patterns and means of analysis of the Western philosophical tradition that Arias Montano did not seem to realize he was replicating them in his "new" natural philosophy. (For one thing, he never referred to them as metaphysical principles.) As a consequence, the Montanian conception of divine agency was reduced to a set of intelligible components analogous to those of other Western philosophical systems and defined in such a way as to be metaphysically sound. Thus all aspects of divine action could be explained by resorting to first principles. As in other systems of natural philosophy, these first principles were the point of departure for finer points of a philosophical system derived through deductive logic. Finally, these principles undergirded an explanatory system that accounted for all natural phenomena.

Arias Montano summed up this aspect of the exegetical exercise on Creation this way: "Therefore there are four principles or origins of things: CAUSA, IEHI, the spirit ELOHIM, and MAIM, or dual liquids, from which all corporeal things proceed and into which they return when unbound."[10] That he found exactly four first principles was no coincidence. They map loosely onto the four Aristotelian causes: CAUSA: Final; IEHI: Formal; ELOHIM: Efficient; and MAIM: Material. (And this despite his pointed dismissal of Scholastic Aristotelianism.) Yet it is telling that the previous quotation continues, "for we affirm they are better able to bring about the forms and species of things, than causes." So Arias Montano, while dismissing the causal system of the Peripatetics, still needed their functional equivalent to explain how forms and species came into being. He looked in Genesis and found there the original explanatory principles. He had an explanation for why the four Aristotelian causes were problematic. In his sermons on Ecclesiastes he explained, "The philosophers of our days—not as learned or as attentive in proper speech—say there are four types of causes, which they call formal, material, efficient, and final. Yet the ancients—and this is so—did not give the name of cause to any but only the final."[11]

Arias Montano was making the case that we need not concern ourselves with the first three causes. They were fabrications and lacked intrinsic purpose. Only the final cause was needed to explain why something mattered, by asking Why? To what end? For what reason? In contrast, the Montanian

10. "Quatuor itaque rerum primordia sive principia sunt, CAUSA, IEHI, SPIRITUS ELOHIM, MAIM, sive liquores duo: ex quibus corporea quaeque constant, et in quae, cùm resolvuntur, abeunt. Nam rerum formas et species causarum potius effecta, quàm causas dicimus.": *Corpus*, 151; *HistNat*, 253.

11. Arias Montano, *Discursos sobre el Eclesiastés de Salomón*, 87.

four principles were purposefully engaged in the operation of the world. All the Montanian principles, however, elucidate the ontology of their respective effects. CAUSA is an emanative consequence of a loving God made manifest in the act of creation. As long as God loves, the world enjoys a final purpose—to exist in this love. With this Arias Montano was getting close to Augustinian voluntarism, but with the caveat that direct agency in the world was the predetermined role of ELOHIM. The world comes into being through a verbal command—IEHI!—that orders things to take a particular form and remain so in perpetuity. This form defines what is and what it will come to be, so that all creation serves God's purpose. These two first principles/causes remain constant and alone would have produced a static—if much loved—world. But the world as we experience it is one of change and cycles, many predictable, but others, like the actions of humanity, highly unpredictable. So Arias Montano conceived of ELOHIM as an active agent in the creation and subsequent continual regeneration of the world, but one whose actions were not sporadic or capricious but followed the divine plan established at Creation. For example, ELOHIM must repeatedly act on MAIM to maintain the seemingly perpetual cycles of nature. MAIM is far from an inert material substrate for all things; in the process of separation and differentiation that took place during Creation, it was divided into different types of matter, each with its own nature, utility, place, and GHOLAM in nature.

Yet the Montanian first principles also share some characteristics with the Neoplatonic notions Arias Montano was familiar with. They constitute a clear hierarchy of dignity and power, with CAUSA as the apex, followed by IEHI, ELOHIM, and MAIM at the bottom. This hierarchical relation resembles the hypostasis of the Neoplatonists and can be loosely mapped onto the Plotinian notions of the One, Intellect, and Soul. The One shares with the Montanian CAUSA the key attribute of being that through which everything takes place and has goodness as its essence.[12] But there is one important distinction between the Plotinian One and the Montanian CAUSA; the One is a being, while CAUSA is the action of a godhead who stands apart from this hierarchy of first principles, although acting through them. In Plotinus, the Intellect or Nous is the first emanation of the One, and in this way it parallels the Montanian IEHI! as the first word spoken by God to begin Creation. Like the Intellect, the command IEHI is the keeper and giver of the Forms that will ultimately define reality. The active role of the spirit ELOHIM is reminiscent of the Neoplatonic Soul that permeates creation while organizing and imparting life. In Montan-

---

12. The discussion of Neoplatonic hypostases follows Remes, *Neoplatonism*, 49–59.

ian philosophy, ELOHIM initiates and regulates a veritable torrent of celestial and terrestrial fluxions, which by means of lesser agents (or spirits) act upon everything on earth, bringing it to life and defining its course. MAIM is not a direct analogue to the Aristotelian prime matter (*hyle*), especially because *hyle* cannot exist distinct from form, as MAIM does in the Montanian scheme. Yet throughout his exposition Arias Montano described MAIM as a very real and tangible thing, not at all like the unintelligible Neoplatonic matter—the bottom rung of the hypostases. Within the Neoplatonists' broad spectrum of ideas about matter, MAIM as a first principle is somewhat akin to the concept of intelligible matter as a substratum for all bodies.

Arias Montano's CAUSA, IEHI, ELOHIM, and MAIM can be thought of as hypostasized causes—in other words, agentive causes. In his scheme the explanatory power of the Aristotelian four causes was augmented by giving their analogues agency to act according to their metaphysical stations. This made them a direct conduit for the divine will and design but also provided a clear agent that actualized all the potential observed in nature. (Not that Arias Montano ever used such Thomist language.) The four Montanian principles bridged the otherworldliness of the Neoplatonic philosophy with the explanatory power of the Aristotelian system. They did so by identifying the metaphysical principles that produced a self-sustaining world, yet one where God could still intercede at will. In doing so Arias Montano trod the difficult path between a theology arguing for a complete Creation—where the world was created once, out of nothing—and another based on the Neoplatonic notion of a world continuously created through emanations of the One. This was clearly on his mind when he explained that some of the Hebrew names for God indicate, among other things, the times and places where this particular expression of God acts. When the Bible used the ineffable name of God IEHVEH (יהוה, *yhwh*) it was referring to the all-being, infinite, and eternal God, perhaps the closest Arias Montano comes to the notion of the Neoplatonic One. As IEHVEH, God does not act in time but exists eternally in his perfection. God as ELOHIM, however, dispatches his efficacies at opportune times to keep the cycles of the world going. Furthermore, the transcendent and immanent aspects of God were manifested in the duties and actions of ELOHIM. In his analogy of God as "center"—more on this later—Arias Montano suggested that God as ELOHIM was immanent, radiating to all confines of the world and throughout everything in it; but because God was infinite and eternal, and therefore not of this world, he was also a transcendent God, in line with Catholic dogma. Arias Montano argued against the notion that God resided within the outermost celestial sphere, as the medievals thought, because this suggested to him that the divine presence was somehow bound.

Some have found Neoplatonic influences in other aspects of Arias Montano's natural philosophy, such as his concept of the biblical language as loaded with hidden meanings whose secrets it will yield only to someone initiated in the art of deciphering it. His repeated, urgent call for studying nature in order to understand the nature of God and thus "approach" the Divinity has also been seen to situate him within the esoteric culture of the late Renaissance that developed around the *Hermetic Corpus* and the *Orphic Hymns*.[13] In my estimation these assessments tend to mischaracterize the scope and purpose of Arias Montano's true project. Despite the unconventional nature of some of his natural philosophical articulations, he never attempts to bar nonadepts from this knowledge, nor does he present the hypostasized causes as esoteric knowledge. In fact he urged his readers to study Hebrew so they could understand the causes at the same level he had achieved. (Exotic as learning Hebrew might have been to some segments of European society, it was more straightforward than learning to mimic theurgist rites or practicing natural magic.) His devotion to biblical Hebrew as an arcane language is more akin to the enterprise by earlier humanists—Christian Hebraists in particular—to rediscover the forgotten philological richness of classical languages and, through studying the ancient civilizations that spawned them, get a better understanding of their cultural production.

MONTANIAN HERMENEUTICS OF NATURE

Once Arias Montano had found a set of metaphysical principles derived from the Genesis account of Creation, he could begin to construct a comprehensive natural philosophy that was also in concert with the historical premises discussed in the previous chapter. He was aware that simply deciphering the arcane language of the Bible was not enough to recover the essential nature of things. The *rerum natura* thus uncovered had to be interpreted through a philosophical lens that situated this knowledge within a comprehensive natural philosophy. It is possible to identify in his work a hermeneutics, or theory of interpretation, that undergirds the whole project of the *Magnum opus* and is clearly at work there. Distilled into a few key points, the Montanian hermeneutics of nature found that:

13. Gómez Canseco, "Estudio Preliminar," in *HistNat*, 25, 33, 53–54.

- The essential nature of things, the way nature is organized, and its purpose, cannot be altered by any natural process from the way it was designed by IEHI at the moment of the Creation.
- Nature was designed and created to serve humanity; therefore utility is inherent in all things.
- Nature depends on the ever-present spirit ELOHIM to initiate movement and thus begin the divinely designed cycle or GHOLAM of things.
- All things in nature are made of the same primal matter: MAIM.
- God wants humanity to learn the nature of things; therefore humans are capable of knowing what they need to know about nature in order to live in the world. The careful observation of nature can inform knowledge of the nature of things, since everything has a number, weight, and measure.

The judicious application of this Montanian hermeneutics of nature informed both the exegetical exercise of deciphering the arcane language of the Bible and the way observations and experience would be interpreted in the Montanian natural philosophy. Since nature was understood as having a purpose (to serve humanity) that could be realized only by humanity (through its domination), Arias Montano understood the natural world through a teleological lens that asked, What is it for? Why is it the way it is? His hermeneutics brought to the interpretation of nature a utilitarian dimension and, by necessity, a reliance on observation, the technical arts, and history. This hermeneutics also supported Arias Montano's notion that God had revealed in the Bible most of what humanity needed to know about nature. Therefore it was reasonable to conclude that the Bible also contained information about useful things such as plant and animal taxonomy, the composition of the heavens and how it affects life on earth, and even the composition of matter.

For Arias Montano the book of nature was written during the Creation, only once and very clearly: humanity had just lost the ability to read it effortlessly to uncover the true nature of things. As James J. Bono explains with regard to the broader context of early modern natural theology, recapturing this knowledge took a process of exegesis that entailed reading both the Word of God and the book of nature and called for a variety of hermeneutic strategies, including some that viewed nature as a highly symbolic and poetic text.[14] In contrast to these approaches, the Montanian hermeneutics of nature did not posit a symbolic or associative relation between different natures of things as signatures God had left for humans to unravel. There is no recourse to doctrines of

14. Bono, *Word of God*, 73–84.

signatures or magical associations, so common in early modern natural philo-sophical proposals. Instead, it presumed a lawlike order in nature—persistent and entirely contingent on God's design.

By the time Arias Montano wrote the *Magnum opus* his hyperliteral exeget-ical methodology served as the first rung on a steep ladder of interpretation. The meanings of words were now dependent not only on their textual and his-torical context, but also on the metaphysical level they were associated with. In other words, he deliberately let his hermeneutics of nature inform his exegesis. Take, for example, his explanation of GHOLAM (עוֹלָם, *ʿôlām* or *saeculum*), a key term discussed in the previous chapter. He relied on it to assert his determinis-tic view of nature. He noted that it had been translated into Latin as "eternity," "eternal," "perpetual," and "perpetuity." To augment this translation Arias Montano first resorted to the etymology of the word. He explained that its root is CHALAM (צָלַל, *ṣalal*), meaning *abscondere, celare, velare, operire*, and *opertum redere*. These suggested to him a process through time that involved uncovering something previously hidden. Thus whenever GHOLAM was used in the Bible it suggested the trajectory (*decursus*) of a thing through its "pro-gression, transit, final consummation, end and revolutions."[15] However, the appropriate meaning of the word also depended on its context and the nature of the thing. For example, when referring to God, GHOLAM meant "eternity" and a going forth without end. If it was used in relation to the sphere of the heavens it described its steady progression, while in relation to animals or the history of humanity it meant the time during which their life or story unfolded or, in Latin, *aetas*. Therefore GHOLAM stood for the duration of a thing while its purpose unfolds according to its nature as designed at the Creation.[16]

God as "Center," the World as a Sphere

Montanian natural philosophy was also open to applying interpretive tools beyond lexical exercises and empirical observations. The tools of mathemat-ics and geometry could help elucidate not only the nature of physical things (since everything was made according to number, weight, and measure), but

15. "Progressiones, evasiones, ac demum consummationes, fines vel revolutions.": *Corpus*, 164; *HistNat*, 269.

16. "Itaque praescriptam unicuique vel naturae, vel rei, ac divino consilio decretam tum ex-ercitationem et ministrationem: tum etiam durationem, seculum sive GHOLAM dicemus, eritque seculum praescripta unicuique rerum decursio ad consummationem usque, hoc est, ad per-fectum ac praestitutum proprium munus, unde fiet, ut Dei seculum nullum habeat finem.": *Corpus*, 164; *HistNat*, 270.

also the nature of more abstract things such as God. Even explanations about these types of things were contained in and followed from the precepts of "this our discipline" (*nostra haec disciplina*).[17] As we will see below, he chose to illustrate this in an exercise that establishes analogies between the nature of God and the geometric center of an infinite space. The mathematical analogy then allowed him to elucidate the nature of God and explain how he could be said to permeate the world.

Key to his reasoning was the notion of a geometric center as the indivisible midpoint of what it is a center of—say, the center of a surface. A center is also axiomatically immobile; thus it is also the center of any forces emanating from it.[18] Therefore one should first seek to understand the center of something, because that is where the nature of the thing—its virtue and its capacity—reside and radiate from. Arias Montano then suggested we imagine a center of a spirit or soul (*spiritus vel anima*). The center, because of its geometrical properties, transmits its very nature to the rest of the body the way lines radiating from a center traverse the surface they define. Should the center or spirit cease to exist, so does the body. Furthermore, Arias Montano explained, if nature were infinite, we could just as well say that it was all center.[19]

In the geometrical properties of centers, Arias Montano found a suitable analogy for God. Since God is infinite, immobile, incorporeal, and unable to be contained, one could imagine him as a sphere of infinite size by virtue of God's being all center.[20] Furthermore, by this virtue God permeates all things, also being the center of all things.[21] Given that the world is the corporeal man-

17. *Corpus*, 158; *HistNat*, 263.

18. "Est itaque centrum indivisum quiddam mediam obtinens eius cuius est vim, eandem-que immotam, in qua, vel circa quam motus omnis sit, et quies firmitudoque consistit. Ex his consequitur amoto centro moveri aut agi, vel agere nihil posse, nihilque quiescere quàm diu suum centrum, cui innitatur, non repererit.": *Corpus*, 160; *HistNat*, 265.

19. *Corpus*, 161; *HistNat*, 266.

20. *Corpus*, 168; *HistNat*, 273. Seneca expressed a similar notion in letter 113 to Lucilius: Seneca, *Ad Lucilium epistulae morales*, 3:292. This idea reappears embedded in the second sentence in the medieval *Liber XXIV philosophorum* attributed to Marius Victorinus (ca. fourth century AD): "Deus est sphaera infinita cuius centrum est ubique, circumferentia vero nusquam.": Hudry, *Livre des vingt-quatre philosophes*, 35–42. The metaphor of God as center of an infinite sphere was further developed by Nicholas of Cusa in *De docta ingnorantia* (book 1, 23): Hopkins, *Nicholas of Cusa on Learned Ignorance*, 78–79. Also on the subject see Brient, "How Can the Infinite Be the Measure of the Finite?," 210–25.

21. "Hoc est, omnis virtute sua et efficacitate ubique constet, nullusque vel cogitando locus existimari possit, in quo infiniti huius, de quo loquimur, spiritus centrum non sit.": *Corpus*, 163; *HistNat*, 268.

ifestation of all the properties of God, albeit only a shadow of the divine reality, by analogy its creator must also be spherical.[22] Furthermore, Arias Montano noted, the senses tell us that large things in the world are spherical, as are earth, air, fire, water, and all things liquid (as is clear from observing droplets). Smaller things on earth, such as trees and flowers, also tend to have rounded shapes. Thus, things we cannot perceive through our senses, but that we understand through intellect and reason—especially if we pay attention to their forces, composition, movements, and efficacies—are also likely to be spherical or to move in this manner.[23] Therefore only a sphere could describe the shape of the *divinum spacium* that contains everything created and is the most beautiful shape.

Arias Montano's God/center analogy is a prime example of how his hermeneutics informed his exegetical approach whether or not it involved the etymology of Hebrew words. According to him, interpreting biblical language required that all possible meanings of a term be in concert with the nature of the thing. Thus a center means one thing mathematically, but when its attributes—the nature of a center—happen to coincide with the nature of God, they also completely explain his nature.[24]

MONTANIAN CREATION

Arias Montano's cosmological scheme was a direct consequence of his historical reading of Genesis and his hermeneutics of nature. The historical sequence of the events of the Creation reaffirmed his historical premises and revealed the metaphysical principles; these same principles were also active agents of the Creation and determined the sequence of events. The circularity of his reasoning seems to have completely escaped him, clearly owing to the certainty he ascribed to the Sacred Scriptures as an infallible starting point for any kind of philosophical reasoning.

Arias Montano understood the Creation as being ex nihilo, with the spirit ELOHIM and MAIM as the first two things that came into existence through the *verbum Dei* IEHI. He explained that in the beginning there existed a shapeless mass of heaven and earth so that none of the things we now recognize in the

---

22. *Corpus*, 159; *HistNat*, 264.

23. *Corpus*, 171; *HistNat*, 277.

24. "Ex his verò consequitur Deum infinitum spatium, infinitam sphaeram, infinitumque centrum ese, atque omnium horum nominum non modò notiones, verùm rationes, vires, efficacitésque unicè ac singulariter omni termino semoto possidere.": *Corpus*, 168; *HistNat*, 273.

world were visible. Nothing could come to be until this mass broke apart. Moses described this in Genesis 1:4 as the separation of light from darkness. This is why light was the first species created through the Word of God IEHI, so that the rest of nature could be seen and its parts distinguished. This act also established, in Arias Montano's view, the means through which God would communicate with humanity—as a light in man's heart. On the second day God ordered the distinction and disunion of bodies when MAIM, the primordial dual waters, agitated by ELOHIM's action, were sent to occupy two different places.[25] The lighter and purer waters went to occupy the superior region, and the less pure, heavier part, better suited to solidification, occupied the inferior region. These heavier waters formed the center of the space where creation would take place (*fabricae spacium*). Arias Montano explained that this command was necessary because in the beginning MAIM had existed as a compound fluid, with the oily part of its nature mixed with the salty part. Subsequently they had been filtered (*excolati*) into two parts because they had different weights (*moles*).[26]

The two liquids were (and are) kept separated by an intermediate body, the result of the first evaporation (*evaporatio prima*). This intermediate body was lighter and more permeable than the inferior bodies but less pure than the superior ones. God called this intermediate region RAKIAGH (רְקִיעַ, *rāqiyaʿ* or firmament), a word that for Arias Montano denoted a region of contraction and expansion (*fisio et expansio*) that functions like a sieve, allowing bodies in the superior region occupied by the more tenuous MAIM to communicate with the inferior ones without permitting the effluvia of the inferior bodies to contaminate the superior ones.[27] It is where regular tempests and mutations originate, and thus it initiates and perpetuates the changes and cycles that happen on the earth below. It is well suited to gather fluids and return them to their proper parts and can also endure changes in color, smell, taste, and temperature. The region occupied by this liquid and the more tenuous ones above sometimes share the name SAMAIIM (שָׁמַיִם, *šāmayim* or *caelum*) in the Bible. As Arias Montano explained, the Hebrew etymology of SAMAIIM implied separation

25. *Corpus*, 152; *HistNat*, 255.

26. *Corpus*, 152; *HistNat*, 255.

27. Here we have echoes of Saint Basil's interpretation of the firmament. It is also interesting that the notion of the RAKIAGH as separating a three-part heaven appears in chapter 4.3 of the *Midrash rabbah*: Freedman and Simon, *Midrash rabbah*, 1:28. Arias Montano also follows interpretations by Ibn Ezra and Christian writers who suggested that the term denoted something stretched or hammered out: Zlotowitz and Scherman, *Bereishis/Genesis*, 1:45, and Ibn Ezra, *Ibn Ezra's Commentary on the Pentateuch*, 1:34.

of two bodies, noting that the ending indicated a dual number. In fact, the Sacred Scriptures sometimes referred to the more tenuous liquids above by the compound name SEMEI SAMAIIM, or *caeli caelorum* (heavens of heavens).[28] All those differentiated natures learned to preserve immediately and forever the rules and norms (*ratio et norma*) they had been given when created. Arias Montano sought clues in the Creation narrative for the subsequent behavior of these celestial regions and took these to be natural laws established by God at the Creation. For example, the separation of the MAIM established that lighter things would always move upward and away from denser things. The Creation narrative clearly described this type of motion, and thus it stood in the Montanian natural philosophy as an established axiom or law of nature. That this coincided with Aristotle's doctrine of natural place might have been perceived by some readers as a happy coincidence, but Arias Montano does not mention the coincidence. Arias Montano returned to these axioms as he developed the finer points of his natural philosophy.

From the heavier (*crassiores*)[29] liquids of the MAIM, two other "bodies" of distinct species were created; because of its weight (*pondus*), the one that was denser and harder had come to occupy the lower space while the other body, more fluid and softer, could more easily be divided, enabling it to flow. (At this point Arias Montano began to refer to the heavier MAIM as bodies with mass.) Because of their weight, they occupied the middle space among all bodies — like the yolk of an egg, Arias Montano explained. The oily, heavier liquids came together to make up land, and the lighter, saltier ones became terrestrial water. Separated yet held together, they formed one body, large and round — like a sphere or an apple, he wrote — with higher, drier elevations prepared

28. *Corpus*, 152, 179–80; *HistNat*, 255, 285. His analysis of the word "heaven" or SAMAIIM differed from interpretations that appear in chapter 4.7 of the *Midrash rabbah*, where the word is explained in several ways: as denoting something compounded from *esh* (אֵשׁ, *ʾēš*, fire) and *mayim* (מַיִם, *mayim*, water), as water or as *sammim* or chemicals of different colors, and finally, as *samayim*, or "laden with water" and congealed: Freedman and Simon, *Midrash rabbah*, 1:32–33. For Maimonides, שָׁמַיִם (*šāmayim*) was a compound word that meant "the [new] name of water" because the waters, now having acquired a new position, merited a new name. Rashi (Shlomo Yitzchaki, 1040–1105) lists "the waters are there" as one of the possible interpretations: Zlotowitz and Scherman, *Bereishis/Genesis*, 1:49. Here, however, I believe Arias Montano was following Pagnino/Kimhi, since in the *Apparatus* of the *Biblia Regia* we read only the explanation that the word is composed from *sham* (there) and *maim* (water): Raphelengius, "Thesauri hebraicae linguae," 6:133.

29. The Spanish translation prefers *fértil* (fertile) as the definition of *crasso*. Here, however, Arias Montano was discussing the arrangement of the sphere of earth and water and thus referring to the heaviness or solidity of the matter that composed these spheres, not their fecundity.

and adorned by God to sustain humanity.[30] The words of the Hebrew Bible in Genesis 1:9–10 indicated the constitution of this sphere. The arid part was initially made of pure salt and called IABASSA (יַבָּשָׁה, *yabbāšâ, arida*), but once it came in contact with the fertile fluids, it became ARETS (אֶרֶץ, *'ereṣ, terra*), or earth, a welcoming place for living things. Arias Montano found this change very meaningful, since the word ARETS denoted a "masculine force" as well as efficacy and strength, firmness and immobility, and above all fecundity.

The third day did not come to a close until a verdant landscape gave testimony to the fertility of this ARETS. For Arias Montano the sequence of events at this point of the creation narrative is particularly important. He noted that this first generation of plants came to be solely by the power of the divine Word, without assistance from "the heavens," since these had not yet been created. Furthermore, God had spoken here in a tense that suggested to Arias Montano that the earth accepted this command and was to maintain it in perpetuity.[31] He did not explain further this aspect of his exegesis. The quotation from Genesis 1:10–13 that appears in the *History of Nature* on which he based his interpretation uses the verb "to germinate" in the future active indicative (*germinabit*), which is, in fact, in concert with the Hebrew Bible (תַּדְשֵׁא, *tadĕšē'*, the leading *tav* indicating the future tense). The Vulgate and the Pagnino version included in the Antwerp Polyglot, however, translate the word in the present active subjunctive (*germinet*). This kind of unacknowledged switch would have been noticed by an attentive reader, especially given the unconventional interpretation that resulted from this translation.

This fruitful earth needed only more adequate shelter (*tectior*) and inhabitants. A properly disposed heaven would provide this shelter, keep the earth temperate, and maintain the cycles needed for the maturation of the fruits of the earth necessary to nourish humans. Thus, on the fourth day the celestial bodies (*astra et sidera*) came to be, not on their own (*sponte*), but through the agency of the same divine verb IEHI and with the same nature as the rest of world. The continued agitation by ELOHIM in the region occupied by the more rarefied MAIM caused the salty part of MAIM to break into like parts (*congeneres*) and form the stars. Each of these then occupied its proper zone

---

30. *Corpus*, 153; *HistNat*, 255.

31. "Atque in hac prima plantarum editione nihil à caelis terram mutuasse aut accepisse, sed quidquid edidit ex verbi divini iussu futuri temporis forma enunciato: ideóque constanter et perpetuum retinendo mandato accepit et reddidit. 'Et vidit Deus quòd esset bonum, et ait: Germinabit terra herbam virentem, et facientem semen: et lignum pomiferum faciens fructum iuxta genus suum: cuius semen in semetipso sit super terram. Et factus est ita. Et vidit Deus quòd esset bonum. Et factum est vespere et mane dies tertius.": *Corpus*, 153; *HistNat*, 256.

in the heavens, all having obtained their proper species and forms. Since they were made of purer MAIM, they remained largely impervious to change or temporal decay.

The stars were assigned their duties: to light the day and illuminate the night, and to keep their precise courses so that time could be kept and the seasons tracked. The sun would rule the day, and the moon would light the night and provide various humors and exhalations (*humores adspirationèsque suppeditare variae*). The duties of the stars, sun, and moon are to generate the conditions that perpetuate the cycle of generation and decay on the earth. The earth needs to be subjected to hot, cold, and temperate weather; to cooling, expansion, and contraction; and to dryness and humidity. It also needs time as defined by the cycles of day and night. These actions and their timing were the final purposes of the celestial bodies and of the RAKIAGH. Arias Montano summarized the relationship as follows: "The author of all these actions was the word IEHI. The spirit ELOHIM was indeed the designer, but the times and the bringing together of matter created for this were two notable ministers."[32] One minister was the tenacious sun, which provided its own light and heat; the other was the capacious but changeable moon. They were joined by several other helpers (*administri*) that, while always obeying the duties assigned to them at the Creation, created the increments and changes (*incrementa ac mutationes*) over the land and sea with their rising and setting. At this point in the commentary, Arias Montano pointed out that no other duties or offices were given to the stars in the Genesis account, in what was clearly his attempt to preclude any other astral duties that might suggest a justification for a judiciary astrology that the Catholic authorities did not consider legitimate.

A world so perfectly disposed, with a fruitful earth sheltered by heavens that communicated the opportune times and climate changes needed for life, still needed permanent inhabitants. Thus, on the fifth day God populated the two regions of the air and water with various species and multiple forms of life and sentient beings. Arias Montano observed that the origin of all these living things was the fertile waters that surrounded the land. There the heaviest bodies, which were oily and milky (*pinguis et lacteus*), condensed to form bones, meat, feathers, and such. The first creatures were created to inhabit the first two regions delimited by the second separation of the heavier MAIM; the fish would inhabit the inferior and heavier waters and the birds the lighter

---

32. "Quarum omnium actionum auctor idem verbum IEHI fuit. Formarum verò concinnator spiritus ELOHIM, oportunitatum autem et materiae condiendae ad eam rem creati, praecipuè insignes ministri duo.": *Corpus*, 153; *HistNat*, 255.

waters that make up the firmament. These were followed on the sixth day by the inhabitants of the land, created and ordered into several groups (*multiplex genus*): BEHEMA, REMES, HHAIAH (cattle), creeping things, and beasts of the earth (*jumenta et reptilia et bestiae terrae*) (Gen. 1:24 VUL). Arias Montano noted that among all these species there was not one that lacked a singular nature, approved by God and with a specific use for humanity. Even those of the most inferior nature had their own names and some utility, so that even the lowly reptile gave testament to the perfect harmony of nature.

Finally, as the culmination of the act of creation, on the sixth day God created man, so that the animals would recognize him as their powerful master and prince. His spot in the order of creation also dictated that man would behave as the supreme commander of the rest of nature and fulfill his role as the cause for the creation of the world. Arias Montano justified this view by pointing to the distinct words and actions God used when creating man (Gen. 2:7). Man, for example, was not formed out of the common earth ARETS from which the animals had sprung; man was made from special earth, GHAPHAR ADAMAH (עָפָר אֲדָמָה, *āpār ʾădāmâ*), the most flexible, fertile, and vivifying soil.[33] This primacy was implied by the nature of the dust used to create him; GHAPHAR (עָפָר, *ʿāpār, pulvis*) is the purest and cleanest thing that can be separated from other substances and is suitable for engendering and shaping the natures of other things. In the *De arcano sermone*, Arias Montano defined GHAPHAR as a type of fine, pure matter owing to its weight, gravity, and inertness.[34] Yet this GHAPHAR dust is lively, has heat, and is barely corruptible; thus it can proffer a suitable proportion of matter and juices to beget and give form to other natures. He explained that the name GHOPHRITH (גָּפְרִית, *gāprît*) "is derived from this source and means the 'juice' of that common earth called ARETS and so is the first principle of all metals and minerals to which our philosophers gave the name *sulphur vivum*."[35] So just as sulfur enlivens the earth, the prin-

---

33. "Pinguissima ac mollissima, et maximè vitali humi selecta portione constaret": *Corpus*, 156; *HistNat*, 258.

34. "Cuius rationes tantum ex pondere et gravitate atque inertia petuntur, appelari solet עָפָר pulvis.": Arias Montano, *Libro de José*, 122, 419, [11]. See also *Corpus*, 350–51; *HistNat*, 467.

35. "Est autem GHAPHAR in quod in terra purissimum defecatissimumque secerni, et ad naturas alias rerum gignendas et informandas aptae, oprtunaeque materiae succos et commoditates praebere potest. Ex quo fonte nomen GHOPHRITH deducitur, quòd in illa communi terra ARETS dicta, succum significat metallorum ac mineralium omnium primum caput, cui Sulphur vivum nostrates philosophi nomen fecerunt. Ut igitur in ARETS natura primam principii laudem GHOPHRITH, ita in ADAMAH, illud quod egregiè secretum GHAPHAR dicitur, ad bonae frugis copiam singulariter commendatur.": *Corpus*, 156; *HistNat*, 258–59.

ciple GHAPHAR supplies ADAMAH with fruitfulness. Arias Montano found the etymological origins of these cognates fascinating in what they could reveal about the nature of man; he advised that those inclined toward philosophizing about the relation between human flesh, things, and their names could use this interpretation of the language to advance wonderful arguments about the nature of man.

God also fashioned man in two parts—a heavy, dense part or body, and another part lacking any density or mass and called soul. This two-part constitution imitated the constitution of the world. Just as the world has an inferior and a superior nature, earth and heaven, so man has a part that can be seen and touched and another, superior part "that cannot be grasped by the senses and the vision."[36] This superior part was also of two parts, a superior one called RUAHH or "spirit" (רוּחַ, *rûaḥ, spiritus*) and an inferior one called NEPHES or "soul" (נֶפֶשׁ, *nepeš, anima*). This permits man to know and command those things whose nature he partly shares, but also to know divine things privy only to spirit. Yet man was also given a superior gift that allows him to comprehend what he perceives by the senses. It is a capacity to "understand and know" that comes by virtue of having partaken of prelapsarian divine light. In Arias Montano's version of the creation story only God rested on the seventh day. The rest of the world came to maturity and completed its development, so that it would continue in perpetuity to carry out the duties and functions assigned at the Creation.

THE COSMOS

The *History of Nature* presented the completely realized cosmological vision Arias Montano had hinted at in the *De arcano sermone*. His account set forth the general parameters for a comprehensive and non-Aristotelian/Ptolemaic cosmology. It posited a common material nature between the heavens and the earth, did away with the notion of an unbreachable boundary between the different parts of the heavens and between the heavens and the earth, reduced matter to a single liquid with a dual nature, and imagined an orderly, law-abiding universe unfolding through time according to a divine plan set forth at the Creation while also existing within the vivifying spirit ELOHIM. After presenting the worldview that resulted from his interpretation of the Creation narrative, Arias Montano moved on to describe the nature of things. Here he would discuss everything from the structure of the heavens and the commu-

---

36. "Quae visu ac sensu comprehendi non posset.": *Anima*, 29; *LibGen*, 129.

nion between the heavens and earth down to the lowliest of plants. The first matter he settled, however, was the shape of the world.

The Structure of the Heavens

Arias Montano's commitment to the spherical figure continued with his consideration of the shape of the heavens, SAMAIIM. His reasoning wove together theological, philosophical, geometrical, and empirical arguments. For example, he combined the Platonic dictum that heaven is a sphere because it is the most fitting figure for the majesty of space with the empirical notion that this shape explains the movements observed in celestial bodies. These observed celestial motions then led him to infer that heaven is immense, since it must accommodate the trajectory of all that moves within it. But just how big is it? Arias Montano explained that this immensity—although beyond what any human can measure—must have a limit, because everything God created in the world was made according to a "measure."[37] God, Arias Montano pointed out, is in no way bound by heaven but permeates everything in the world.[38] This last observation was clearly aimed at those who continued to maintain the medieval conception of heaven as composed of neatly nested crystalline spheres, the outermost being the one where God "resided." This immense yet bound heaven is the space that contains all the bodies from the surface of the sphere of earth and water through the outer limit of the circular world, encompassing everything endowed with weight, body, and size (*molis, corpus, mensuraque*).[39] It stretches over the earth like a liquid curtain, as Arias Montano found to be unambiguously expressed in the Psalms: "You spread out the heavens like a tent; you raised your palace upon the waters" (Ps. 104:2–3 NAB).

His interpretation of the Genesis narrative and metaphysical principles also led him to conclude that the heavens were incorruptible by virtue of the material they were composed of: the most rarefied MAIM. Although they shared a common nature with the MAIM that later formed the earth, their tenuous nature made them impervious to change or defect caused by passing time. This unchanging nature allows the heavens to follow the order assigned at the Creation and to continue doing so as long as it is God's will, through the

---

37. *Corpus*, 179; *HistNat*, 285.

38. "Itaque praestantia nomine Deo omnia tenenti, omnia complenti, gubernanti omnia, supremum caelum ut regia sedes tribuitur.": *Corpus*, 178; *HistNat*, 283.

39. *Corpus*, 175; *HistNat*, 280.

power of the word IEHI.[40] By assigning to the heavens a nature dictated by the purest version of MAIM, Arias Montano was able to preserve the ancient notion of the incorruptibility of the heavens without resorting to the presence of an Aristotelian fifth element, aether. SAMAIIM shared the same nature as the rest of creation, although its purity was beyond anything found "below" it. Arias Montano's hermeneutics of nature also explained the purpose of the heavens. God had arranged the heavens with "precise orders and balanced laws" to provide the times, functions, and uses of things through an ingenious communication of songs, virtues, and spirits.[41] This was the basis for his understanding of the extent and purpose of astrological influences. I will return to this topic later.

The word heaven, especially in the phrase "heaven of heavens," suggested to Arias Montano that the celestial region comprised at least three distinct parts, each defined by its purpose and the nature of the bodies or spirits inhabiting it. He was unambiguous on this point in his later commentary on Ecclesiastes, proclaiming that "if we think about it, the machine of the world is disposed into three parts."[42] The first part of the celestial realm, called the firmament, extends from the surface of the sphere of the earth and water up to where the superior bodies begin. It is made up of a rarefied matter or RAKIAGH capable of transmitting the influences of the bodies in the heaven above it to the bodies below it. The second region of the heavens extends from the firmament to the farthest stars and is in turn divided into the regions of the celestial bodies, the attending bodies or planets (*ministrantes*), and militant bodies (*militiae*) or fixed stars. Beyond these is a third region of heaven or *tertium caelum* inhabited by the spirits of holy men; it is immobile and without stars. The earth firmly seated below the RAKIAGH remains fixed and unmoved at the center of the world, surrounded by the majestic display of the heavens. The three-part world he described, as we saw earlier, was an unconventional but not unheard-of cosmological interpretation of the Genesis narrative.[43]

40. "Natura autem caelorum quod illis liquoribus puris ac defecatis constat, purissima nitidissimaque est, et pellucida omnis: atque hac ratione nulli mutationi, turbationi, vetustatis aut residentiae vitio nullis rubigini obnoxia, sed ut ab initio dispositis partibus ac regionibus extitit: ita etiam perpetuo constat, constabitque omnis distributasque vires iisdem illis locis quibus primum accepit: ex conditoris voluntate verbique IEHI iussu conservabit.": *Corpus*, 179; *HistNat*, 285.

41. *Corpus*, 181; *HistNat*, 286.

42. Arias Montano, *Discursos sobre el Eclesiastés de Salomón*, 94.

43. See the discussion of Saint Basil and Agostino Steuco in chapter 2.

The RAKIAGH's place within the uppermost part of the first region, acting as a barrier between heaven and earth, suggested to Arias Montano that God could alter the channels (*viae*) through which the influences could travel between the upper and lower realms. This is the case of the *cataractae caeli*, or "cascades from heaven," that God had on occasion "opened," for example, to shower the earth with fire and brimstone in the case of Sodom and Gomorrah or with water in Noah's flood. Two proverbs testify to the nature of the RAKIAGH. "The LORD by wisdom founded the earth, established the heavens by understanding" (Prov. 3:19 NAB) meant that the heavens were established according to certain "arrangement, nature, and laws," while "When he established the heavens I was there, When he marked out the vault over the face of the deep, When he made firm the skies above" (Prov. 8:27–28 NAB)[44] illustrates the RAKIAGH's "series, weight, and mission."[45]

The lower part of the second region of the heavens, adjacent to the RAKIAGH, is where rain, ice, and lightning originate. Arias Montano explained that the Bible refers to it as SAHHAKIM (שְׁחָקִים, *šĕḥāqîm*), but the insufficiency of the Latin language had forced translators to use as synonyms inadequate words such as clouds, heaven, ether, or air.[46] Above it and within the second region is the home of the planets, the seven wandering stars. Just beyond this region was the home of the fixed stars, whose positions had been assigned at the Creation and that kept to their stations like soldiers (TSEBAOTH or *militiae*). Both parts of the second heaven move by their own will and complete their trajectory in a day, in the manner of a wheel, carrying with them the stars.[47] (This is the equivalent of the Ptolemaic daily motion.) There are other orderly movements in the heavens, such as those of the sun ("365 and almost a quarter of a day"), Moon (29 days), and all the way to the planets with the longest circuits, so that they eventually return to the same spot they occupied at the Creation. Celestial motion can be easily observed and predicted with a bit of mathematics by anyone not too stupid, given that God designed the celestial bodies according to "number, weight, and measure" so that their motion could be understood, Arias Montano observed.

44. The Vulgate version of Proverbs 8:27–28 evokes Arias Montano's interpretation much better that the English translation: "Quando praeparabat caelos aderam quando certa lege et gyro vallabat abyssos: quando aethera firmabat sursum et librabat fontes aquarum."

45. "Caelorum superiorum institutionem naturam, et leges priori versu expansionis, sive RAKIAGH, sive firmamentu situm, seriem, pondus, vel officium posteriore indicavit sapiens.": *Corpus*, 178; *HistNat*, 284.

46. *Corpus*, 180; *HistNat*, 285.

47. *Corpus*, 191; *HistNat*, 297.

Arias Montano noted that the ancients noticed that stars maintained the same distance and aspect relation to each other. This led them to believe that the stars in the highest place were fixed and immobile and that the heavens themselves moved separately from the parts below.[48] (He was referring here to the fixed sphere of the stars that, depending on the cosmological system in question, could be the eighth or ninth crystalline sphere.) He found no need to maintain this distinction between the different parts of the heavens, since their very nature, composed as they were of a very light and tenuous (*tenuissimum ac purissimum*) body, could endure the transit of celestial bodies. To explain this he resorted to the well-known contrast between the movements of the stars and those of fish in water: "Stars themselves are moved in their regions, not like fish in the sea or as in the air all the species of animals with wings, but rather keeping their stations and its laws and progressions in perpetuity."[49] He found no biblical justification or natural need for the complicated system of nested crystalline orbs. Instead he found in the Psalms the justification he needed for an orderly self-movement of the stars: "He assigned them duties forever, gave them tasks that will never change" (Ps. 148:6 NAB). He observed that each star or planet follows two movements, one assigned to the heavens— moving all they encompass in a daily motion—and another that pertains to and is characteristic of each celestial body. Therefore stars and planets move in the heavens not like fish in the sea or birds in the sky—by which he meant not randomly—but rather by keeping "stations, laws, and progressions" in perpetually assigned orbits.[50] However, just as a fish swimming through the water does not cause any change to the water, neither does the transit of celestial bodies affect the heavens. To Arias Montano this sufficed to explain all the movements observed in the heavens without resorting to the opinion of wise pagans who assigned single or multiple orbs to carry each planet.[51] In his

48. "Quamobrem gentium aliarum sapientibus fixae et immobiles putantur esse superiori in loco stellae, et caelum ipsum ab inferiori distinctum, moveri creditum.": *Corpus*, 191–92; *HistNat*, 298.

49. "Moventur verò astra ipsa suis in regionibus, non secus atque in aquis pisces, et in aëre pennigerorum genera omnia: suis tamen stationum intervallis, et progressionum legibus perpetuo obtentis.": *Corpus*, 191; *HistNat*, 298.

50. The idea that the planets traveled in a fluid medium was of biblical origin, as discussed in chapter 3. The notion of fluid orbs was also commonly held among Jewish natural philosophers: see Efron and Fisch, "Astronomical Exegesis," 72–87.

51. "Atque hoc pacto caelis et astris converso motu suos peragentibus gyros, et orbes describentibus ex singulorum praescripto divina sancito lege, aspectus, respectus, comparationes, coniunctiones, oppositiones, et, quas observatas novimus, regressiones ac directiones facilè à

explanation he had resorted to a well-known image of the movement of fish and birds through a fluid medium that Scholastics had used to argue both for and against the self-movement of planets.[52] But he had yet to account for the cause of the motion of the celestial bodies. Here again the principles established at the Creation provided the necessary explanation. The author of the motion was the word IEHI, which had commanded the stars to "be!" moving objects, but it was the agency of the spirit ELOHIM, with its capacity to agitate things, that imparted motion to them.

The Sun, Moon, and Stars

Arias Montano derived the properties of the three principal bodies in heaven—sun, moon, and stars—from the Bible and confirmed them through his own observations. The sun, he explained, was constituted through the agency of ELOHIM from the very pure MAIM. He noted that when describing the events of the fourth day (Gen. 1:16) the Bible uses the Hebrew word MAOR (מָאוֹר, *mā'ôr*, luminary) or, more specifically for the sun, MAORHAGADOR or the large luminary. He noted that the Hebrew etymology of MAOR was VR (אוֹר, *'ôr*, light) and found that it also meant "illuminate," "shine," "inflame," and "heat." Thus MAOR means "light united with heat," and these are observable, intrinsic, and indivisible qualities of the sun. Furthermore, it is given from the first principles that the very tenuous nature of the celestial MAIM ensures that the sun will partake of the very rapid movements associated with penetration and persuasion. Arias Montano explained that this is confirmed by the Bible's use of two other words for the sun: SEMES (שֶׁמֶשׁ, *šemeš*) and HHAMMAH (חַמָּה, *ḥammâ*).[53] The former refers to "the faculty to administer" and is etymologically associated with the word ES (אֵשׁ, *'ēš*), which means fire, while the latter refers to heat as a force (*vis*). The word SEMES, when considered as SEM plus ES (*nomen* or name plus *ignis* or fire), reinforces the notion that the sun was given a name in the divine language that represented its true nature. From this Arias Montano

---

siderum spectatorum peritis collectae ad calculos revocantur: idque longè promptius, quàm si, ut Gentium sapientes opinantur, singulis planetis proprios ac singulares aut multiplices etiam orbes tribuendi sint, et stellis etiam altioribus, quas militiam vocamus, communem orbem adscribere libeat, quo ipsae fixae moveantur proprio, rapiantur verò diurno gyro.": *Corpus*, 193; *HistNat*, 299.

52. The reference to planets perhaps moving like fish in the sea is at least of medieval origin, although the author of the analogy is uncertain: Randles, *Unmaking of the Medieval Christian Cosmos*, 35n15; Grant, *Planets, Stars, and Orbs*, 274n16.

53. *Corpus*, 181–82; *HistNat*, 287–88.

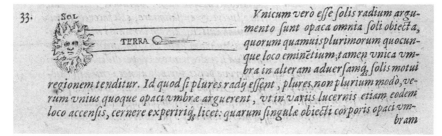

33.

Vnicum verò esse solis radium argumento sunt opaca omnia soli obiecta, quorum quamuisplurimorum quocunque loco eminētium,tamen vnica vmbra in alteram aduersamq̃, solis motui regionem tenditur. Id quod si plures radÿ essent, plures,non plurium modo,verum vnius quoque opaci vmbrae arguerent, vt in varÿs lucernis etiam eodem loco accensis, cernere expeririq̃, licet: quarum singulae obiecti corporis opaci vmbram

F I G U R E   8 . 3 . Rays of the sun. Arias Montano, *Naturae historia*, 182. Biblioteca Histórica de la Universidad Complutense de Madrid (BH FLL 14614).

concluded that the ray of the sun is a concrete form of heat that has the same properties in the heavens as on earth. It is not an accident of the sun but of the same kind as the sun, and thus the ray of sun is essentially the same as the sun. Arias Montano's observations of shadows and sundials confirmed his assertion about the unitary property of the solar ray. Do we not see only one shadow when the sun casts its ray on an object? If there were many distinct rays emanating from the sun, would we not see many shadows, as when many lamps in a room cast their rays on a single object?[54] This single ray is of the same diameter as the sun and encompasses the whole of the inferior orb in the manner of a column or a torrent, delivering to the earth the full substantial qualities of the sun (fig. 8.3). Philological gymnastics and some empirical observations resulted in a natural philosophical explanation for the composition of the solar *radius* reaching the earth. But lest we forget the purpose of Arias Montano's enterprise, he then took this argument that the sun and its ray substantially share their natures to address the divine nature and the mystery of the relation between God the Father and God the Son. It was clear to him that when the prophets referred to God as Sun they were referring to the trinitarian unity of God and employed this metaphor because they knew the relation between the sun and its ray.

The Montanian hermeneutics of nature also determined other properties of the sun. For example, since the sun is made of the oily liquid of the primordial MAIM, it can conserve its capacity to heat without being consumed or diminishing in virtue. This gives it an eternal quality that keeps the sun in its GHOLAM and as one body. As we saw earlier from Arias Montano's interpretation of these two words, this means that the sun carries on the duty prescribed at the Creation that permeates its body. What other virtues or powers were

54. *Corpus*, 182–83; *HistNat*, 288.

assigned to the sun? He returned to the etymology of SEMES (שֶׁמֶשׁ) and found that its three Hebrew consonants are the same as in SAMAS (שְׁמַשׁ, *šĕmaš*), an Aramaic word that means *ministrare*. This additional meaning suggested that the sun is also a minister of nature—that is, it was instructed by IEHI to act on other bodies that are capable of being illuminated and heated and that can in turn heat other bodies.[55]

Arias Montano carried out a similar philological/empirical exercise with the moon. The moon is made of the same MAIM that makes up the heavens, but with a "different appearance of parts and elements" (*diversa partium et elementorum habitudo*).[56] The Bible refers to it first as the small star, MAOR KATON (הַמָּאוֹר הַקָּטֹן, *hammā'ôr haqqāṭōn*), but also as IAREAHH (יָרֵחַ, *yārēaḥ*), a word whose root Arias Montano believed to be RAVAHH (possibly רָבָה, *rābâ*),[57] meaning "dilate," "aspirate," or "expend wind and spirit." Because the word is in the future tense, it suggests that the effects of the moon are changeable and will be different in the future. Therefore the moon must also be made of a very pure substance, but mostly of the salty dual liquid. This is why the moon does not shine with its own light and does not share the sun's fiery nature. Arias Montano remarked that it serves the balance of nature to have another luminary with a nature complementary to the sun's. However, because of its nature and its proximity to the sun, it acts like a glass full of clear water by shining, warming, inflaming in size (waxing and waning), and getting hot.

Finally, Arias Montano found that the rest of the stars were incubated from the dual liquid by the spirit ELOHIM. In a telling analogy, he explained to readers that he imagined this process to be like one he had observed among Belgians and Batavians, who turn churned milk into very different products: cheese, butter, "sur melch," and whey, creating from one liquid different products of the same genus and even the same name, but with different proportions, natures, forces, and uses.[58] He believed the process of the Creation had operated in a similar way. In response to God's command, the continuous agitation caused by ELOHIM broke the salty liquid into similar parts and propelled them into their proper places. These parts—now stars—then obtained their proper species and forms, with each occupying its proper place.

55. *Corpus*, 183–84; *HistNat*, 289–90.

56. *Corpus*, 185; *HistNat*, 291.

57. Arias Montano's source for this root is unclear. It does not appear in the Pagnino/Kimhi thesaurus. Gesenius explains that the root of יָרֵחַ (*yārēaḥ*) or "moon" is יֶרַח (yeraḥ), which means "month."

58. *Corpus*, 186; *HistNat*, 292.

What more does scripture reveal about the nature of stars? In Genesis 2:1 the stars are referred to as TSEBA, in Latin *ornatus* or *militia*. Arias Montano was unhappy with this translation. The Hebrew word could also refer to the actions that resulted when things or persons carried out the duties they were assigned, such as soldiers in war and sailors on ships.[59] This suggested to him that stars must also have duties. What, then, is the duty of a star? Arias Montano explained that in the arcane language celestial bodies are sometimes referred to as CHOCHABIM (כּוֹכָבִים, *kôkābîm*), a word ripe for etymological and phonetic analysis. He first deconstructed the word into two parts, CHOAHH and CHABIM, noting that the second part, CHABIM, comes from KABAH (actually, קָבַל, *qābal*), meaning "to take," and KAB (קָב, *qāb*), meaning "receptacle." The first part of the word he interpreted as CHOAHH (כֹּחַ, *kōaḥ*), explaining that phonetically the final *het* (ה) of that word was typically aspirated when next to *kaf* (כ), so that over the years the ה had been "lost" in the spelling of the word. CHOAHH (כֹּחַ) means *vis, virtus, potestas*. Thus CHOCHABIM, the biblical name for star, stood for something that was a receptacle and had a "capacity of virtue."[60] Therefore all celestial bodies, separated into their classes, must contain and dispense their own virtue, power, and efficacy. This reasoning supplied the biblical basis for his astrological beliefs.

THE QUESTION OF CHANGE

Unity and lawlike constancy were the principal attributes of the Montanian cosmos, but Arias Montano had yet to account for change, mutation, and re-generation of kind (e.g., when an animal dies but leaves behind progeny of the same genus). The answer to the question had to account for the orderly perpetuation of the world while not resorting to any kind of special creation, given the finality he read into the Genesis account. He began by noting that not all things created at the beginning were to remain unchanged; most things under the heavens are subject to death and decay according to their natures. Some suffer less change than others; the earth and sea seem to be exempted (although mutations have been observed in the sea). The answer was found in a process of "addition" (*suffectio*) that God had instituted at Creation. Nothing new needed to appear on earth beyond what had come into being during Crea-

---

59. "Convenit autem nomen TSEBA rebus ac personis quae vel consilio ac ratione, vel sorte, vel etiam lege alicui rei augendae fuerint destinatae.": *Corpus*, 187; *HistNat*, 293.

60. *Corpus*, 188–89; *HistNat*, 294–95.

tion; it just needed to be restored. This restorative process took place under the ministry of ELOHIM and an army of spirits working through the forces and liquids contained in the "receptacle" of the stars. These spirits continuously move and agitate the surface of the liquids they have been assigned to, differentiating them and delivering them to their assigned places below, where they can "continue, perpetuate, and establish" things.[61] Realizing, perhaps, that he was wandering far from what an audience trained in Scholastic natural philosophy would understand, Arias Montano conceded that what was produced by the separation of the liquids might be better understood in Latin as a separation of orders into many distinct forms.[62]

Relying on Psalm 146:4 (VUL), "He numbers the stars, and gives to all of them their names" (Ps. 147:4 NAB), Arias Montano argued that each star has a distinct function and follows the order prescribed to it by God. Therefore it is not licit to think that some stars lack purpose or that they can convey forces different from the order assigned to each. This is because stars are receptacles of everything that exists in nature, separated into their respective species and distinct forms.[63] The stars keep within them pure versions of the liquids that constitute all things and that are used to replenish the world at their assigned, proper, and due times. And although they provide these liquids to the earth at periodic and opportune times, in the end only God knows when this happens, since he determined the GHOLAM of each thing at Creation.

COMMUNION BETWEEN HEAVEN AND EARTH

The extent and timing of the restorative communion between heaven and the earth is governed by the sun, Arias Montano explained. The biblical justification for this was to be found in Psalm 103:19 (VUL), "sol cognovit occasum suum," or Psalm 104:19 (NAB), "the sun that knows the hour of its setting." Arias Montano preferred the original Hebrew word MEBOO (מָבוֹא, *mābô'*) for *occasus*, since the word means *accessus* and thus suggests that the sun knows

---

61. "Continuationis sive perpetuitatis atque instaurationis generum, de quibus nunc agimus, ministrum spiritus ELOHIM sese indefinenter praebet cum legionis ipsi commissae militia, sive ornatu, sive exercitu spirituum quamplurium, de quibus loco proprio opportunè dicemus. Horum enim unusquisque motus agitationesque suas super eas quae sibi creditae sunt, aquarum sive liquorum facies exercet.": *Corpus*, 188; *HistNat*, 293–94.

62. "Igitur ex illis liquorum faciebus, sive, ut Latinè loquamur, ordinibus multiplici forma distinctis, variae et multiplices copiae seiunctae sunt.": *Corpus*, 188; *HistNat*, 294.

63. *Corpus*, 189; *HistNat*, 295.

its duties. This is also evident in Genesis 1:16 (NAB): "God made the two great lights, the greater one to govern the day, and the lesser one to govern the night; and he made the stars." Therefore, Arias Montano reasoned, the sun knows how to proceed so that the earth enjoys the sun's ministrations during the day. Its heat collects various virtues from the stars according to their stations (*positiones*) during its passage across the heavens, drawing them down onto the RAKIAGH.[64] They commingle within the RAKIAGH, generating new virtues, and after gradually growing, they are delivered to the earth as rain and snow, ice and fire, lightning and rays of sunlight, salt, oil, dew, honey, and sulfur. Meanwhile, during the night the moon brings forth the watery and airy spirits from the earth and mixes them in the RAKIAGH with those spirits from the stars.[65] The moon's obligation is to diffuse the watery and airy spirits drawn from the RAKIAGH through rain, wind, and turbulence on the sea. It also moistens and diffuses all the forces the sun deposited on the earth during the day. From this collaboration, all things that were created in the beginning can be nourished, grow, and complete their cycles.[66]

Arias Montano finds that the arcane language uses the word TAL (טַל, *ṭal*), dew, to denote the commingling of these celestial influences and terrestrial exhalations that are ruled by the passage of the sun and moon. The word can mean water or a very tenuous liquid, also abundant rain, snow, or hail, or else thick fluxes resembling honeylike, milky, or oily liquids. The liquids that nourish the earth are called SEMEN (שֶׁמֶן, *šemen*), meaning *pinguedo* or richness, fatness, and they consist of all other fluids (except for salty ones).[67] All the "adornments" and plenitude of the world continue in perpetuity when both liquids (TAL and SEMEN) mix in a "legitimate, pure, and ritual manner." This "matrimony" begets all things needed for life, as indicated in Genesis 27:28 (NAB): "May God give to you of the dew of the heavens, and of the fertility of the earth abundance of grain and wine." The timing of all generation and maturation proceeding from this union is signaled by the word MOGHEDOTH (מוֹעֵד, *môʿēd*, *tempora, dies, opportunitates*), which also means "conditions" and "constructions." The word was used in Genesis 1:14 when the sun and moon were assigned their duties. This suggested to him that all the products of the union of TAL and SEMEM are related to the stations of the sun and moon.

64. *Corpus*, 212; *HistNat*, 319.
65. *Corpus*, 211–12; *HistNat*, 319–20.
66. *Corpus*, 211; *HistNat*, 318.
67. *Corpus*, 214-15; *HistNat*, 322.

Astrological Determinism

Arias Montano endorsed a notion of astrological influences that was conservative and orthodox, if unusual in its formulation and justification. The church recognized as within the bounds of licit astrology formulations that understood the arrangements of the stars and planets as affecting the climate of the earth. Since human constitution and health were understood through the Galenic humoral theory, any disruption to the climate could affect humoral balance and could be seen as a cause of disease or as affecting a person's temperament. Furthermore, the climate could also affect the success or failure of crops, the arrival of plagues, and the general well-being or dissatisfaction of the populace, with concomitant repercussions on a king's well-being. This licit astrology excluded any formulation implying that celestial arrangements directly influenced an individual's behavior, especially if it suggested the person was unable to overcome these influences. Self-determination or free will had been one of God's greatest gifts to humanity. The often-repeated phrase "the stars incline, they do not determine" (*astra inclinant, non necessitant*) summed it up nicely.

In Arias Montano's world of celestial torrents and terrestrial fluxions, the stars' effects on the earth and its inhabitants were beyond question. In fact, it was the way God had ensured that the fertility of the earth and sea would be preserved and the cycle of generation and regeneration would persist through time. Humanity could not control or draw down influences from the stars. His explanation of celestial influence on the earth is permeated with statements voicing skepticism about humans' capacity to learn much more about these complex processes.[68] The task was immense, given the size of the heavens, the number of stars, and humans' inability to count them or even see them clearly. Still, understanding the virtues of each star might be possible, although challenging. First it was necessary to understand that the sun was the only source of light in the heavens; other celestial bodies like the moon simply acted as mirrors reflecting the sun's light. The effect of the stars depended on their being vessels that reflect what they contain. The color of a star reveals something about the "juices, spirits, or any other nature" they contain, Arias Montano explained, and could be studied to determine its influence.[69] So, although it

---

68. "Verùm singula haec nominatim ac distinctè non omnibus hominibus quamvis maximam eruditionem ac doctrinam professis in promptu est cognoscere.": *Corpus*, 212; *HistNat*, 319.

69. "Alia verò pellucidi vasis ex vitro aut crystallo confecti instar habent; uti stellae omnes, quarum corpora pellucent, et pro contenti vel succi vel spiritus, aut cuiuspiam naturae facie

was difficult to determine the specific virtue of each star, it was possible to discern their order in the heavens, the order of the three heavens, and how the celestial bodies move according to their stations. He did not explain how knowing the position of the heavens could be used to determine the virtues of each particular star or how and when its influence would reach the earth. Although his views suggest that astrological interpretation was possible, it was limited by how little could really be known about the specific virtues of a given star, as he alleged that many biblical authors had testified.

Arias Montano skirted the doctrinal problems associated with astrological determinism by falling back on one of his fundamental premises: God designed and created the world so that it will continue to function according to his design as long as he wills it to be so. The stars follow this divine—if obscure—plan and constantly obey his will. In this predetermined universe, where all celestial motions and related effects had been carefully designated at the Creation, there was no opportunity for humanity to manipulate astral influences. Furthermore, if only the learned, after much toil, could hope to understand even a fraction of when and how the virtues of the stars would come to bear, Arias Montano—while not saying it in so many words—seems to be taking a stand against some conceptions of popular natural magic, astrology, and hermetism advertising that such manipulation was possible.

STOIC INFLUENCE IN MONTANIAN NATURAL PHILOSOPHY

Seneca and Horace played important roles in the revival of Neo-Stoic moral philosophy in seventeenth-century Spain, and historians have often noted Arias Montano's inclination toward the philosophy, particularly the aspects that called for a quietist attitude toward worldly matters.[70] His friendship and correspondence with Justus Lipsius testify to his inclination toward the phi-

---

ac ratione, tum viribus tum claritate distinguuntur, ita ut quaedam ignis, aliae sanguinis, aliae lactis colorem referant: aliae purissimi crystalli, aliae ferri, aeris aliae aut calybis speciem praeferant: nonnullae virorem imitantur; pars etiam pallent: ut vel ex coloribus ac tincturis ipsis, qui diligenter atque attentè huic studio operam dare statuerint, vires etiam et efficientias coniicere possint, earum, quas numerare et consequi liceat: Namque omnes, nec Abrahamus doctissimus, ut fertur in Chaldaeis disciplinis, assequi potuit: cui in patentem producto locum dictum est: Suspice caelum, et numera stellas, si potes. Ordines verò stellarum munia et officia uni Deo auctori cognita, et pro temporum oportunitatibus imperata, ex quamplurimis apud sacros scriptores, observatis locis, indicare non erit difficile.": *Corpus*, 190; *HistNat*, 296.

70. Blüher, *Séneca en España*, 96–97.

losophy.[71] Ramiro Flórez ascribes to him a "calculated stoicism" that ran parallel to a desire for Christian aesthetic contemplation of the world. While Arias Montano's Christianity led him to reject the Stoics' belief in the role of fate in human lives, he did sympathize with Stoic ethics and believe that natural laws derive from divine law and individual self-knowledge.[72] Given his engagement with Stoic thought, it is important to ascertain how far he might have derived any of his natural philosophical ideas from Stoicism.

The fraught transmission of the natural philosophical aspects, specifically the physics of Stoic philosophy, proved challenging for humanist scholars interested in reconstituting the Stoic understanding of the natural world as an alternative to Scholastic Aristotelianism. Piecing together a description of the world from a variety of Stoic material—principally Epictetus, Diogenes Laërtius, Cicero's *De natura Deorum*, and Seneca's *Letters* and *Naturales quaestiones*—proved challenging to early modern authors who tangled with them.[73] Lipsius's *Physiologiae stoicorum* (1604) was the first attempt to compile the principles of Stoic physics into the broader project of Stoic moral philosophy, and even to Christianize the pagan philosophy.[74] It is unclear how much Arias Montano and Lipsius discussed their respective positions on natural philosophy. Lipsius certainly would have known the *De arcano sermone* and the *Anima* before he completed his *Physiologiae stoicorum* about 1597, but we do not have correspondence testifying to any substantive discussions between the two on the subject. Yet we find Lipsius emphasizing some aspects of Montanian philosophy beyond what was suggested by the Stoic corpus, among them the primacy of water. How far Lipsius adapted the Stoic notion of a world soul and seminal reasons in a manner that might parallel Arias Montano's concept of ELOHIM merits further investigation.[75]

Yet Montanian natural philosophy and Stoic philosophy shared a similar commitment to understanding the natural world as an important prerequisite

71. Lipsius, *Epistolario de Justo Lipsio y los españoles*, 53–54; Landtsheer, "Benito Arias Montano and the Friends from His Antwerp Sojourn," 46–50.

72. Flórez and Balsinde, *Escorial y Arias Montano*, 367.

73. Sellars, *Stoicism*, 135–44.

74. For an example, see Lipsius's attempt to untangle whether the Stoics thought the world was a sentient being in 1.8 and again in 2.10, in Lipsius, *Iusti Lipsi Physiologiae stoicorum libri tres*, 16–20, 92–96. On Lipsius's interpretation of Stoic physics, see Lipsius, *Juste Lipse et la restauration du stoïcisme*, 63–65, 206–53, and Saunders, *Justus Lipsius*, 117–27, 165–201, 215–17.

75. Lipsius, *Physiologiae stoicorum*, 1.7–8 and 2.8. See Saunders, *Justus Lipsius*, 162–64, 191–94.

to understanding one's place in nature.[76] In fact, one of the objectives of Stoic moral philosophy was achieving a course of personal behavior that was in rational concert with nature, thus making the study of the natural world essential to leading a Stoic life. Although this was not its ultimate objective, Montanian natural philosophy also made understanding the natural world a central aspect of realizing an individual's purpose. Thus, for both philosophies an understanding of the natural world was perceived as the foundation for a virtuous life, despite their vastly different conceptions of what a "virtuous life" entailed. The philosophies also shared a static view of nature as operating according to laws ordained by God for humanity's sustenance. But whereas for the Stoics this unyielding aspect of nature was an argument for the Stoic *ataraxia* and acceptance of fate, for the Christian Arias Montano it was only the manifest testimony to a benevolent God. Both philosophies relied on natural theological and intelligent design arguments to prove the existence of God. Cicero's articulation of these Stoic ideas in the *De natura Deorum* (book 2) were very influential in the revival of early modern Stoicism. But here again the substance of the arguments supporting Stoic and Montanian notions of design differed markedly.[77] For Cicero, the Stoic concept of an animated world proved the existence of gods, while for Arias Montano the design of nature itself, created as it was to shelter and sustain God's greatest creation—humanity—was enough testimony to the existence of the Divinity.

Two aspects of Montanian metaphysics resonate in their similarity to Stoic thought: the first principle CAUSA and the conceptualizing of the role of ELOHIM as a pervasive spirit that transmits life and rules the cycle of exchange between heaven and earth. The first cause (along with matter) in Stoicism also serves as the ontological basis for everything in the world; as Seneca explained in his epistle 65, "Now, however, I am searching for the first, the general cause; this must be simple, inasmuch as matter, too, is simple. Do we ask what cause is? It is surely Creative Reason [*ratio faciens*]—in other words, God." And later, "And that which creates, in other words God, is more powerful and precious than matter, which is acted upon by God. God's place in the universe corresponds to the soul's relation to man."[78] Arias Montano also identified his principle CAUSA with a persistent creative agency, "on account of which

76. For the Stoic perspective, particularly as articulated by Seneca, see Inwood, "Why Physics?," 201–23.

77. Salles, *God and Cosmos in Stoicism*, 1.

78. Seneca, *Ad Lucilium epistulae morales*, 1:451, 457, 459. For Lipsius's interpretations, see Saunders, *Justus Lipsius*, 123–25.

each type is, or may be" (*quamobrem unumquodque genus est, aut sit*) and as an all-encompassing, vivifying force "in which all, from which all, and through which all exists" (*in quo omnia, à quo omnia, et per quem omnia*), mysteries he sees encapsulated in the word PRINCIPIO.[79]

How similar is ELOHIM as principle to the notion of the Stoic *pneuma*?[80] For Stoics, the sole two principles in nature were God and matter (though not conceived as fully individuated entities). The essential nature of all matter was fire, from which all originates and into which all will resolve in the great conflagration. Fire in its most fundamental form resides in heaven and gives the celestial bodies their light and heat. Like terrestrial fire, celestial fire needs fuel, which is provided by a derivative substance, *pneuma*, a material substance (fiery and airy) that penetrates bodies, holding together everything in the world and animating it as its soul.[81] This *pneuma* has been interpreted alternatively as the material means through which God is immanent in everything in the world, ordering and, in the process, enlivening the world, or as a god with a material body.[82] In Stoic philosophy *pneuma* pervades everything that is beneath the fiery region, swirls into the earth conveying celestial influences downward, and returns to the heavens replenished.[83] Eventually this cycle would exhaust the earth, and the *pneuma* of the heavens would be absorbed by the sun while the universe reverted to pure "fire" in a giant conflagration. Furthermore, since the celestial bodies also partake of the universal *pneuma*, they have an intelligence that directs their motions and thus are comparable to living things. While the Stoic *pneuma* is an entirely material substance that transmits the divine will to everything and is therefore immanent in the world, the only immanent thing in the Montanian cosmos was the spirit ELOHIM. Equating ELOHIM with the Stoic *pneuma*, however, would require a significant caveat; Arias Montano considered ELOHIM to be an agentive, incorporeal spirit and definitely not a material substance.

As in both Stoic and Aristotelian philosophy, the Montanian cosmos is completely full and abhors a vacuum. Along with the Platonists and Aristotelians—as with most ancient natural philosophies—Arias Montano shared the conviction

79. *Corpus*, 149; *HistNat*, 252.

80. Sellars, *Stoicism*, 96–104; Barker, "Stoic Contributions to Early Modern Science," 138–40.

81. For a synthesis of Chrysippus's ideas concerning *pneuma* as preserved by Cicero, see Hahm, *Origins of Stoic Cosmology*, 156–66.

82. Salles, *God and Cosmos in Stoicism*, 6–7.

83. On Stoic notions of *pneuma* or "air," see the first twelve chapters of book 2 in Seneca, *Natural Questions*, 163–69.

that the world was spherical. But to make his cosmos analogous to the Stoic view, the Montanian world would have had to be filled with *pneuma* and admit an extracosmic void; instead, it is filled with a rarefied derivative of MAIM and is definitely bound. Yet like the Stoics, and in a drastic departure from Aristotelian cosmology, Arias Montano drew no distinction between the essential material natures of the supra- and sublunar realms—they all derived from the primordial MAIM. In the Montanian scheme—and somewhat analogous to the degree of *pneuma* accepted by Stoics—what differentiates the regions is the purity and tenuousness of the material that occupies them.

Early modern Stoic proposals presented yet another viable alternative to Aristotelian, Platonist, and atomistic natural philosophies. A young Roberto Bellarmino, for example, discussed Stoic cosmological notions in lectures he delivered at the University of Louvain from 1570 to 1572.[84] In Louvain, during the same years when Arias Montano was in contact with prominent members of its faculty, Bellarmino found an environment where these ideas were avidly discussed, most notably by Cornelius Valerius (1512–78), who advocated the elemental nature of the heavens in his *Physicae*, published by Plantin in 1568.[85] Bellarmino shared several of Arias Montano's theories, including the idea that corruption of the heavens was possible owing to their elemental nature.[86] Furthermore, and in line with Stoic thought, Bellarmino maintained that the substance of the heavens was fire and that planets move within this fiery—but possibly also airy—medium by their own motion rather than being carried about by the celestial spheres. He even used the analogy of the planets moving in the heavens as birds move through the air and fish through water.[87]

Bellarmino's views also coincided with Montanian cosmology on two important questions: What is the firmament? and How many heavens are there?

84. Baldini and Coyne, "Louvaine Lectures"; Blackwell, *Galileo, Bellarmine, and the Bible*, 40–43.

85. Valerius, *Physicae, seu de naturae philosophia*, 16, 21; Kelter, "Reading the Book of God as the Book of Nature," 176–80.

86. Barker, "Stoic Contributions to Early Modern Science," 144.

87. As Bellarmino explained, "But be that argument valid or not, if we wish to hold that the heaven of the stars is one only and formed of an igneous or airy substance, an hypothesis which we have declared more than once to be more accord with the scriptures, we must then of necessity say that the stars are not transported with the movements of the sky, but they move of themselves like the birds of the air and the fish of the water. In fact, it is known that the motion of the planets is diverse: one is faster, the other slower, and it is clear to everyone that one same heaven cannot move at the same time with diverse velocities.": Baldini and Coyne, "Louvaine Lectures," 18–20.

Bellarmino maintained that the firmament could act as a bulkhead separating the waters and that, based on scripture and the evidence astronomers had provided, there are only three heavens.[88] Like Arias Montano, he preferred to let the Bible adjudicate many of these cosmological issues.[89] Stoic cosmology, with all its vagueness when considered by agile minds, could open avenues that plainly contradicted Scholastic natural philosophy yet stayed within the bounds of scripture.[90] My using Bellarmino as an example of the kind of theoretical speculation possible during this time is no coincidence; in fact it is rather self-serving. For as we will see in chapter 12, it would be Bellarmino who, more than thirty years after delivering these lectures, supervised the expurgation of Arias Montano's works for the Roman Congregation of the Index in 1607, roughly a decade before the Galileo affair.

*

The intent of the Montanian natural philosophical program was to build the apparatus of human knowledge anew using as its foundation the revealed Word. In the prefaces to both the *Anima* and the *Corpus*, Arias Montano was aware that perhaps some aspects of what he argued in the book might resemble what earlier philosophical schools had taught. Yet he insisted that he sought guidance and authority solely from the Sacred Scriptures, inquired about things and causes with rectitude of mind and spirit, and submitted only opinions (*sententiae*).[91] The hermeneutics of nature that resulted from his historical premises and metaphysics yielded an unconventional cosmology for which he found complete support in the words of Genesis. His conclusions testify to the kind of terminus that the critical spirit developed in the hexameral commentary tradition, the methodology provided by hyperliteral biblical exegesis, and the desire for new natural philosophical proposals could achieve during the late sixteenth century.

---

88. Ibid., 14–16.

89. Ibid., 40n92.

90. Tycho Brahe and Johannes Kepler sided with the view of an orbless, fluid heavens, although they differed on the nature of the mechanism that moved the planets: Barker, "Stoic Contributions to Early Modern Science," 146; Boner, "Life in the Liquid Fields," 275–97.

91. *Anima*, Praefatio, fol. †5v; *LibGen*, 97.

CHAPTER 9

Meteorology, Matter Theory, and Mechanics

Having laid down the first principles undergirding his new natural philosophy, Arias Montano undertook a systematic, but far from exhaustive, exploration of the nature of things through the hermeneutic lens he found so clear in the Bible. Among the topics he addressed were what the modern language of science calls meteorology, matter theory, mechanics, botany, and zoology. Significant portions of the *Magnum opus*, however, are devoted to anthropology, or what Arias Montano would have considered the corporeal and moral history of humanity. As we saw in chapter 8, to explain the corporeal aspects of the human body he turned to an etymological analysis of the work ADAMAH. Yet he devoted thousands of words to rich meditations on humanity's moral condition and how it related to his natural philosophy; these deserve a far more expansive treatment than I can provide here.

In chapter 8 we saw how Arias Montano conceived of the relationship between heaven and earth as a constant communion and exchange of materials that replenished the earth and kept it a suitable dwelling place for humanity. A similar relationship existed between the land and sea, as well as between the land and sea and the RAKIAGH. The communion operated in perfect harmony or, as he would have it, in accordance to the command IEHI, with the constant oversight and action of ELOHIM, and with each natural thing obeying the mission it had been assigned at Creation. He thought of these communions as circulating, rejuvenating cycles that ensured the fertility of the earth.[1] He

---

1. "Quandoquidem praecipuae ac maximae mundi partes ita à Conditore comparatae sunt, ut quamquam locis ac sedibus regionibusque discretae, caelum, abyssus, ac terra: tamen effici-

found evidence of this communion in climatic events, sometimes terrifying but nonetheless vital, that kept the earth welcoming for humanity. Here again the communion takes place by the exchange of fluids, with the spirit ELOHIM keeping it in perfect order and equilibrium. Arias Montano was keen to show that, although ELOHIM oversaw these exchanges, they took place through natural processes that are part of the *rerum natura* of each thing involved. He made frequent asides to elucidate the principles of mechanics and properties of materials that explained how the processes took place. For example, he devoted substantial passages to the physical properties of water, explaining how it flows, how siphons work, and how compound mixtures occur in nature. The discussions of matter theory and mechanics, in particular, do not take place as systematic expositions within the traditional disciplinary categories of Aristotelian physics and Plinian natural history but seem to be placed haphazardly throughout the text. In fact Arias Montano deliberately avoided organizing his natural philosophy following the ancient canons that dominated the field. He chose to explore each subject as warranted by particular natural events or things. For example, he discusses mechanics—an old Aristotelian category and the cornerstone of that philosophy's physical theory—as a truly ancillary topic, brought up when needed to support a claim about the constitution of the world or to explain a particular type of observed motion. He does the same with matter theory, though here his approach is more in keeping with the conventions of the time. During the sixteenth century, matter theory was not a disciplinary category per se but was usually discussed as a constituent part of physical theory or in alchemical treatises. Thus his approach requires us to collect scattered bits of his discussions on these subjects throughout the *History of Nature* to reconstitute his understanding of matter theory and mechanics, acknowledging that this approach does some violence to the historical style he chose.

THE NATURE OF THE EARTH AND OF THE SEA

To Arias Montano the earth was the immovable center of the world, a sort of base and support the world rested on.[2] The psalmist stated as much—"You

---

ente communicatione, coniunctissimae semper maneant, et superiores non sua, sed inferiorum causa in actione ac motu perpetuo sint, ut testimoniis citatis inferius intendemus. Prius videndum quemadmodum inferiori corpora Abyssus ac Terra, et horum corporum spiritus habeant, et agant.": *Corpus*, 205; *HistNat*, 312.

2. *Corpus*, 194; *HistNat*, 300.

fixed the earth on its foundation, never to be moved" (Ps. 104:5 NAB)—and Isaiah wrote, "Thus says the LORD: The heavens are my throne, the earth is my footstool" (Isa. 66:1 NAB). Arias Montano interpreted the reference to the earth as God's footstool to mean that the whole of the world rested on the immobile earth. Characteristically, Arias Montano did not pause here to address alternative theories, choosing instead to derive his cosmology solely based on biblical readings. So, as expected, there are no references to ancient heliocentric notions such as those of Hipparchus or even Copernicus or any of a handful of contemporaries who considered movement of the earth a possibility. The epistemic stakes around this question were particularly high for Arias Montano, since earth's immobility and central position mirrored the concept of God as center. His hermeneutics of nature dictated that the earth's purpose—to shelter humanity—made it a reflection of the divine presence and will. The earth was prepared for generating and supporting life by having been covered with the fertile primordial waters, then shaped at the Creation with various geological features.[3] Take its round shape yet irregular surface. This lumpiness provided dry areas that let the earth fulfill its purpose: housing humanity and providing it with sustenance. This designed landscape, furthermore, remained unchanged in its essential and fundamental form. Among all geological features, it is not surprising that Arias Montano considered mountains—his beloved namesakes—the most important and prominent. They were made of various types of stones that were the product of "juices of the earth" and humors responding to cycles of heat and cold. Veins of oily liquids produced a variety of metals and minerals in some mountains, while others flowed with water and salt, producing sustenance for humans and animals.[4] Thus these observations confirmed for him that the biblical words for mountain—HARIM (הָרִים, *hārîm*) and HAROTH (הָרוֹת, *hārôt*)—derived from HARAH (הָרָה, *hārâ*), "to be pregnant."[5]

3. *Corpus*, 195; *HistNat*, 302.

4. "Montes qui prorie HARIM et HAROTH dicuntur, saxosi plerumque sunt, variosque tenent, creant, et cohibent, aut etiam produnt lapides pro vario terrae et inclusi, aut accepti humoris succo, quos subeuntis caloris aut frigoris vicissitudo excoquit ac durat. Inter has etiam venae sunt pinguioris liquoris refertae, ex quibus eiusdem vicissitudinis oportunitate mineralia et metalla diversa, fluoresque varii conficiuntur, et formantur. Id quod virtute spiritus ELOHIM, liquoribus per ministros spiritus incubantes efficitur. Sunt praeterea salibus multis, et aquis pleni partim dulcibus, partim sapore, odore, ac vi efficacitateque differentibus. Unde et vario latice distincti fontes manant, quorum fluxu et copia multiplices animantum generi omni, atque homini praecipuè usus capiuntur" (in the margin הרים, הרות). *Corpus*, 196; *HistNat*, 303.

5. Raphelengius, "Thesauri hebraicae linguae," 6:22.

According to Arias Montano, the sea—conceived as the abyss, though safely contained by the surrounding high land—constituted one-seventh of the mass of the earth.[6] He saw this clearly stated in 2 Esdras 6:42: "Upon the third day thou didst command that the waters should be gathered in the seventh part of the earth" (NAB). Arias Montano observed that the depths of the sea remained unexplored by humans and therefore were known only to God, but he uncovered in the Bible a few hints about the nature of the sea, or rather the "seas." The shift in the order of the letters of the word IAMMIM (מַיִם, *yammîm*) used in Genesis 1:10 caught his attention. IAMMIM suggested to him that the seas originated from the dual liquid MAIM, but as the liquids were grouped together to form seas, the result were seas with different material compositions, some saltier than others.[7] Seawater, he observed, is salty by design and remains so because that is its *prima natura*; the salt in the sea is constantly replenished by the water's washing arid and salty coastlines. While rivers and rain deposit greasy fluids into the sea, the continued agitation of ELOHIM keeps the saltiness distributed throughout, although in different proportions, and ensures that the sea does not "coagulate." Water itself, by virtue of its circular nature, also has a particular "spirit" that causes motion, though the tides, Arias Montano explained, result from the continuous action of ELOHIM rather than from the cycles of the moon.[8]

Just like the relationship between the heavens and earth, the "eternal communion" and "mutual collaboration" between land and sea enabled them to share their forces in a measured way by commingling their liquids. Land could not be fecund without water; the sea could not support life without the greasy fluid that flows from the land through springs and rivers. The exchange takes place as water from the sea is drawn into the land and filtered (*excolatus*) as it makes its way through the land back to the sea. This removes salt so the water becomes drinkable as it emerges from springs and rivers. As it flows to the sea, it gathers the "fecund juices" of the earth; the many types of water in rivers and springs give testament to the variety and abundance of the "spirits" that flowing water gathers on its way back to the sea, where it tempers the sea's

6. *Corpus*, 200–203; *HistNat*, 307–10.

7. *Corpus*, 203; *HistNat*, 309.

8. "Atque motionis et agitationis huius, quae dicta est, quaeque in assiduo maris fluxu et refluxu cernitur, minister est ille spiritus ELOHIM, cuius opera superficies liquorum moventur et commiscentur, id natura ad producendarum et alendarum rerum oportunitatem et aptum idoneumque temperamentum exigente.": *Corpus*, 201; *HistNat*, 309.

saltiness and provides nourishment to sea life.[9] This unceasing life cycle, or GHOLAM, in which all living things take part precipitates their return on death to the place of their origin: residents of the land are absorbed back into the land, and those from the sea go back into the sea.[10]

While the spirit ELOHIM initiates the agitation that drives the sea into the land, the cycles necessary for replenishing the earth are perpetuated by the spirit of each thing. For example, the internal spirit of water has a propensity to flow in a circular manner. Although Arias Montano acknowledged that the source of water in rivers can be snow and rain, most of the water that flows on the surface of the land comes from within the earth, pushed into it from the sea. According to Solomon, whom Arias Montano considered "the most expert sage in the whole world,"[11] "All rivers go to the sea, yet never does the sea become full. To the place where they go, the rivers keep on going" (Eccles. 1:7 NAB). Arias Montano was compelled to explain that, just as a pump draws water from a well, so channels or "veins" inside the earth draw water from the abyss into springs and rivers.[12] The entryways of these channels are the biblical *fontes abyssi* that gave way and brought on the deluge (Gen. 7:11). Note that, although in many aspects Arias Montano's meteorology coincided with Aristotle's, his understanding of the nature of springs contradicted what Aristotle set forth in *Meteorology*, book 2.

METEOROLOGICAL EVENTS

Although Arias Montano understood the acts of the seven days as resulting in a complete creation, he allowed for the temporal appearance of aspects of nature that were previously created but as yet unseen. Some things in nature do not always exist in species and form but appear when the Creator, through the agency of ELOHIM, desires it, as do meteorological events—atmospheric fires, winds, clouds, rain, snow, and dew. Some appear at regular intervals and are

9. "Atque hoc pacto terrae cum mari perpetua communicatio est, duplicis liquoris in variam ac multiplicem rationem attemperati et commixti, unde varia rerum origo ac nutritio certis legibus ac terminus constat iuxta perpetuum IEHI imperium omni rerum generi, quod ad naturam pertinet, observatum. Itaque mare liquores suos terrae suggerit: rursusque terra suos mari evestigio rependit; et utraque natura proprios producit et fovet, aut alit fœtus, et primorum generum formas, quibus orbis ornatur, perpetuò sufficit.": *Corpus*, 211; *HistNat*, 318.

10. *Corpus*, 234; *HistNat*, 342.

11. *Corpus*, 208; *HistNat*, 314.

12. *Corpus*, 210; *HistNat*, 317.

called *oportuna*, while others, appearing only occasionally or less predictably, are called *miranda*, wonderful, strange, or prodigious.[13] These phenomena bring comfort to humanity, but they might also be divine punishments.

Arias Montano understood climate events as resulting from either of two types of heat: one that came from the sun and another that permeated the earth, the "spirit of heat."[14] The purpose of the sun's heat was to absorb moisture from the sea and land and carry it to the regions above in the form of vapor. The nonsolar "spirit of heat" consumes airs from within the earth and those it encounters as it leaves the center of the earth by turning them into vapor. As these vapors rise into the RAKIAGH they commingle with similar vapors from the stars, each with its own kind, and remain there, tossed about by the agency of ELOHIM. This is the activity we see in the creation and movement of clouds, their different colors indicating the nature of the vapors they contain. The efficacy of the moon then dispatches these virtues, separated and differentiated into their similar natures, to replenish the earth. Most return to the earth as rain, snow, hail, and dew. Rain is the condensation (*densatur*) of tenuous humid vapors that, having been brought together by wind, gain weight and gravity and become spherical droplets of rain.[15] Once joined with the "juices" of the earth, rain provides each thing with the "natural spirits" it needs to carry out its mission. Arias Montano summarized the process with a mixture of Aristotelian and alchemical terms:

> Thus the author of the rain and downpours is God; matter: vapor drawn up from the earth and sea; the arranger: spirit ELOHIM; the condiment: celestial flux; the place of seasoning: the expansion or firmament; the inciter: winds; the collector: mostly the earth; the means: irrigation; the events: commingling of liquors; the effect: production of subterranean bodies and plants; the cause: at last, the renewal of the fruits and nutrition of living things and animals.[16]

13. *Corpus*, 224; *HistNat*, 332.

14. *Corpus*, 213–14, 225; *HistNat*, 321, 332.

15. "Namque vaga ac tenuis humidi vaporis natura à vento vel paulatim, vel celeriter coacta densatur, densataque gravitatem pondusque accipit, et in rotundam sive sphericam figuram efformatis guttis diffiso aëre, utpote leviore ac rariore descendit, à quo rursus coëunte impellitur, et à vento etiam protruditur.:" *Corpus*, 225; *HistNat*, 333.

16. "Itaque pluviae et imbris auctor Deus est, materies vapor è terra atque mari sublatus, concinnator spiritus ELOHIM, condimentum caelestis influxio, condiendi locus expansio sive firmamentum; impulsor, ventus; exceptrix, terra praecipuè; efficientia, irrigatio; eventus, liquorum commixtio; effectus, subterraneorum corporum plantarúmque productio, causa demùm fructuum instauratio, et viventium animantiumque nutrition.": *Corpus*, 225; *HistNat*, 333.

The greasy liquids that formed part of the primordial MAIM provided Arias Montano with explanations for fiery atmospheric phenomena such as lightning bolts, comets, shooting stars, and "other things of the superior fire." These greasy liquids were also drawn up from the land or sea by the power of the sun or drawn down from the stars and mixed in the RAKIAGH. Once there, they stood separated into zones in the sky depending on the density of the resulting mixture. Even in a mixture, he added, they obey the nature of the pure original spirit and thus persist uncorrupted, like the soul in the body of an animal.[17] But because they are all of a greasy nature, they are very susceptible to the "seed" of fire that is present in all greasy liquids. When these liquids are irritated and agitated by the wind or by being surrounded and compressed by salty liquids, they spark, igniting the sulfuric part of the liquid first because it is more spirited.[18] Any adjacent salty, watery, cold liquid will react violently and push the fire away in any direction, causing it to extend a great distance until it is extinguished.[19] This we witness first as the light and then as the lightning bolt. The sound of thunder is the result of the sudden and rapid disruption of the air following the fire's ignition.[20]

He explained that wind was a wave (*unda*) of air and shares the same spirit as air, although it has much more power and efficiency.[21] In nature, the moon acts as the "mechanic" (*opifex*) of the wind, enlivening the spirits drawn up by the sun as it warms the surrounding air, expanding it and causing it to move. Earthquakes are caused by wind trapped in caves inside the earth. When this trapped wind is heated by sulfur, carbon, and vinegar coming from the earth, it pushes forth violently against the nitrous substance in the surrounding air and causes the agitation of subterranean "storms," which make the land tremble and can bring down mountains. Arias Montano found etymological evidence for this interpretation in Psalm 18:7 (NAB), "The earth rocked and shook; the foundations of the mountains trembled; they shook as his wrath flared up."

17. *Corpus*, 220; *HistNat*, 328.

18. Like Aristotle, Arias Montano made sulfur the cause of lightning, but where the Greek philosopher was vague as to the mechanism behind the phenomenon, our biblist explained it with luxury of detail. His description, in fact, is similar to Seneca's (2.21–24): Seneca, *Natural Questions*, 172–74.

19. *Corpus*, 220; *HistNat*, 328.

20. "Repentina ac celerrima aëris diruptione et rarefactione ab igne.": *Corpus*, 221; *HistNat*, 328.

21. "Ventus est aëris fluens unda vel à sui vel ab alterius generis vi implusa.": *Corpus*, 221; *HistNat*, 329. For another explanation of wind along the same lines, see Arias Montano, *Discursos sobre el Eclesiastés de Salomón*, 96–98.

He takes exception, however, to the translation of HARAH (חָרָה, *ḥārâ*) as *irasci*, "to be enraged," preferring to work with HHAMAH (חָמַם, *ḥāmam*), "to heat" or "ignite," given that "all ire kindles from inflamed air near the heart."[22]

Early modern meteorology, as Craig Martin explains, "was by no means static. It evolved as authors of meteorological treatises incorporated new evidence and theoretical models."[23] Among the disciplines addressed by the Aristotelian corpus, meteorology was considered the most empirical and conjectural, and therefore most accessible to revision. For Aristotle, as for Arias Montano, meteorological phenomena resulted from the movement of "exhalations" between the surface of the earth and the sky. These were of two types, a vaporous one and one that was hot and dry, that could act in unison as imperfect elemental mixtures. Natural phenomena originating within the earth, such as earthquakes and hot springs, resulted from two analogous exhalations working within the earth in the presence of a persistent internal fire.

Breaking with Aristotle on this point, Arias Montano attributed these natural phenomena not to the existence of a subterranean fire but rather to a spirit of fire residing in the interior of the earth. This allowed him to explain that earthquakes were infrequent because of the occasional and "unreasonable" increase in heat caused by the spirit of fire acting on "sulfur, carbon, and vinegar."[24] An earthquake took place when this overheated, inflamed air, seeking to escape from within, caused the earth to tremble. Aristotle also saw the immediate cause of earthquakes as overheated air and the resulting powerful wind, though he seems to prefer Seneca's description of wind as "a wave of flowing air pushed either by itself or by another type of force."[25] Yet, lest

22. "Quo loco verbum 'iratus est,' à Latino interprete pro verbo HARAH redditur. At verò HARAH voce ac significatione affine est verbo HHAMAH, quod calescere et incendi significat. Namque ira omnis ex accenso circa praecordia aëre excandescit. Et Jeremias, 'Ab indignatione eius commovebitur terra.'": *Corpus*, 235–36; *HistNat*, 342–43. This may be a reference to Saint Thomas's *Summa Theologica*, part 2, 1, question 48, in a reference to Saint John of Damascus: "On the contrary, Damascene says (*De fide orth.* 2, 16) that 'anger is fervor of the blood around the heart, resulting from an exhalation of the bile.'"

23. In this section I follow Martin's explanation of Aristotelian meteorology: Martin, *Renaissance Meteorology*, 2, 6–9, 62–64, 80–105.

24. *Corpus*, 234–35; *HistNat*, 342.

25. "Ventus est aëris fluens unda vel à sui vel ab alterius generis vi impulsa.": *Corpus*, 221; *HistNat*, 328. He subsumes into the first sentence of the section on the wind two aspects Seneca develops in the first chapter of his writing on the same subject, "Wind is flowing air" ("*Ventus est fluens aer*"), followed by a discussion of how wind moves like the waves of the sea (5.1): Seneca, *Natural Questions*, 73.

his readers think he was borrowing too much from the ancient philosophers, Arias Montano buttressed his interpretation of the meteorological cycle with a dozen biblical citations testifying to his interpretations, Job—whose exposure to inclement weather made him an eloquent bard of weather—being his favorite source.

MATTER THEORY

The Montanian world of commingling celestial and terrestrial virtues and constant, predetermined regeneration rested on a matter theory that ensured the orderly transference of influences and forms between the heavens and the earth. This theory relied on the demarcation between the offices and duties of corporeal matter and of incorporeal spirits. As we saw for the stars, Arias Montano also conceived of corporeal bodies as vessels or receptacles made of matter. These bodies lay inert until infused by an incorporeal spirit that either gave them motion and life, as with plants, animals, and man, or assigned them duties, as with inert matter. To explain how this worked Arias Montano had to lay down three fundamental precepts: the difference between corporeal and incorporeal things; the nature of elemental matter—its origin and constitution; and the behavior of mixtures.

Corporeal and Incorporeal

For Arias Montano everything is made of MAIM and even the most tenuous bodies are corporeal—including those sometimes referred to as "spirits." A true spirit—one that can penetrate bodies, give them form, and move them—is incorporeal and eternal, although it may at times imitate bodies. The wind, terrestrial exhalations, vapors, even hotness and coldness are examples of the corporeal kind, while the spirit that is within man or lies within or adjacent to animals, plants, and other natural things is incorporeal. (This latter kind is far more crass than the spirit of man.)[26] Note that he had not abandoned the division between spirit and body established in the *De arcano sermone*. But whereas in that instance he was describing semantic relationships, now he was describing ontological ones. Unburdened by having to adhere to or challenge notions such as universal hylomorphism (which would have had him posit some sort of spiritual matter to explain angels) or having to explain the nature

26. *Corpus*, 166–67; *HistNat*, 272.

of angels in Thomistic terms, he could simply draw predicates concerning the nature of angels from the Bible to substantiate his claims.[27]

In Montanian natural philosophy the existence of incorporeal spirits goes unquestioned, largely on the strength of biblical evidence testifying to their existence. Interestingly, though, he does not draw a sharp distinction between them but suggests they exist on a continuum, with the incorporeal ones separated from corporeal ones by an immense difference in their tenuousness (*subtilitas*).[28] He deduced the properties of incorporeal spirits from the properties of very tenuous corporeal ones.

He also drew his explanation for the behavior of corporeal spirits from observation. For example, the most subtle bodies can permeate denser ones yet not be contained by them, as when a ship on the sea seems to part the water without fracturing it or becoming one with it. Likewise, a corporeal spirit can penetrate a denser body, as when a person is affected by vapors, wine, or the smell of vinegar. The spirits of heat and cold also penetrate denser bodies, moved as they are by their tenuous spirits. This is why a cool body gets warm when placed next to a hot one. Among the corporeal spirits, the more subtle and acute (*subtilior et acutior*) of these, the spirit of the sun, can penetrate to the core of the earth and to the bottom of the abyss, while the denser spirits do not penetrate as deeply. Recall that Arias Montano considered the spirit of the sun an inseparable combination of light and heat. This conception of the sun as a corporeal spirit suggests he thought of it as having a material basis. Although he places hotness among the corporeal spirits, he makes no mention of light, nor does he concern himself with the mechanics of corporeal spirits' transmission other than to say they "penetrate."

Using the behavior of tenuous corporeal spirits as his model, Arias Montano was able to posit that a true spirit, a wholly incorporeal one, can easily penetrate a corporeal body, as with angels and evil spirits. This is also how God infused spirit into the first human. For Arias Montano it is clear that a tenuous body can penetrate a denser one, but not the other way around. This is why God and his angels cannot be affected by lesser spirits (those understood to have limits or bodies), so there are no actions humans can take to influence the behavior of incorporeal spirits—a clear dismissal of the theurgist's rites described by Iamblichus and other Neoplatonists. Again using as an example the way tenuous corporeal spirits affect the bodies they pervade, he explained that by analogy this was how the spirit ELOHIM causes his influence (*efficien-*

27. On the incorporeity of angels, see *Corpus*, 137–39; *HistNat*, 239–41.

28. *Corpus*, 168–69, 213; *HistNat*, 274, 321.

tia) to be transmitted to the world. ELOHIM can permeate corporeal things, agitating them and imparting motion to them; in turn this action, carried out through the agency of lesser "ministers" (such as wind), can cause changes on the earth.[29]

MAIM as a Fluid with a Dual Nature

The principle MAIM allowed Arias Montano to fashion a matter theory that supported his idea of a common essential nature between the heavens and the earth. He rejected the notion of the four Empedoclean elements—earth, water, air, and fire—and of the Aristotelian fifth element, aether, as elemental matter. It is likely, although he does not state so explicitly, that he also rejected the atomistic notion of hard, indivisible, and impenetrable particles as the constituents of matter; likewise, he does not engage with the vague Aristotelian and Platonic notions of prime matter. Arias Montano's universe was a fluid one, where everything could be reduced to the two liquids that came about when the spirit ELOHIM agitated the primordial MAIM, dividing and recombining them in different proportions and order into everything in the world. This is how the first constituents of the world were defined, and they in turn, under the continued agitation of ELOHIM, became the component parts of things.

The exegetical exercise that gave rise to the matter theory is vintage Arias Montano and clearly something he had been thinking about for a long time. We saw the first hints of it in his editing of the second verse of Genesis in the Pagnino Bible of the Antwerp Polyglot. He explained in the *History of Nature* that he based this notion on the interpretation of the phrase GHAL PENE HAMAIM (עַל־פְּנֵי הַמָּיִם, *ʿal-pĕnê hammāyim*), which had been translated into Latin as *super facies aquarum* (Gen. 1:2 VUL). Working with an alternative meaning of *facies* as "multiple aspects," Arias Montano paraphrased the Vulgate as *iuxta, circum, supra gemini liquoris aspectus et species*, or "near, around, above the appearance and the type" of the dual liquids. (Later Arias Montano cited the phrase as "spiritus ELOHIM ferebatur super PENEI HAMAIM.")[30] This meant that the dual liquids ELOHIM was agitating had different aspects or appeared

29. "[I]n eáque re instar viseretur totius orbis cuius corpora membraque omnia unius spiritus ELOHIM efficientia per inductos immissosque ministros, corporeos inquam, spiritus moventur, aguntur, et exercentur: sic enim, uti iam diximus, mare nunc aurarum, nunc ventorum opera pro varia causa movetur: id quod ad omnem totius qui caeli infimo ambitu cingitur, orbis partem referendum esse, iam naturae contemplatione declaratum est. Ubi illud pro certo statuimus, citra spiritus agitationem nullum corpus moveri.": *Corpus*, 359; *HistNat*, 476.

30. *Corpus*, 176, 187; *HistNat*, 282, 293.

different in the primordial world, so that MAIM, once divided, takes on a variety of appearances in composite matter (both terrestrial and celestial) yet remains the essential and singular basis of all things. Arias Montano indicated that this dual liquid would last in perpetuity, or a long as God wills it. This endowed large, preeminent bodies, whether in the heavens or on earth, with perpetual duration (*duratio perpetua*), so they may continue to replenish smaller, individual bodies created to have shorter durations.

Fire

Arias Montano realized that his theory of MAIM explained most corporeal things, yet he found fire intriguing. Its uniqueness stemmed from its being the only created thing whose nature was endowed with a unique type of spirit.[31] It serves to purify and was created chiefly for its utility to humanity. Yet fire does not generate anything but only destroys, as indicated by one of its Hebrew names, ESOCHELETH (אֵשׁ אוֹכֵלָה, *ʾēš ʾôkēlâ*), "devouring fire" (Isa. 29:6). Fire feeds on greasy liquids, demonstrating some affinity with them, but it completely rejects water or anything salty, hissing loudly to show its rejection. Meanwhile it silently and smoothly consumes things that are acid and rough (*acre et asperum*) or that are continuous (*continuum*), even if they do not appear to be so. Examples of these are sulfurous things, such as *sulphura viva* (especially once it has been liquefied and condensed) and other liquids that have had the salts removed through fire or human art and have become very acidic (*acerrimi*), such as *aqua vitae* or quintessence.[32]

Arias Montano observed that he did not assign fire a place in either the superior or the inferior part the world as ancient philosophers had done. This was because parts of fire or its force can be detected in everything except water. Fire, recall, was created to serve the "inferior world" and so existed within every thing that has some measure of sulfuric grease. Fire generally lies inert until subjected to some violent motion or compression. Then it can spark and, by feeding on nearby air, advance and consume a great many things. Its heat and voracity exist in proportion to the greasiness of the burning thing. Because fire burns by agitating and pushing away the salty components and consuming the greasy parts, burned things become desiccated and brittle.[33] Yet the purer and

31. "Inter cetera quae ad inferioris orbis usum condita fuerunt rerum communium genera, ignis numeratur singularis natura suo etiam spiritu praedita.": *Corpus*, 216; *HistNat*, 324.

32. *Corpus*, 217; *HistNat*, 324.

33. *Corpus*, 217; *HistNat*, 325.

more tenuous the greasy liquid that fuels the fire, the gentler the fire's nature. He mentioned that a fire lit with what the *chymici* call *aqua vitae* (alcohol) or quintessence will not even burn linen. In contrast, things like carbon that burn very hot must contain a heavier and stronger type of grease.

Arias Montano also noted the different nature of the heat from fire and heat from the sun. The sun's heat was intended to sustain the natural cycles such as procreation and maturation on earth, while common fire was meant simply to burn and exhaust what it consumes. Neither is the light fire gives off the same as the sun's light. The reason is easily observed: the color and type of flames vary depending on what is being burned. (The light of comets and shooting stars, for example, is somewhat sulfuric, having been produced by carbon mixed with a purer sulfuric matter found in heaven.)[34] However, despite these very different natures, humanity has devised ways to use the heat given off by fire to simulate the beneficial heat of the sun in useful ways.

Composite Matter and Alchemical Decomposition

Having discovered in the arcane language of the Bible the existence of the primordial dual matter, MAIM, Arias Montano was eager to discover traces of it in things. According to his matter theory, vestiges of it persisted in composite bodies under many different appearances; therefore it should be possible to see, feel, or discover them through "skill of art/artistic skill, or investigation" (*artis peritia atque indago*). Arias Montano explained how this could be done using alchemical processes. The two principal natures of MAIM, when mixed with celestial influences and earthy exhalations, can still be recognized in the world as two distinct liquids—the greasy SEMEN and the salty TAL—and can be distinguished by how they react to fire.[35] The greasy part of the mixture or composite body easily catches fire, while the salty part is "anathema" to fire. Of the salty kind, the purest is a flavorless, colorless, and odorless water, *salis fundamentum*, that might not exist near the earth. At the other end of the density range of the salty liquids is mercury—as the *chymici* call it, he added—or in Latin, *argentum vivum*. Between these two are numerous other salty liquids of different densities that Arias Montano excuses himself from describing, considering it beyond the scope of his book and his skills. The

34. "Cometarum verò et crinitae stellae fulgor propter sulphuream, qua producitur, tametsi puriorem, materiam, colorem admixtum carbonibus consimilem habet.": *Corpus*, 219; *Hist-Nat*, 327.

35. *Corpus*, 215–16; *HistNat*, 322.

greasy (*pingues*) liquids that are part of MAIM are akin to olive oil, but far lighter and purer. Sulfur is at the other end of the density range of the greasy liquids. As the most rarefied of them, sulfur "governs" other liquids of its genera and can sometimes be smelled in the air (a reference to the unmistakable stench of anything sulfuric).[36] For Arias Montano, sulfur serves as the principle of combustibility, since it is found in some measure in all greasy liquids that burn easily. Like Seneca, he calls on it to explain all sorts of fiery phenomena.

Liquid solutions play an important role in Montanian matter theory, since they redistribute essential fluids throughout the earth. He noted, for example, that fluids containing a more tenuous nature in solution, such as wine or vinegar, flow more quickly than water or greasy fluids, while potable water and other tenuous liquids mix easily with other substances.[37] Just as the MAIM could be identified in compounds by their material similarity to salty or oily liquids, by extension different parts of a compound could be seen as having distinct forces, powers, and uses, since they retain in different proportions the nature of the dual liquids. He suggested readers consider the differences among the salts commonly used in food and medicine: *sal gemmae, sal terrenum, tartar, talcum*, or salts made from sea or freshwater (*nitrum*).[38] Each has distinct forces and uses, yet they share in the common nature of salt. Arias Montano also pointed out the importance of the cleansing property of flowing water. Notice, he urged, how when we want to clean a vessel we agitate and spin the water in it. Flowing water collects impurities and cleans as it carries away oily liquids suspended in or mixed with it. Greasy liquids, on the other hand, are responsible for composite matter.[39] Mixtures where the greasy MAIM predominates, such as all types of oils, are good for transporting medication into the body, since they penetrate the body slowly and to the marrow. Yet very pure oils, such as olive oils, are not suited for human nutrition because they lack a base (*fundus sive basis*), making it more difficult for the vital heat of the body to consume them.[40]

When Arias Montano turned to metals as another example of compounds, he left some further tantalizing clues about his knowledge of alchemical matter theory. All metals, he explained, derive from liquid sulfur found on earth, but they also contain some mixture of the salty liquid according to the efficacy of

36. *Corpus*, 176–77; *HistNat*, 282.

37. *Corpus*, 204; *HistNat*, 311.

38. *Corpus*, 186; *HistNat*, 292.

39. "[C]ui praecipuam definitionum in compositis corporibus praerogativam adscripsimus.": *Corpus*, 266; *HistNat*, 375.

40. *Corpus*, 268; *HistNat*, 377.

the spirits that incubated them.[41] He repeatedly identified sulfur, GHOPHRITH (גָּפְרִית, *gāprît*), as the vivifying juice of earth and as the first principle of all metals and minerals, "to which our philosophers gave the name *sulphur vivum*."[42] The primacy he ascribed to sulfur as constituent to all matter that burns was an idea common in alchemical thought. Given that Arias Montano placed mercury as the densest salty liquid and made sulfur its counterpart among the oily substances, his matter theory suggests that he was attempting to map his explanation of metals onto some version of a mercury/sulfur theory, such as the dyad theory of the Jābirian corpus.[43]

Arias Montano "saw" metals through alchemical spectacles and assessed the relation between greasy and salty liquids by carefully observing what happens in the alchemical furnace. Clearly his observations of the behavior of various substances brought to the fire testify that the author of the *History of Nature* was familiar with contemporary interpretations of matter theory and with alchemical practices. He noted, for example, that as a very fine powder GHAPHAR (עָפָר, *'āpār, pulvis*) can be used in the fire to supply other things with essential qualities, and he observed that the purer the substance, the more capable it is of imparting its essence to something else. In addition to understanding the role of fine particulates in chemical reactions, he also knew what to expect from subjecting mixtures to fire. Given that even the simplest organic distillations readily yield what appear to be oily liquids, while heating can reduce watery ones to salts, I think we are witnessing the work of someone well informed in the practices of alchemy.[44] This should come as no surprise: he had ready access to the largest and perhaps most sophisticated distillation laboratory of the sixteenth century in the *botica* of El Escorial, where in the years preceding the completion of the *Magnum opus* he had worked as librarian and royal chaplain.[45] Yet it is key to remember that his objective, more than following the precepts of the alchemists, was to bring in laboratory observations as empirical evidence to support his biblical natural philosophy.

41. "Id quod virtute spiritus ELOHIM, liquoribus per ministros spiritus incubantes efficitur. Sunt praeterea salibus multis, et aquis pleni partim dulcibus, partim sapore, odore, ac vi efficacitateque differentibus.": *Corpus*, 196; *HistNat*, 303.

42. *Corpus*, 350–51; *HistNat*, 258–59.

43. Principe, *Secrets of Alchemy*, 33–37.

44. As Debus explained, "The working chemist actually saw vaporous, inflammable and ashy portions every time he performed an organic distillation.": Debus, *Chemical Philosophy*, 113.

45. Rey Bueno, *Señores del fuego*; Rey Bueno and Alegre Pérez, "Renovación de la terapéutica real."

Although Montanian matter theory does not map directly onto the Paracelsian *tria prima*, which considered salt along with sulfur and mercury as the constituent elements of all material entities in the world, it does coincide with some aspects of Paracelsian thought. Like the Montanian version of Creation, the *Mysterium magnum* of Paracelsus (1493–1541) begins with the separation of a primordial matter to yield a world that is later sustained by the cyclical exchange of fluids between the higher and lower realms.[46] Later Paracelsians built on their preceptor's chemical understanding of Creation, with adepts such as Heinrich Khunrath (1560–1605), in his *Amphitheatrum sapientiae aeternae Christiano-Kabalisticum* (1609), giving interpretations that also relied on the Hebrew Bible for guidance.[47] But notwithstanding an inclination toward literal exegesis, their versions of Creation are substantially different from Arias Montano's earlier interpretation, although in Duschene, for example, there are references to dividing the waters of Genesis into fluids, one airy/mercurial and the other oily/sulfurous.[48] Despite the alchemical clues scattered in the *History of Nature*, the source of Arias Montano's ideas about a primordial matter with a dual nature remains a bit of a mystery (that is, only if we deny the originality of his ideas and his route to them purely through biblical interpretation of Hebrew etymologies).

Let us further consider two aspects of Montanian MAIM: its role as primordial matter and its role as matter with a dual nature, one nature being oily. As Arias Montano presented it, MAIM is not analogous to the Aristotelian prime matter, or *hyle*, in itself a vague concept much written about since the Middle Ages yet still an essential topic of discussion in any Aristotelian commentary.[49] What was clear from Aristotle's somewhat sparse writing on the subject was that his prime matter did not exist in the world separate from form and had resolved into five elements: earth, water, air, fire, and the supralunary aether. In Arias Montano, MAIM eventually does give rise to earth, water, air, and fire, but these are clearly compounds of MAIM, not elements. By contrast, in Stoic

---

46. Debus, *Chemical Philosophy*, 87–94, 111–13; Hedesan, "Mystery of the *Mysterium Magnum*."

47. Forshaw, "Vitriolic Reactions," 122. The same is true for the chemical interpretation of Genesis by Robert Fludd: Walton, *Genesis and the Chemical Philosophy*, 82–90.

48. Joseph Duchesne's *Liber de priscorum philosophorum* (1603), as cited in Debus, *French Paracelsians*, 54.

49. Diego de Zuñiga in his *Philosophiae, Prima pars* (1597) argued for the existence of a prime matter shared between the heavens and the earth. Although he noted the existence of a perfect celestial substance akin to the Aristotelian aether, it was the same type of matter present elsewhere in the universe: Zuñiga, *Física*, 187–88.

philosophy, Seneca (following Thales of Miletus) posited that although fire was the element the world resolved itself into at the great conflagration, water was the starting point of the world as we know it.[50] As we saw in an earlier chapter, in the hexameral tradition of the Christian Latin West, particularly in literal exegetical exercises, water was rarely considered the sole primordial element. Patristic interpretations generally considered Genesis 1:1 as indicating that God had created some type of matter to act on, whose nature was then differentiated into the four Empedoclean elements in subsequent days. Saint Basil, though varying somewhat from this notion, explained, "Thus, although there is no mention of the elements, fire, water and air, imagine that they were all compounded together, and you will find water, air and fire, in the earth."[51]

In the Jewish tradition, the work of Philo of Alexandria (ca. 20 BC to ca. AD 50), *On the Creation of the Cosmos according to Moses*, entirely skips the discussion of the separation of the waters on the second day.[52] Among the exegetical Midrashim commentaries, the verse-by-verse commentary on Genesis of the *Bereshith rabbah* (post-400 CE),[53] explains the elemental composition of the world as a refutation of those who argued that God had worked with some eternal matter signified by the *tohu va-bohu* (תֹהוּ וָבֹהוּ, *tōhû va-bōhû*), to which others added "darkness, water, wind, and the deep."[54] By the twelfth century, Maimonides (1135–1204) was advocating for Aristotelian physics as an exegetical aid for interpreting Genesis. He thus found in the first two verses a clear description of the creation of the four elements. Although he noted the division of the waters, this simply indicated to him that "there has been one common element called water, which has been afterwards distinguished by three different forms; one part forms the seas, another the firmament, and a third part is over the firmament, and all this is separate from the earth."[55] Another camp of Jewish scholars, principally kabbalists, preferred to retain the

50. As Seneca explained in book 3, chapter 13.1, "[Thales] thinks that it was the first element, and everything arose from it. We too hold the same opinion, or something close to it: for we say that it is fire that seizes control of the world and turns everything into itself; then it becomes faint and weak and dies down, and when the fire is extinguished, nothing else is left in nature except moisture. The hope of a future world lies hidden in it. So fire is the end of the world, and moisture is its starting-point.": Seneca, *Natural Questions*, 33.

51. Homily 1.7 of the Hexameron, in *St. Basil, Letters and Select Works: Hexameron*, 8: 1395–96.

52. Philo, *On the Creation of the Cosmos according to Moses*, 55.

53. Strack and Stemberger, *Introduction to the Talmud and Midrash*, 304.

54. Freedman and Simon, *Midrash rabbah*, 2, 8.

55. Book 2, chapter 30; Maimonides, *Guide for the Perplexed*, 214.

Platonic and Neoplatonic interpretation of Genesis based on the belief that the Torah contained all there was to know about the world. Genesis was thought not to describe the physical process but to explain the coming into being of the world through the implanting of forms by the action of the ten emanations from the godhead or *sefirot*. This line of thinking extended, for Abraham bar Hiyya among others, to interpreting *tohu va-bohu* as representing the metaphysical principles by which matter (*tohu*) took on form. (*Tohu* in this interpretation is sometimes identified as the *hyle* of the Greeks or described as a chaotic mass of matter.)[56] Just as with Christian exegesis of Genesis, the several Hebrew explanations of the book varied according to natural philosophical commitments. Paul Fagius collected a number of these views in his *Exegesis sive expositio dictionum hebraicarum literalis et simplex* (1542).[57] None of these interpretations posit that the primordial waters were an element or that they had a dual nature.

As for primordial matter having a double nature, sources for this type of conceptualization have proved difficult to find. The closest to Arias Montano's ideas, in my estimation, are some medieval alchemical theories that relied on the distinction between aqueous liquids and fatty or unctuous ones.[58] These theories drew on Aristotle's explanation that moisture held earthy substances together, and that when this moisture contained oily liquids it resisted evaporation and allowed the cohesive substance to persist. By the Middle Ages and through the influence of Arabic sources, the notion of unctuous moisture became a suitable explanation for things such as the persistence of metals in fire and the occurrence of sulfur in all things that burn. This tradition, however, does not hold unctuous moisture to be primordial matter. Again, it is worth noting that almost half a century after Arias Montano, Jean Baptiste van Helmont (1579–1644) saw the waters of Genesis 1:2 as the primordial element and the *archeus* (or "chief workman of the efficient cause") as an enlivening force not unlike ELOHIM.[59]

56. Tirosh-Samuelson, "Kabbalah and Science in the Middle Ages," 505.

57. According to Rabbi David Kimhi in his book of roots, "Thohu est res cui non est similitudo ac figura, sic tamen disposita et praeparata, ut quamvis recipere possit similitudinem et formam; id quod graeci hylin vocant. Ceterum Bohu, est ipsa forma, hoc est, res cui inest vis et potentia, ipsum Thohu induendi similitudine et forma.": Fagius, *Exegesis sive expositio dictionum hebraicarum literalis et simplex*, 9.

58. Martin, *Renaissance Meteorology*, 84–85; Freudenthal, "The Problem of Cohesion between Alchemy and Natural Philosophy," 107–12; Freudenthal, *Aristotle's Theory of Material Substance*, 161–64, 200–207.

59. Newman and Principe, *Alchemy Tried in the Fire*, 62–63; Helmont, *Oriatrike, or Physick Refined*, 35, 48.

MECHANICS

Arias Montano never addressed motion or mechanics as a distinct topic in the *History of Nature*. Instead, he integrated them into the general discussion of specific divine and natural entities. Since he dwelled only on those aspects he could use to support a particular argument or interpretation, he did not address many topics that would typically be discussed in, for example, a commentary on Aristotle's *Physics*, such as projectile and accelerated motion, magnetism, or other topics with particular valence in mechanics. He took other aspects as unquestioned principles, generally those that could be corroborated by everyday experience—for example, that the natural motion of light bodies is to rise, while heavier bodies fall. This type of motion is implicit in Arias Montano's description of the separation of the waters during Creation.

The Transmission of Forces

We find an example of Arias Montano's piecemeal approach to the mechanics of motion in his discussion of God as the central locus of motion, discussed in chapter 8. To sustain his argument about how a center conveys its nature to all that lies within the periphery it defines, he had to establish what he understood to be the mechanical properties of spherical bodies according geometric principles as well as the principles of weights and measures.[60] This led him to consider the mechanics of linear and circular motion, of natural and violent motion, and of change in general. He then used those conclusions as his point of departure for a discussion on the transmittal of incorporeal virtues (*virtutes*) and forces (*vires*) between the center of a body and its periphery—that is, between God and the world.

Arias Montano defined "center" as the middle of any one thing, be it a figure or a body, but also as virtues of the forces (*virtutes virium*) or the efficacies of motion, stillness, or stability. This center has the same properties as the geometrical point: it is indivisible and cannot be penetrated. It is essential to motion, since "nothing can move unless on its center."[61] Likewise, a center is necessary for anything to remain still or come to a stop. It can also be said that a center is what something moves or is moved on, what it may be displaced from,

---

60. "Atque haec ad id quod suscepimus indicandum argumentum ex figurarum ponderum ac mensurarum ratione mutuata deprompta fuerint nunc satis, quae depictis ad sensum characteribus sive notis clariora evident.": *Corpus*, 160; *HistNat*, 265.

61. "Nihil enim moveri nisi super centrum potest.": *Corpus*, 160; *HistNat*, 264–65.

or what it remains stable on. A center defines the middle of the force of whatever it is the center of, either while still or while in motion. Therefore the whole of the thing partakes of the virtue and efficacy of the center, while the center, in turn, partakes of every area and part of the thing, whether near the center or far from it. He used as an example an "impetus" applied anywhere on a body; the center ensures that some of the stimulation is transmitted throughout the body by means of a straight line.[62] He was convinced that understanding the center of something reveals not only its shape, but also its forces, weight, efficacy, and the way it moves.[63]

Ultimately, Arias Montano's objective in the discussion of God as center was to construct an argument that supported divine immanence in nature; in doing so he left evidence of being well informed in geometry and mechanics. For example, he understood the notion of transmitted force as something that took place when bodies came into contact, as with an impetus. But far from being a mechanist, he also understood incorporeal virtues as being transmitted the same way, by means of lines of force emanating from any center and throughout the surface of a body. His explanation seems to presuppose a completely filled world through which the forces of the center are transmitted, but he does not say this explicitly.

The "Actions of Things" and Motion

Arias Montano's ideas about motion were based on a strict distinction between "things" (*res*) and "actions of things" (*actiones rerum*). He explained the difference between these two concepts from a grammatical perspective in *Adam, sive De humani sensus interprete lengua*.[64] He had first broached the subject in the *De arcano sermone*, but in the *Adam* he simplified the definition considerably.[65] Recall that he believed everything existing in the world could be divided into two categories: "things" (described by nouns) and the "actions of things" (verbs). "Things" may include the unseen and spiritual, while the

---

62. Arias Montano based the explanation for this on the geometrical principle that an entity cannot be divided into anything other than what makes it up. A point cannot be penetrated or divided by a line or a line by a surface. The opposite is true: a line can be penetrated or divided by a point and a surface by a line. Therefore, a line cannot penetrate the center, but it can penetrate a surface and transmit the virtue of the center to the surface: *Corpus*, 160; *HistNat*, 265.

63. *Corpus*, 172; *HistNat*, 278.

64. The surviving copy of the *Adam* manuscript made by Pedro de Valencia likely was copied in the early 1590s; Arias Montano, *Tratado sobre la fe*, 75.

65. Arias Montano, *Libro de José*, 93–94, 404, [1v].

"actions of things" are what a thing is moved by (*quo quicquam exerceatur*).[66] Both notions rely on the concept of time. A "thing" is something that "is" and persists unchanged through time. Arias Montano associated it with the immutability of the created world, its image persisting unchanged in our minds. "Actions of things," in contrast, are lawlike behaviors that take place in time.[67] His definition suggests that the potential for an action also falls into this category. The distinction between "things" and "actions of things" was in harmony with the Montanian hermeneutics of nature, which assigned permanence of type to created "things" and—not unlike Aristotle—considered change or "actions of things" as a type of motion. In the Montanian scheme, however, motion could originate at any time by the agitation of the spirit ELOHIM or at a predetermined time from the internal spirit of something guided by its GHOLAM, which had been determined at Creation. Thus the one aspect of mechanics that Arias Montano could not avoid addressing was motion.

He understood motion in corporeal things as originating from two possible sources: one was an internal source he described as the virtue and efficiency of the spirit of the thing, the other was an external impulse acting on the body. He defined the incorporeal spirit of a body as what gives shape to, drives, and moves the body.[68] Although it is finite, the incorporeal spirit can move what contains it as well as other things, all according to the extent of its force and weight.[69] Although the corporeal spirit can penetrate other bodies, it cannot be penetrated by denser bodies, but only sliced. Arias Montano's explanation of how an internal spirit moves bodies does not go much further than this. However, in what he did write on the subject, it seems he conceived of natural motion as initiated by the corporeal spirit of a body, while violent motion was the consequence of an external "impetus," a reference to an impressed force of some kind.

His clearest example of such motions originates not from a biblical passage but from personal observation. While gazing at the freshwater spring that flows to this day at the Peña de Alájar he surely must have marveled at how water rose through the mountain to ultimately feed the fountains he had installed at

66. Arias Montano, *Tratado sobre la fe*, 98, 99.

67. "Actiones ea, quae artibus iussis proficiscuntur, et cum tempore incipiunt et promoventur et absolventur, atque ita tempore coniunctae sunt, ut sine ipso expediri non possint, ideoque animo cum mente hominum, cum temporis ratione adiuncta plerumque cogitantur, tametsi separatim a tempore cognosci possint.": Arias Montano, *Tratado sobre la fe*, 99.

68. "Corpora informatur, aguntur, et moventur.": *Corpus*, 166; *HistNat*, 272.

69. "Itaque corporeus spiritus quamquam finitus sese ipsum movet, atque alia pro virium ponderúmque modo.": *Corpus*, 169; *HistNat*, 274.

his retreat. Given that he understood subterranean water to be largely seawater drawn into the earth through the *fontes abyssi*, his challenge was to explain how water (a type of heavy matter) could flow uphill, against its natural motion. He began by describing two types of motion, one "spontaneous" (*sponte*) and the other "coaxed" (*coacta*).[70] (Thus he avoided tangling with Scholastic distinctions between things that exist *per se* or *per accidens*.) He established that observing how a part of a homogeneous body behaves tells one about the behavior of the whole, both its magnitude and force and its efficacy and movement.

Spontaneous motion occurs as bodies continuously seek their place, moving in that direction unless they are impeded by another body. They move this way without an external impulse, driven by an innate spirit that never wanes in its desire to find either rest or its center.[71] Thus spontaneously moving water flows in a manner consistent with the Aristotelian doctrine of proper place.[72] For biblical examples of this motion, Arias Montano noted the way the spirit ELOHIM agitated the waters as it constantly hovered over them and how celestial bodies move in the heavens. This spontaneous movement is perpetual, that is, until it reaches its end (*finis suus*) as assigned by God. Coaxed motion, on the other hand, begins and continues because another force has been impressed upon a body. It always resists motion so far as its own internal force is capable of doing so. The resulting motion wanes as the impressed force wanes. While spontaneous motion is perpetual unless completely impeded by something else, coaxed motion cannot be perpetual by its nature, but may continue as long as whatever force moves it does not wane and until the thing moved withstands the movement.[73] When coaxed motion is added to spontaneous motion, both motions increase, but they also may "impede" each other, and the coaxed motion eventually falls off or desists. Yet when a coaxed motion yields to a spontaneous motion, motion can persist if the spontaneous motion is natural to the body.

To illustrate these two types of motions, he used "an instrument from the mechanical discipline," a pump or *machina ctesbica* (fig. 9.1, left). First he

70. *Corpus*, 208; *HistNat*, 315.

71. "Quae moventur corpora vel sponte vel coacta moventur. Ea moveri sponte dicimus, quae locum suum petentia, nisi aliàs impediantur aut prohibeantur, quo usque nacta fuerint, moveri non desinunt: idque nullius impulsu externo agunt, sed inveniendae propriae quietis, hoc est, centri sui cupiditate insitoque spiritu aguntur. Hoc motus genere aqua in Abyssum rotata voluitur.": *Corpus*, 208; *HistNat*, 315.

72. Aristotle, *Physics*, 4.8 and 8.4–5.

73. *Corpus*, 208; *HistNat*, 315.

needed to explain how a siphon works. He pointed out that a straw standing in a vessel of water will fill to the same level as the water in the container, with the rest of the straw filled with air. Removing the air in the straw by sucking with the mouth or a machine will drag water into the straw by means of the "impulse to drag" (*protractionis impulsus*).[74] Why does this happen? The answer was entirely predictable: because it is not possible to completely vacate a place, since a vacuum cannot exist in nature. The specific answer explained that under these conditions the water seeks to avoid any greater and common discomfort (*quando maioris & communis incommodi vitandi causa*) and will swiftly move into the place vacated by the air. Water will remain in the straw as long as air is kept out, but the slightest intrusion of air will cause the water to recede to its original level. If the siphon is extended so that the end of the straw is below the bottom of the vessel, that is, closer to the "center," the water that had been coaxed into the tube will now flow out of it with perpetual motion (*perpetuus motus*) as long as there is liquid in the vessel. With the siphon experiment, Arias Montano found a way to explain the combined case of coaxed motion—impelling the water into the tube by sucking—followed by a spontaneous motion, as the water continues to move to a "better place" as long as the end of the straw is below the bottom of the vessel. The coerced force is now joined with the spontaneous force of the water seeking its proper place. These same principles applied to the water pump (which he noted he had frequently seen used in Flanders). In this case the water intake of a pump functions like an animal's larynx, opening and closing the entrance to the passageway that leads the water upward. When the lever "hits" the water in the tube, it impels it upward with each stroke. Once the water reaches the opening at the top of the pump, it can flow freely (fig. 9.1, right).

The explanation of the siphon at work in a pump provided all the theoretical material he needed to fulfill his main objective of explaining how water from the sea enters the land and supplies springs. In fact, Arias Montano maintained that a pump is a mechanical contrivance or work of artifice that simply mimics natural springs, because the natural communion between land and sea requires both spontaneous and coaxed motions. The coaxed motion in this case is provided by ELOHIM's agitating the water and and impelling it into the *fontes abyssi*. Once inside the earth's passageways, the water cannot escape back out because of the weight (*moles*) of the sea behind it, nor can it overcome the rush of more incoming water. Under this coaxed motion it will travel inside the earth until it finds an opening. At that point the spontaneous

---

74. *Corpus*, 208–10; *HistNat*, 315–17.

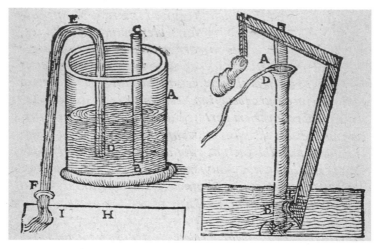

FIGURE 9.1. *Machina ctesbica* (left) and pump (right). Arias Montano, *Naturae historia*, 209, 210. Biblioteca Histórica de la Universidad Complutense de Madrid (BH FLL 14614).

motion natural to the spirit of water takes over, and the water flows over the land collecting the juices necessary to make the land and sea fertile and maintaining the divinely designed balance between land and sea.

This section of the *History of Nature* has not gone unnoticed by historians. Some have argued that Arias Montano was the first to develop the modern concept of atmospheric pressure based on the weight of air, preceding by decades the works of Isaac Beeckman, Evangelista Torricelli, and Blaise Pascal.[75] They find that because he gives air "a fundamental role" in explaining the limitations of pumps in raising water above a certain height, his work "became a precursor of the idea of atmospheric pressure."[76] Let us examine carefully Arias Montano's presentation of the pump problem. When discussing how the water, having been impelled up the straw by sucking, returns to the vessel once a small intrusion of air is permitted, he wrote, "The water, which had climbed on account of the sucking strength, ceding with the sudden admittance of air,

75. This was an opinion among some exponents of an older, often positivistic historiography of Spanish science, such as Felipe Picatoste, Acisclo Fernández Vallín, and Eduardo Lozano y Ponce de León, but it is also shared by some contemporary ones: see Cobos Bueno and Vaquero Martínez, "Benito Arias Montano y el estudio de los fluidos," 89. Cobos and Vaquero's study is inclined to overinterpret Arias Montano, often seeking modern definitions where none exist and, most important, neglecting in their analysis the action of ELOHIM, which was central to Arias Montano's explanation.

76. Ibid., 92.

subsides to its proper region and level."[77] There is no reference in the quotation to the air exerting its weight on the water. Arias Montano described the action as water ceding (*cedens*) to the air rushing in, which suggests that this violent action pushes the water down. Furthermore, water is said to be returning to its proper place (*propria regio*), a clear reference to Aristotelian doctrine. Finally, when it is time to show how these mechanical examples support his view about the movement of water that supplies mountaintop springs, he calls on the agitation of the spirit ELOHIM, *not* the weight of the air, to supply the force that impels the water into the earth. This, along with the explicit statement about the impossibility of a vacuum, in my view does not support the assertion that Arias Montano developed some notion of atmospheric pressure. He was operating well within the Aristotelian doctrine of proper place for bodies subjected to a force opposing their natural movement.

Another example of Arias Montano's piecemeal exploration of mechanics occurs in his consideration of the flow of water as it redistributes life-giving juices to the land. His explanation hinges on the physical shape of water's constituent matter. However, he supports his argument without making corpuscularian claims, simply basing it on the strength of biblical citations and Aristotelian principles of natural place. He begins with the premise that certain characteristics (*ratio*) are shared among bodies such as earth, heaven, and liquids, while other characteristics are particular to an individual body.[78] One of these latter characteristics is weight; it varies in particular bodies, from heavy to light. We know that heavier things settle in lower places while lighter things occupy higher places, but neither the characteristics bodies have in common (*communis utrique generi ratio*) nor their respective weights determine the particular motion of each body (*proprius singulorum motus*). The way a body moves is determined more by its shape than by its weight. Round things "love to roll and spin," while bodies with other shapes prefer to move in a straight line. A careful observation of water droplets shows that the proper nature of

77. "Aqua, quae propter illatam vim ascenderat, admisso repentè aëri cedens, propria regione ad libellam residet.": *Corpus*, 209; *HistNat*, 316.

78. "Alia terrae, caeli, liquorum sive aquarum ratio est qua corpora sunt; alia verò qua singularibus ac propriis definiuntur naturis. Corpora omnia vel gravia, vel levia sunt, vel medium quiddam inter grave et leve habent: quod non tam ad motum proprium, quam ad situm facit. Nam gravia omnia eo quòd gravia sunt, commodius in inferiore loco sidunt: contrà verò, levia superiorem occupant locum. Verùm haec communis utrique generi ratio non indicat proprium singulorum motum et exercitium; hoc enim à figura potiùs quàm à pondere aut levitate proficiscitur. Quae natura rotunda sunt, ea rotari et volvi amant, aliumque motum quemcumque non sponte, sed invita exercent.": *Corpus*, 203–4; *HistNat*, 310.

water is spherical; therefore, given that the movement of a sphere is circular, it is proper for the movement of water to also be circular. Thus water and other liquids exert their nature in a circular motion. This is why we observe water moving gently over flat terrain but starting to roll and tumble violently when it meets an obstacle and begins rotating to overcome the impediment. The prophet Amos testified to this: "But let judgment run down as waters, and righteousness as a mighty stream" (Amos 5:24 KJV). Arias Montano preferred to translate the Hebrew IIGALGEL (from גָּלַל, *gālal*) in the quotation as "to roll" (*volvo*)—as he did in the Pagnino edition in the Antwerp Polyglot—rather than the Latin *revelare*, "to uncover" used in the Vulgate.[79]

No less fascinating is his description of motion in a medium. This discussion focuses on the flight of birds and how fish can swim.[80] There are three factors to consider when exploring movement through the length, width, and depth of a medium, he explained. Downward vertical transit is easier than moving upward because the body is helped by its weight. Upward movement requires greater efficiency of spirit, yet the farther removed the body is from its center, the easier it becomes to fly or swim.[81] This is because the body is in turn sustained by the mass beneath it and pressed on less by the one above.[82] Horizontal movement in a straight line, always keeping the same distance from the "bottom," whether along the length or the breadth of the medium, requires sustained and steady effort. Again we see that Arias Montano is comfortable with some aspects of the Aristotelian ideas about natural place, but with the caveat that some special conditions apply when a body transits through the middle of the medium. He seems to suggest that a body in motion can find a point equidistant along the vertical axis that balances the heaviness of the medium above it with the support provided by the medium below it. This point allows the body to move through the medium more easily in a straight horizontal motion. Although Arias Montano does not use the terms, it suggests an understanding of buoyancy.

Thus, he explained, flying and swimming obey these rules because they entail cutting or penetrating a medium by the *impetus* or thrust of the animal in a particular direction, the only difference being the density of the medium. Since cutting—of any kind—needs strength and acuity, it must be that birds and fish

79. AntPoly, 7:144.

80. *Corpus*, 297–98; *HistNat*, 407–8.

81. "Quo magis à sua regionis fundo sublatum est, eò levius habet.": *Corpus*, 298; *HistNat*, 408.

82. "Namque et ab inferiore partis mole crassiore sustinetur, et minus idem à superiore premitur.": *Corpus*, 298; *HistNat*, 408.

have been given a particular force of spirit or wind (*spiritus vis*) in their wings and fins that allows them to cut through the medium. With this brief physics lesson as his preamble, Arias Montano could explain the preeminence and dignity that high-flying birds such as eagles have in the Bible. They must have the greatest force of spirit in their wings, allowing them to soar to great heights. This is why Solomon found the eagle's flight wondrous (Prov. 30:18–19).

*

Although Arias Montano's treatment of meteorology addressed the themes that had become associated with the topic's early modern formulations, we can attribute this piecemeal treatment of mechanics and matter theory to the literary approach he chose for the *Corpus*. He never forgot that he was writing a *history* of nature during the seven days of Creation as dictated to Moses. One may suppose that, had Moses discussed projectile motion in the account of Creation, our author would have been compelled to discuss that as well. In all fairness, Arias Montano did not set out to write a systematic commentary or to refute ancient natural philosophy; thus he did not have to engage with subjects he ultimately found to be "fabrications" of the philosophers. In the Montanian hermeneutics of nature the Bible's description of the behavior of bodies and its corroboration of his own experience and observation were sufficient reason to take certain aspects of mechanical behavior as axiomatic.

CHAPTER 10

A Biblical Natural History

A quarter of the *History of Nature* is devoted to describing plants and animals. By this point in his text, Arias Montano had reiterated that these were created primarily to be useful to humanity. Yet plants' and animals' usefulness went beyond simply providing sustenance: they were also *exemplaria* of natural theological arguments, since God had placed some animals on earth so their intricate design would testify to the Creator's power. This was true of his precious mollusks, which he collected avidly.

The natural history sections of the *History of Nature*, perhaps even more than those on cosmology or meteorology, show that Arias Montano was a careful observer of the plant and animal kingdoms. Given the purpose he ascribed to them, it is not unexpected to find a plant or animal's utility highlighted in his descriptions. Recall that he considered a thing's purpose the key to finding out its true nature, which in turn was one of his objectives in writing a history of a plant or an animal. Yet the *History of Nature* displays little of the learned empiricism typical of the natural histories of his day. There are no long exercises in collecting and comparing descriptions from ancient and modern authors, nor is there a single explicit citation of Pliny or any contemporary naturalist. Arias Montano had little room and even less patience for a Pliny or a Dioscorides. Such was his adherence to the biblical text that when he discussed animals he preferred to follow King Solomon's classification rather than any alternative proposed by the ancient philosophers.[1] So it remains unclear how far he drew on natural historical reference works, as surely he must

1. *Corpus*, 313; *HistNat*, 424.

have done. Arias Montano's short descriptions of distinct plant and animal species are largely the result of a close study of intertextual references within the Bible, his usual etymological deconstruction of the Hebrew name of the plant or animal, and personal experience. Each section begins with a general description of how the plant or animal came into being during Creation, followed by a discussion of where it fits into the order of nature relative to other species. The descriptions of particular species try to identify analogous counterparts that would have been familiar to readers or, if this was not possible, to clarify the characteristics of the plant or animal as mentioned in the Bible. This approach made the *History of Nature* particularly useful as a reference tool for biblical interpretation but perhaps unsatisfying for contemporaries used to the encyclopedic approach of natural histories of the late Renaissance.[2]

During the first half of the sixteenth century, botany had undergone a significant transformation, largely through humanist naturalists' engagement with the ancient works of Theophrastus (ca. fourth to third century BC) and Pedanius Dioscorides (ca. first century AD). While revisiting the work of Theophrastus, they had to consider two main problems: the challenges presented by his plant classifications and how to deploy the empirical tools this student of Aristotle brought to bear in making these classifications. In Dioscorides they encountered what had become the model of books on materia medica and a plant taxonomy organized around plants' medicinal properties. Yet as sourcebooks of factual information about particular plants, the works of both ancient Mediterranean authors proved increasingly inadequate, particularly to northern European naturalists. To correct or amend these works, some resorted to the humanist's usual tool kit, mostly textual comparison and philological analysis. But by midcentury, after publication of the works of Otto Brunfels (ca. 1489–1534) and Leonard Fuchs (1501–66), a new epistemic standard began to prevail for natural histories, particularly herbals. For, whereas learned empiricism was often still the norm, in the form of reliance on borrowed firsthand observations of a trusted authority, increasingly the epistemic basis of botany became empirical, centered on direct observation of nature, personal experience with a plant's properties, and realistic illustrations of a specimen.

Arias Montano was well acquainted with a generation of naturalists who flourished in the Low Countries, among them Rembert Dodoens (1517–85) and Carolus Clusius (Charles de l'Escluse, 1526–1609). Through the work of

---

2. One of the few discussions of Arias Montano's contributions as a botanist can be found in Cobos Bueno, Oyola Fabián, and Vallejo, "Dimensión botánica del humanista extremeño Benito Arias Montano," 127–65.

Dodoens, he became aware that the principal preoccupation of the botany of his day was finding a taxonomic scheme to account for the variety observed in nature. This is likely the reason he made this the central concern of his section on plants. Arias Montano's descriptions of plants and animals were an attempt to unravel and bring some order to what he perceived as the confusion caused by the profusion of species God created. Learned men, he explained, had attempted to study plants and trees by grouping them according to their particular shapes (*singulis forma*), something he himself had done while traveling and inquiring, until the careful examination of the Sacred Scriptures showed him a shorter way to accomplish this study.[3] This was one of the insights he considered a divine gift he had reaped from studying scripture. It was reason enough for him to put forth a biblical taxonomy and a new method of identifying plants to situate them within it.

He considered the systematic study of plants and their classification a virtuous pursuit, one he had practiced during his travels and that gave him great benefit. No one but Solomon, he explained, had mastered the discipline and found God's favor in doing so. The wisest of kings had not neglected any aspect of Creation, studying the cedars of Lebanon with the same zeal as he devoted to the humble hyssop growing out of the wall. While Arias Montano acknowledged the work his many botanist friends had done over the years in identifying plants and attempting to classify them, he believed that the divine grand plan had not yet been discerned and that confusion reigned when it came time to not just classify plants, but also identify and name them. In the *History of Nature* he was putting forth God's taxonomy, and he explained that he would rather leave it to his "most studious and skilled" (*studiosissimi et peritissimi*) botanist friends to complete the work. The printed edition of the book mentions in a single passage some of the most important naturalists of the day whom he considered to be doing this work—Dodoens, Clusius, and Matthaeus Lobelius (Matthias de l'Obel, 1538–1616)—but he had intended to add a few other names to the list: Bernardus Paludanus (Berend ten Broecke, 1550–1633), Jacob Monaw (1545–1603), and Lorenz Scholz von Rosenau (1552–99).[4] The original list also included two friends from his circle of close

3. "Id quod nobis quoque studium peregrinantibus & inquirentibus frequentabatur magnum fecit negotium, donec divinorum scriptorum observationi attentis compendii ac methodi in hac etiam parte inveniendi ratio indicata fuit. Quod etiam beneficium divinis quoque chartis referimus acceptum.": *Corpus*, 238; *HistNat*, 345–46.

4. *Corpus*, 241; *HistNat*, 349. In a letter to Moretus on 16 January 1596, Arias Montano mentions wanting to add the other three naturalists to the list of those in the manuscript. Yet the names of the last three—Paludanus, Monaw, and Scholz—never appeared in the published

associates in Seville, Simón de Tovar and Francisco Sánchez Oropesa. But he did not want his call for further study of botany along the lines of his own taxonomic scheme to go unnoticed. In a letter to Clusius written some twenty years after they parted ways in the Low Countries, he mentioned that the first part of the *History of Nature* included a brief treatise on how to identify plants in his "sacred philosophy" and inviting him, Tovar, and other friends to take on the challenge of enhancing it. Clusius, he hoped, would lead the enterprise "for the honor of God our creator."[5]

Arias Montano considered the diligent study of plants and animals in themselves a necessary and virtuous occupation whose ultimate purpose was to explain divine Creation, but he saw his contribution as limited to pointing out that the Bible offered a comprehensive—even divinely inspired—taxonomy. And though he did not consider himself well suited for the arduous task of plant classification and study, he encouraged others to carry on with the project:

> Yet it is necessary for the diligent and studious man of this agreeable and gratifying and, in addition, most useful knowledge of plants to carry out his work with diligence, so that he does not become weary working and dealing with very common and most vulgar things and is not ashamed of investigating, searching, and following strange things, and even very rare and singular ones. Only then will he master this discipline with great reward for his effort and as one who for his diligence and learning the sacred history recommends Solomon as a leader, guide, and model.[6]

He intentionally relegated the specifics to another project, mentioning on several occasions that certain aspects of nomenclature were to be discussed in the third part of the *Magnum opus*, the never-published *Vestis*.[7]

GENERATION AND PERPETUITY OF TYPE

Within the Montanian hermeneutics of nature the question of perpetuity of type was answered by the stability of genera implied by the acts of the third day

---

book: Dávila Pérez, *Correspondencia en el Museo Plantin–Moretus*, 2:836, and Morales Lara, "Otras tres cartas," 225–26.

5. Letter from Arias Montano to Carolus Clusius, 19 February 1596, Seville; translated in Clusius, *Correspondencia de Carolus Clusius con los científicos españoles*, 108–11.

6. *Corpus*, 279; *HistNat*, 389.

7. *Corpus*, 243; *HistNat*, 351.

of Creation and by the phrase *juxta genus suum*, "according to its own kind" (Gen. 1:11–12 VUL).[8] This ensured that a genus retained its true nature and passed it on to the next generation in perpetuity. Yet this did not imply that the species of plants had appeared all at once but rather, Arias Montano believed, meant that they had been ordered during Creation to appear at opportune times and according to their own innate quality (*ingenium*). The arcane language expressed this intent with the phrase "and it was so" (*et factum est ita*) or VAIHI CHEN (וַיְהִי־כֵן, *wayĕhî-kēn*), which suggested to our interpreter that the action was done in a way that was "undeviating, suitable, and according to the rule" (*rectè et aptè ad amussim*).[9] For plants, the seeds or sprouts were responsible for perpetuating the species, so Arias Montano based his method of plant identification (if not his taxonomy) on seeds. In Montanian natural philosophy, perpetuation depended on the active participation of the dual liquids MAIM and also the spirit ELOHIM, both following the command IEHI that gave genera their essential character. For animals the problem was more complex, since the exegetical exercise also had to account for spontaneous generation, sterile species (e.g., mules), and the different mechanisms suggested by live births and generation from eggs.

In explaining the generation of plants from seeds, Arias Montano perhaps unconsciously turned to the Aristotelian notion of change as a form of motion. He therefore needed a source of motion to initiate germination, and he identified it as a seminal spirit (*seminalis spiritus*) contained in seeds. Germination, he wrote, needed three things: the two types of primordial liquids MAIM in a given proportion, and a seminal spirit. The spirit ELOHIM moved these and heated the dual liquids until they were reduced and transformed into the two main parts of a plant: root and stem. Arias Montano found in the etymology of the word SORES (שֶׁרֶשׁ, *šereš, radix*) the clues to the structure and function of roots. One element of the word, SR, meant principal or perfect, while the other, ES, stood for force and nature of fire. Therefore the root of the plant was really its "head," since that was its most efficient part and the seat of the plant's seminal force.[10] The principal function of this force, hot by nature, was to heat the two liquids inside the seed, joining them with other liquids surrounding the seed, identified as the fecundity of the earth.

8. "Et protulit terra herbam virentem et adferentem semen iuxta genus suum lignumque faciens fructum et habens unumquodque sementem secundum speciem suam." (Gen. 1:12 VUL).

9. *Corpus*, 238; *HistNat*, 346.

10. "Est itaque radix certum et verum planta caput, principe, id est, efficacissimo ingenuoque igne praeditum, in quo nativa seminalis spiritus vis praecipuam sedem efficientiamque obtinet.": *Corpus*, 244; *HistNat*, 353.

Once germination took place, the seminal juices of the seed continued to provide force and efficiency. Because of their active functions, he identified the seminal juices as masculine and the fecundity of the surrounding earth as the feminine receptacle where nutrition and gestation took place. The result of this "embrace" was a tiny plant that contained in its "genital field" (*genitale arvum*) all the parts of the grown plant it would eventually become, including its future seeds. The seminal spirit housed in the plant's roots continued to oversee the mixture of the plant's liquids with those of the earth, ensuring the plant's growth.[11] To Arias Montano, this mechanism of plant germination and growth explained perfectly the parable of the sower found in the New Testament, especially the verse, "Other seed fell on rocky ground where it had little soil. It sprang up at once because the soil was not deep. And when the sun rose, it was scorched and it withered for lack of roots" (Mark 4:5–6 NAB). The plant withered, he concluded, because the rocky ground at the root of a seed—its "head" or principal part—could not find enough earthy juices to nourish its growth.

Although the germination of a seed was brought about by agents solely contained in the seed—and nourished by a feminine embrace—it was perfectly clear to Arias Montano that only the animals present at Creation could reproduce through the union of male and female.[12] He returned to the word MIN (מִין, *mîn*), which as we will see below was essential to his discussion of the generation of plants, but with regard to animals he found that the arrangements of the letters in MIN signaled the union of the masculine letter *mem* (מ) and the feminine letter *nun* (נ) united by the letter *iod* (י). The iod also stood for the number ten, which was the sign of perfection and the absolute and perfect number "beyond which nothing can exist still" (*ultra quem nihil aliud superesse potest*).[13] Therefore the word MIN, taken as the union of the male and female, meant "perfect genus." Further, the *iod*, as the very beginning of a pen stroke, signals the offspring or beginning of something that will later become larger. It is worth noting here that Arias Montano was not calling on the asso-

11. "Itaque seminis vis et spiritus praecipuam in radice sedem habet, efficacitatemque exercet suam; indéque totius plantae rationes et gubernat, et oportunos membris omnibus suppeditat succos.": *Corpus*, 246; *HistNat*, 354.

12. "Huiusmodi animalis formam terra parens à se editam non agnoscit, ideoque tamquam spuriam repudiatam, et infelici sorte addictam intelleximus. Singulis enim quadrupedum generibus mundi initio editis, suum genus proprium quo sese propagaret, quoque terrarum tractus frequentaret divini verbi vis et indulgentia concessit, et generis etiam sive MIN nomen tribuit.": *Corpus*, 341; *HistNat*, 453.

13. *Corpus*, 342; *HistNat*, 454.

ciation between the *iod* and the *sefirot* made by the kabbalists but relied instead on the idea of the number ten as signaling both completion and beginning.[14]

All generation, whether of plants, fish, birds, or "animals of the land," resulted from the comingling of the two salty and oily liquids. For oviparous animals this took place inside the egg through the agitating and warming action of ELOHIM. But there was an exception. Flies could be generated from their own seed or "from the corrupt and excremental oily liquid and the heat of the earth."[15] Flies born this way tended to come in swarms, corrupting the earth and overpowering other living things. The Bible had no shortage of examples of such plagues. A particularly disgusting (*foedissima*) type of fly was the ZEBUB (זְבוּב, *zĕbûb*), which was bred in, born in, and fed on blood. The etymology of its name, according to Arias Montano's idiosyncratic analysis, derives from ZABAH, which means to flow (possibly זוּב, *zûb*). Not coincidentally, he added, the inhabitants of Ekron derived the name of the prince of demons from it, BAGHAL-ZEBUB (בַּעַל זְבוּב, *baʿal zĕbûb*) or Lord of the Flies (2 Kings 1:2).

Although Arias Montano did not dedicate a separate section to the generation of mammals, he nonetheless noted that the mule presented a problem for the claim that all created animals had been instructed to reproduce their own kind, as indicated by the word MIN, or *genus*.[16] He stated categorically that Adam never saw a mule, let alone named the wretched beast. Mules could not possibly have been part of the first creation. When, then, was the mule created, and by whom? Arias Montano was convinced that the mule was a work of artifice—of human intelligence—and the result of a particular historical circumstance. Moses had said as much when he attributed the "experiment" to Anah: "This was that Anah that found the mules in the wilderness, as he fed the asses of Zibeon his father" (Gen. 36:24 KJV).[17] Arias Montano followed this with an extended philological analysis of the Hebrew word for mule, IEMIM (יְמִם, *yēmim*), which again relied on the shape of the letters and their position in the word. At the core of the word, he explained, were two letters *mem* (מ and ם); the first was an "open" letter, but the final one was a

14. The Jesuit Juan de Pineda, for example, found this type of explanation "kabbalistic." See below, chapter 12, note 51.

15. "Ex corrupto et foeculento pingui liquore, ac terrae calore.": *Corpus*, 306; *HistNat*, 417.

16. *Corpus*, 341–45; *HistNat*, 453–58.

17. The Hebrew HAIEMIM was mistranslated in the Vulgate as *aquas calidas*, an error Arias Montano was quick to point out. He had many options to choose from; the Septuagint translates it as *qui invenit Eamin in deserto*, while the Latin translation of the Chaldean paraphrase reads *qui invenit gigantes in solitudine*. Saint Jerome argued that the word had a Punic origin and meant "hot springs."

closed letter. This indicated that the male and female mules might try to mate, but their union would be sterile. The *iod* at the beginning of the Hebrew word for mule lay tellingly outside the word, suggesting that the animal indeed exists but that it cannot culminate the union of male and female because it does not contain the principle of reproduction *within* it, as in the case of MIN, where the *iod* is nestled between the two letters *mem* and *nun*. The mule's sterility confirmed the beast's artificial origin. This was indicated by its Hebrew name PERED and its root PARAD (פָּרַד, *pārad*), which mean to divide or disperse.[18] It was a characteristic that ancient Hebrew observers and biblical authors had noted when giving it such a meaningful name.

PLANT TAXONOMY AND IDENTIFICATION

For Arias Montano understanding the plan behind the apparent complexity of the plant kingdom was a prerequisite for understanding the grander plan of Creation. Yet what seemed an immensely complex task was in his view rather simple. In this instance he returned to what he interpreted as the most synthetic description of God's plan: all things were created according to "number, weight, and measure." Arias Montano found a sign in Moses's use of the word *genus*, or MIN (מִין, *mîn*), when describing the generation of plants; it suggested that the number of plant types was finite and integral. This meant all plants were created with distinct characteristics. The word, Arias Montano explained, derives from MANAH (מָנָה, *mānâ*), meaning, among other things, "to count" and "to distribute" (*numerare, censere, manere, partiri, dotare*), so MIN could also be taken to mean "definition" and "description." MIN also referred to the Solomonic "number" and suggested that "all genera are established with a number of forms, which may be called 'species.'"[19] This exegesis had important implications; it meant that plants resulted from a deliberately ordered creation, but also that they could be grouped with similar plants into natural genera.

His plant classification was based on groups he identified as mentioned in the Bible and defined by their utility. He recalled the two names used in the Bible when plants were first created, considering these the *summa genera* of all plants that germinate: herbage, or GHESEB (עֵשֶׂב, *ʿēseb, herba*), and trees, or GHETS (עֵץ, *ʿēs, arbor*).[20] The rest of the classificatory scheme derives from

18. *Corpus*, 341–45; *HistNat*, 453–59.
19. *Corpus*, 240; *HistNat*, 348.
20. *Corpus*, 238–40; *HistNat*, 345–48.

TABLE 10.1

DESSE
(דֶּשֶׁא)
Plants

GHESEB (*herbae* or herbaceous plants)
- *Herbae virentes* (herbage that grows)
  - *Faenum* (eaten by animals): hay
  - *Genus medium*: *mentham* (mint), *cuminum* (cumin), *rutan* (rue), *omne oleras*
  - IEREK (eaten by man)
    - *Olera* (vegetables): a great variety
    - *Herba servituti hominum* (plants at service of man)
      - medicinal plants
      - poisonous plants
      - odorous plants
      - beautiful plants
- *Herba faciens semen* (seed-making herbage)
  - KETANITH (*legumina*, beans): a great variety
  - DAGAN (*frumenta*, grains)
    - HHITIM (*triticum*, wheat)
    - SIGHVRIM (*ordeum*, barley)
    - CVZZEMET (*secale*, rye)
    - SIBOLETH SUGHAL (*spelta*, spelt)
    - SIPHON (Latin name not known)

GHETS (*arbores*, trees)
- GHETS PERI (*ligna faciens fructum*, fruit-bearing trees)
  - *TAPVAHH* (*mala*, tender fruits): apples, pears
  - *bacchae*: olives, palm trees, grapevines
  - *AGUZ* (*nuces*, nuts)
  - *HHARUB* (*siliquae*, pods of legumes): tamarinds, cinnamon, cane
- ASER ZAR GHO BOLEMINO ("*semen in ipso*," self-propagating trees)

(*Asperiora*, rougher plants)
- KOTS (*spinae*, woody thorns)
- DARDAR (*cardui*, herbaceous thorns)

these, determined first by the order of creation and then by their usefulness to man. Table 10.1 shows my interpretation of Arias Montano's plant classification.[21] The subdivisions of the GHESEB were again mentioned in Genesis 1:12. Arias Montano interpreted the passage as meaning there there were two

21. Readers interested in identifying plants mentioned by Arias Montano and their modern counterparts should consult Cobos Bueno, Oyola Fabián, and Vallejo, "Dimensión botánica del humanista extremeño Benito Arias Montano," 162–65.

distinct types of herbage: an "herbage that grows," or *herba virens*, that is useful to humans (and animals) because its greens become edible before it goes to seed, and another, "herbage that makes seeds," or *herba faciens semen*, that is useful for its seed. Among the *herba virens* meant to be eaten, he identified several groups: those meant to be eaten by animals (*faenum*), such as hay, those to be eaten by humans IEREK (קֶרֶ, *yereq*), and a *generis media*. Within the type eaten by humans there were three kinds: vegetables (*olus*), plants at the service of humans *(herba servituti hominum)*, and seed-making plants (*faciens semen*), which he does not identify with a Hebrew word. Humans first availed themselves of the vegetables and later learned by art how to process the grains. The *faciens semen* category was divided into two kinds, DAGAN (דָּגָן, *dāgān*, *frumentum*), grains, and KETANITH—possibly a reference to קִטְנִיּוֹת (*qiṭniyyôt*, *legumen*), beans. He found that the Bible refers to five types of grain, while the legumes were so well known that studying them should not be particularly difficult.

Later in the text Arias Montano returned to the problem of plant classification to account for plants that were "not completely useless." These plants, because they often feature in biblical accounts, had to be situated within his classification scheme yet did not fit the utility paradigm that characterized the major groups.[22] These were plants considered troublesome or rougher than others (*asperioris*) and seemed by design to be the opposite of the GHETS and GHESEB. They included KOTS or woody thorns (*spina*) and DARDAR or herbaceous thorns (*carduus*).

Although Arias Montano had found the names used in the order of Creation and throughout the Bible helpful in developing a plant taxonomy, he realized that identifying individual plant species and grouping them with similar species was not as straightforward as setting up larger genera. In a section titled *De cognitione ac definitione herbarum*, Arias Montano explained that plant identification was best done by choosing as the principal descriptor that part of the plant that was useful to humanity and therefore indicated its essential nature—in most cases the seed. Therefore if, like Adam, we were to bestow names that described the essence of a plant, we would use the seed and the plant's utility as a guide. If the genus of a plant was unclear it could be discerned from the seeds or by observing its uses, Arias Montano believed, given that these had been made by the Creator according to distinct "number, weight, and measure." Furthermore, he found in the shape and design of seeds and seedpods another example of providence; God took care to design proper

22. *Corpus*, 272–73; *HistNat*, 382–83.

shelters for the seeds and fashioned them to fulfill their diverse and specific capacities.[23] As we saw earlier, his overall taxonomic approach was not morphological but was based on his hermeneutics of nature: the most useful part of anything created was what served humanity. (He did, however, show interest in plant morphology, particularly when describing specific plants and finding helpful links among them.)

The purpose of one type of herbaceous plants was to produce seed; that is, they fulfilled their GHOLAM by producing seed. Given this purpose, it was reasonable to group them into subcategories according to the shape of their seeds—spire, or pointed (*spica*), pod (*siliqua*), or shell (*theca*)—since each seed type reflected a type of plant.[24] For example, the seeds of most legumes, such as peas and beans, were in pods. Other herbaceous plants fell into the category of being useful to humanity because parts of them were edible either by humans or by animals, because they were useful for making medicines, or because of their aroma or beauty. He noted four distinct types among those whose roots were edible and characterized them by the shape of the edible part: bulbous, tuberous, meaty, and woody.

In grouping and describing similar plants one should analyze at least three "orders" of the seed: material composition, shape, and color. The first order of inspection of each plant should be the nature of the juices (*succus*) or liquids that make it up, since the dual liquids, greasy and salty, combined in proper proportion to give the seed its nature and force. The greasy liquids, however, were found not only in the seed, but also throughout the plant. The slightest variation of the combination of the two liquids might yield variation in the seed and plant, but the plant still belonged to the same genus.[25] The second order involved the seed's shape. Here Arias Montano displayed the experience he gained during his many years of botanizing.[26] Seeds can be shaped like spheres or cylinders, the surface can be honeycombed or shaped so it is clear where the seed will sprout. Others are rhombic, flattened, or elongated; some even are shaped like an egg or a kidney. The third order is to describe the seed's color. Arias Montano also suggested paying attention to the roots of plants, especially cultivated ones, since the capacity for generation, growth, and governance resided in the roots.[27] How could these orders be discerned? He noted

23. "*[F]igurae, diversa capacitatis ratione concinnatae.*": *Corpus*, 242; *HistNat*, 350.
24. *Corpus*, 241; *HistNat*, 349.
25. *Corpus*, 243; *HistNat*, 352.
26. *Corpus*, 243; *HistNat*, 351.
27. *Corpus*, 246–48; *HistNat*, 355–56.

that humanity had been given five senses, varying in the trustworthiness of the information they yielded. For classifying plants, smell and taste were the most important, permitting us to tell the difference between salty and greasy and even degrees of saltiness and greasiness. It is very important to determine the proportion of saltiness of an otherwise greasy liquid, he noted, since salt conserves the greasy liquid's liveliness (as opposed to letting it go putrid), force, and weight.[28]

In the discussion of plant classification, Arias Montano postponed presenting his solution to the problem of plant nomenclature until the promised, but likely never written, third part of the *Magnum opus*, the *Vestis*.[29] But he nonetheless proposed that once the seed had been carefully examined, taking into consideration its material nature (whether it is salty or greasy), its shape, and its use, it should be easy to reach agreement about the plant's name. This consensus had so far eluded the lovers of the botanical discipline, he noted, yet by following the classification scheme revealed in the Bible and his own method of identifying plants according to their seeds, the problem was bound to be quickly resolved.

In the final chapter of the section on plants, he left some hints about the form this system of nomenclature would take. Most nations, he commented, have simply given each plant a particular name, so that within the designation "nuts" they include almonds, chestnuts, and hazelnuts, while others have suggested comparing the uses of the plants or naming them after their discoverers or the places where they can be found. He returned to a comment made earlier in the classification of trees and suggested using numbers to designate orders of plants within a genus, for example, "first genus of pears, second, third, and so on," noting that this required observing the plant carefully during its life cycle and writing detailed descriptions of the plants and the changes they underwent.[30] A more precise way of identifying and naming plants would take into account a number of characteristics to distinguish plants in the same category.[31] These would include the plant's roots, leaves, branches, color, seed, taste, efficacy, and force (such as the qualities of being hot, cold, toxic, or nontoxic, and how it feels to the touch); the name of the person who discovered it; and the place where it grows. This last property, he noted, was the way most commonly used to identify plants in the Bible.

28. *Corpus*, 244; *HistNat*, 352.
29. *Corpus*, 242–43; *HistNat*, 351.
30. *Corpus*, 277; *HistNat*, 387.
31. *Corpus*, 277; *HistNat*, 387.

Arias Montano demonstrated his system of plant classification and description in a section on flowers. He answered the question, What is a flower? in the following way: a flower is the efflorescence resulting from the plant's concentrating its greasiest liquids in preparation for producing its seed. Flowers also indicate the nature of a plant, since they too are made of the greasy and salty liquids that can be separated by fire and chemical operations to make perfumes and scented oils.[32] Flowers are divided into four types: *lilium, rosa, viola,* and *herbaceus flos.* He gave three rules to help in classifying flowering plants. The first stated that flowers of similar shapes have similarly shaped seeds, although they might differ in size.[33] This is true, for example, for plants with pods, such as beans and alfalfa, whose flowers resemble the seedpod, or for berries (*bacae*), whose capsules (*vascula*) also resembles the seed. Therefore, given one known shape, the names can be based on this and distinguished by adding other names to indicate differences. Arias Montano's discussion of flowers is not exclusively morphological, however. He dwelled amply on the significance of flowers in general and especially on what the rose and the lily signified in the Bible.

The second rule called for describing the plant. Once the genus of the plant was identified, he argued, it was easy to observe it diligently and add to its description. New observations should be indicated in notes or by amending notes and descriptions from ancient and contemporary authors after examining the plant according to this art and method. He promised to demonstrate how this was to be done in the third part of this book, likely a reference to either a yet unwritten section of the *History of Nature* or even the *Vestis.* But in the meantime he invited anyone to begin describing particular flowers, first under the heading *Solanum* and then using numbers for the other species.[34]

The third rule stated that, although every flower seemed to have a given color and scent that the plant tries mightily to preserve unchanged, the color could be affected by several factors, including the degeneration of the seed, the type of soil, and the climate where it grows. Arias Montano noted that while

32. "Pinguissimus in planta ad seminis concretionem lactescens et fervens succus florem, tamquam spumam, praemittit, naturae suae certum agnoscendumque indicem et antesignanum, duplici etiam parte instructum, oleo videlicet et sale: quae ignis vi et chymica arte separari ac dignosci possunt.": *Corpus,* 248; *HistNat,* 357.

33. "Similes florum figurae simile praesignant seminis vas, et semen demum, quanquam magnitudine dispar, forma tamen non fore dissimile. Rursus etiam, similia seminis vascula similes florum figuras praecessisse docent.": *Corpus,* 250; *HistNat,* 359.

34. *Corpus,* 250; *HistNat,* 359.

traveling in England during his perilous first voyage to Antwerp, he noticed that a plant with a flower of any color, when brought to the island of England, once planted there would eventually bloom white. He ventured to speculate that this was because of the salty nature of the soil—evident from the white vapor it emitted, which he had witnessed while on his travels in that land. This had led ancient geographers and mariners to name the island Albion, and it was also the reason, he added, that the land was so rich in very pure tin (*stannum*).[35] Although some might argue that the forces inherent in the plant might change under these conditions and therefore cause the plant's nature to change, Arias Montano remained steadfast: the change in the color of the flower did not change its species. He conceded that the strengths (*vires*) of the plant might change, but nonetheless the mixture of liquids within the plant assigned at the Creation persisted as established by *et factum est ita*, guaranteeing that the original nature of the plant would not change.[36]

The discussion of trees that follows was also written with the purpose of establishing a suitable classificatory scheme that revealed the abundance of Creation. He explained that there were two genera of trees designated in Genesis 1:12: GHETS PERI (fruit-bearing trees) and ASER ZAR GHO BOLEMINO (self-propagating trees that reproduce by shoots or their own seed). Although the Bible seemed to imply that the first trees created were all fruit-bearing, a careful reading of Genesis 1:12 suggested to Arias Montano that there were others as well. In his opinion, the Hebrew phrase conveyed better than the Latin the idea that the second type of trees, described as "with its seed in it," meant a second genus of trees that can propagate without the need for fruit. But Arias Montano also pointed out that some of the Psalms indicated different classifications: trees that need human care in order to bear fruit (SADEH) and trees that do not bear fruit but are useful for other reasons (ARAZIM), such as for their wood or for shade.[37]

Thus the classification of trees should proceed according to their seeds and their fruit—the two forms of reproduction—rather than through their roots, branches, or flowers as with other plants. This, he reminded readers, is be-

35. The reference to Albion and tin appears in book 4, chapter 41 of Pliny's *Natural History*.

36. "Floris mutato colore non variari speciem, quamquam vires nonnihil immutari siquis contendat nil contra simus conaturi. Nam cùm omnia eodem se habent modo quo ab initio vel ab antiquo habuisse accepimus, vires quoque integras conservari censendum est. Hic enim succus pinguis cum hoc salso temperatus id perpetuo ex se praestabit, quod hactenus praestiterat.": *Corpus*, 251; *HistNat*, 359.

37. *Corpus*, 252; *HistNat*, 361.

cause God referred to them in Genesis 1:12 by their purpose, or in Montanian language, by how they fulfilled their GHOLAM.[38] Arias Montano then divided the first genus, the fruit-bearing trees, into four "very broad groups" based on the physical constitution of their fruit. The hard-skinned, meaty fruits were in the *malus* group. Trees whose fruits has softer skins and were juicier were *bacchae* (*sic*). Then he considered trees that bore fruits with very hard shells that were suitable to eat (*nux*), and finally the *siliqua*, whose seeds were harbored in long segmented and reedy pods.[39] Yet again he postponed discussing nomenclature until the third part of this work and even cut somewhat short the discussion of trees not bearing fruit.

The section in the *History of Nature* dedicated to plants ends with an extended reflection on the uses of fruits and trees for nutrition, perfumes, oils, shade, and wood and includes the discussion of the precious wood of Ophir in chapter 6. In the rest of the section, Arias Montano selected names of plants and trees from the Bible and tried to establish, when possible, a description and an analogy to a known plant. He paid particular attention to human nutrition. An acknowledged vegetarian for most of his adult life—at a time when this was rare, if not suspicious—he devoted a section titled *Annonae et fructibus parabilis ratio* (An easily accessible system of provisions) to show that humans can obtain and sustain adequate nutrition from only the products of seed-bearing plants and fruit-making trees, in particular those products described in the Bible as DAGAN (seeds to make bread), THIROS (juices extracted from fruits), and IITSHAR (oils).[40]

The biblically inspired taxonomic nomenclature schemes Arias Montano proposed were two among a number of approaches suggested toward the end of the sixteenth century. While naturalists among his contemporaries continued to use the classification of Theophrastus, dividing plants into trees, shrubs, half-shrubs, and herbaceous plants, they also devised idiosyncratic systems of their own, many of them based on a plant's medicinal properties or utility, or they simply listed plants alphabetically. Dodoens, while basing on utility the five-part division of the plant kingdom he proposed in his *Cruydeboeck* (1554), later refined his classification scheme in his *Stirpium historiae pemptades sex* (1583) by dividing plants into twenty-five subgroups, again based on their uses. Yet historians of botany have detected within these subgroups an

38. *Corpus*, 251–52; *HistNat*, 360.

39. "Malum dicimus quemcunque arborum fructum qui pelle tenaciore, . . . durescit, et asservatur.": *Corpus*, 253–54; *HistNat*, 362.

40. *Corpus*, 257; *HistNat*, 366.

inclination to group plants with similar morphology, or in Dodoens's words, *forma et figura*.[41] A year earlier, Andrea Cesalpino (1519–1603) had published his *De plantis libri XVI*, which also took up plant classification. An avowed Aristotelian, Cesalpino sought to group plants according to their natural "affinities" (*similitudo*) and differences. These affinities were most apparent in their flowers and fruits, and particularly in their seeds.[42] The essential affinity, however, resided in the plants' "vegetative soul," which was responsible for the conservation of the species.[43] He therefore focused on the parts of the plants responsible for reproduction: their seeds and fruits. Yet these affinities and differences had to be substantial as opposed to accidental, that is, essential and necessary to the plant itself and not contingent on its medicinal applications or where it grew. Arias Montano does not seem to have been too preoccupied by the hazards of using accidental qualities in his classification, so certain was he that careful observation and perhaps the alchemical analysis of a plant's oily and salty liquids could reveal its true genus. It is perhaps Arias Montano's and Cesalpino's shared Aristotelian commitment to essential and substantial natures that makes their systems resonate, although there is no evidence Arias Montano drew on the Italian botanist's work.[44] For, whereas the Italian limited his study of nature to plants, for Arias Montano studying plants was an incidental exercise on the way to unraveling the much grander plan God had imprinted in nature.

ANIMAL CLASSIFICATION

On starting the discussion of animals, Arias Montano wrestled openly with determining the best way to present the material: he could either follow Solomon's example and start with the most dignified animals or do as Moses did and follow the order of the Creation. He decided on a hybrid of the two approaches. First he followed the order of Creation, beginning with animals created on the fifth day—fish and birds—then took up the land animals cre-

41. Visser, "Dodonaeus and the Herbal Tradition," 54.

42. For a synthesis of Cesalpino's classification system and underpinning philosophy, see Sachs, *History of Botany (1530–1860)*, 37–58; Greene, *Landmarks of Botanical History*, 2:808; and Ogilvie, *Science of Describing*, 223–26. On the possible influence of Cesalpino on Arias Montano, see Cobos Bueno, Oyola Fabián, and Vallejo, "Dimensión botánica del humanista extremeño Benito Arias Montano," 137.

43. Repici, "Andrea Cesalpino e la botanica antica," 72.

44. The similarity of their systems was noted in Navarro Antolín, Gómez Canseco, and Macías Rosendo, "Fronteras del humanismo," 120n75.

ated on the sixth day.[45] He interpreted the order of Moses's exposition as also indicating the different origins of the animals, from the waters and from the earth. In the watery category were all the animals that lived in a fluid medium—fish and birds—while the land category included beasts of the earth, beasts of burden, and reptiles. Within each group, however, he decided to imitate the order that he believed "Solomon had taught." Solomon had believed, according to Arias Montano, that to teach something so that its reason for being was understood (*cognoscendi ratio*), it was essential to present the material in the proper order so the student could discern the overall plan. He lamented the loss of the books in which Solomon had recoded his great knowledge about animals, but he presumed that the king had discussed the most "dignified" animals first and proceeded in order until he reached the lowest creature. For it was written that Solomon "discussed plants, from the cedar of Lebanon to the hyssop growing out of the wall, and he spoke about beasts, birds, reptiles, and fishes" (1 Kings 4:33 NAB). This way of explaining nature was clearly what divine instruction required. Instead, Arias Montano complained, some "wise men" had decided it was best to teach about the simplest things first, so that even the "rudest intellect" could understand.[46] He limited his discussion to animals mentioned in the Bible, although a personal interest in certain topics led him to digress at length about other animals such as his beloved mollusks. Generally, for each category—fish, birds, and land animals—he first discussed the most appropriate method of classification, drawing on biblical references whenever possible, then provided shorter descriptions of the most important members of each group.

It was no coincidence, Arias Montano argued, that the inhabitants of the sea and air shared the nature of the primordial material they were made from. Fish and birds were made from the greasy liquid of the MAIM that makes up water and air, all according to "measure, number, and weight," but in different degrees to suit the singular nature of each.[47] Their spirits were also designed to suit the places where they had been instructed to live. Aquatic animals had a spirit that could be replenished by the cold of the liquid they inhabited, while flying animals were given a more generous portion of spirit so they could easily move through the air.[48] The similarity between aquatic animals and flying animals extended to the way they moved through the medium they had been

45. *Corpus*, 280, 313–14; *HistNat*, 390, 424.

46. *Corpus*, 313; *HistNat*, 424.

47. *Corpus*, 281; *HistNat*, 390.

48. *Corpus*, 293–94; *HistNat*, 403–4.

instructed to inhabit, as I discussed in the section on mechanics in chapter 9. But more important, their "watery" nature determined the place on the earthly sphere they were meant to inhabit, as stated in Genesis 1:20. Arias Montano interpreted this to mean that this assigned place limited the species' movements. The flying species were confined between the heaven and the surface of the earth; that is, they could not venture under the surface of earth or above or beyond the RAKIAGH. There were therefore no flying animals that could go with impunity beyond the firmament or the face of the earth and no fish that could live out of the water or move on the surface of the earth.[49] Subsequent generations of fish and birds, after their perfect and complete creation, resulted from the divine blessing and the commandment, "Be fertile, multiply, and fill the water of the seas; and let the birds multiply on the earth" (Gen. 1:22 NAB). This clearly indicated to our interpreter that fish would outnumber birds in nature; they had been instructed to *fill* the water, so much so that their variety could not be counted or even known.[50]

Fish and birds also reproduced in similar ways, by means of eggs, where the egg acted as a receptacle or membranous sac filled with the salty and oily liquids. Once heated and agitated by the spirit ELOHIM, the liquids developed a life form similar to the animal's parents. The time of gestation was preordained and overseen by ELOHIM as decreed in Ecclesiastes 3:2 by the words, "A time to be born, and a time to die" (KJV). The number of progeny was determined by the care the parents gave to their offspring. Fish, for example, have innumerable offspring because they promptly abandon them, whereas birds lay only a few eggs because they were made to tend their offspring and supply them with food. This essential behavior of birds gave Arias Montano an indication that they could be grouped into three genera: those that bring food to their offspring in the nest using their beaks or talons; those that regurgitate the food for their offspring to consume; and still others, such as chickens, that promptly push the offspring out of the nest and teach them to find food on the ground. Arias Montano found further clues for classifying flying animals in their wings; one type had feathered wings, while the wings of the other type consisted of

49. "Quo dicto quis utrique generi locus, quibus metatus terminis ad proprium motum assignaretur, declarabatur, hoc est, huic volatilium generi inter caelum ac terram, non inferius terrae solo, nec ulterius aut superius firmamento caeli, id est, RAKIAGH. cui claritatis causa expansionis nomen tribuebamus. Nullum enim volatilium est quod vel firmamento ulteriora tentare, vel terrae faci profundiora volando contendere impunè possit. Ut nullus est piscium qui curà vitae periculum altius in aquis, quam terrae superficies extollatur: nedum in ipsa terra moveri possit et procedere.": *Corpus*, 294; *HistNat*, 404.

50. *Corpus*, 304; *HistNat*, 414.

TABLE 10.2

SERETS
(שֶׁרֶץ)
Creatures of the "waters"

DAG, DAAG, or DAGAH (*pisces* or animals of the waters)
- THANINIM (*Cete grandia, dracones*): whales and dolphins
- REMES HAROME SETH: fish with scales and fins
- SERETS:
  - ○ Mobile
  - ○ Nonmobile
    - ▪ Tenuous body: sponges, slugs, and sea worms
    - ▪ Hard-shelled: lobsters, turtles, crabs
    - ▪ Construct their homes: shells, oysters and mollusks (*testacea*)

GHOPH CHOL CHANAPH (*volatiliae*, flying animals)
- SERETS HAGOPHEPH (*reptile volans*, flying reptile): bees, flies, crickets
- TSIPHOR CHANAPH (*aves alatae*, winged bird)
  - ○ Carnivorous birds: eagles, vultures, crows, ibis, storks, and bats
  - ○ Pure birds that do not eat carrion:
    - ▪ feed offspring by regurgitation
    - ▪ teach offspring to forage

a small membrane (*membranulae*).[51] In fact, he considered the wings the best way of identifying birds (table 10.2).

Arias Montano also found three genera of aquatic animals indicated in Genesis 1:21: large sea creatures, living fish, and fish that move. The Latin of the Vulgate, in his view, made the three-part distinction clearer: "Creavitque Deus cete grandia, et omnem animam viventem atque motabilem, quam produxerant aquæ in species suas, et omne volatile secundum genus suum."[52] The THANINIM genus included sea snakes (*dracones*), the mighty Leviathan (whales), and other aquatic animals larger but without scales (perhaps a reference to dolphins). The aquatic animals of the second genus had "innumerable" forms so that it was difficult to reckon, yet God had given them a common trait (*communis nota*): fins and scales. The third genus of aquatic animals fell into two groups: those that remain immobile, attached to rocks under the water, and those that move, carried either by currents or by their own motion. Those in the latter group could be of three kinds: those with a tenuous body

51. *Corpus*, 295; *HistNat*, 405.

52. "God created the great sea monsters and all kinds of swimming creatures with which the water teems, and all kinds of winged birds" (Gen. 1:21 NAB).

(sponges, slugs, and sea worms), those covered by a hard shell (lobsters, crabs, and turtles), and those that construct their own coverings (shells, oysters, and other mollusks).

Like creatures of the "waters," the animals of the earth were made from the medium they lived in, but in this case the quality of the earth they were made from determined their "dignity." All animals were made from ARETS, with its fertile juices generated from the proper proportions of the dual liquids MAIM.[53] But when these animals were created, the earth had to obey another very specific command, to "produce a living soul" (Gen. 1:24) from this material. NEPHES (נֶפֶשׁ, *nepeš*) indicated not just any soul, but a soul that gives the body the faculty of feeling and desiring and is ruled solely by appetites and the senses. This NEPHES animates creatures but allows them to enjoy only the simple life found on earth. Arias Montano reminded readers that the capacity of the soul in creatures made from the ARETS is very different from that of humanity, whose soul is made from a different type of earth, ADAMAH.[54] Animals were also given a different type of corporeal soul that works only on the appetites and on the senses, called NEPHES HHAIAH (נֶפֶשׁ חַיָּה, *nepeš ḥayyâ*), or *anima* in the singular (Gen. 1:24). Humanity, however, was given a spirit described as NESEMAH HHAIJM (נִשְׁמַת חַיִּים, *nišmat ḥayyîm* (possibly from Gen. 2:7), or *animae* in the dual or plural form (*animus duplicis*), indicating humanity's potential for living two lives, one on earth and another purely spiritual afterlife. *Anima* in the singular referred to the spirit of all other living things.[55] Throughout this discussion of different types of souls in animals and humans, Arias Montano does not use the terminology associated with the three-part division of the soul as intellective, sensitive, and vegetative—pace Mariana's remarks.[56] The land animal's soul has the faculty of feeling and desiring (*sentiendi et appetendi facultatem*), which of course maps neatly onto the sentient soul of Aristotle; further, as Arias Montano points out in his discussion of cattle, land animals lack a capacity akin to the human intellect or the ability to speak.[57]

Arias Montano found the biblical classification of animals straightforward and does not mention the issue of nomenclature that plagued the study of plants. He found that in the Bible animals created from earth or ARETS fell

---

53. *Corpus*, 314; *HistNat*, 425.

54. *Corpus*, 314–15; *HistNat*, 426.

55. *Corpus*, 358–59; *HistNat*, 475–76.

56. On his interpretation of the human intellective and sentient soul, see Arias Montano, *Discursos sobre el Eclesiastés de Salomón*, 143–44.

57. *Corpus*, 314-15, 329; *HistNat*, 426, 441.

TABLE 10.3

NEPHES HAIA
(נֶפֶשׁ חַיָּה)
Animals of the earth

HHAIATH HAARETS (*bestiae terrae*, beasts of the earth)
- HAIAH RAGAH (*ferae malae*, fierce beasts): lions (four types), leopards, tigers, panthers
- HHAIAH HASADEH (*bestiae agri*, beasts of the field): bears, wolves, foxes, dogs
- HHAIAH IAGHAR (*ferae sylvestres*, ~~wild~~beasts of the forest): deèr, goats, gazelles
- *Ferae minores*
  - ARNEBETH: hares
  - SAPHAN: burrowing animals

BEHEMAH (*iumenta*, beasts of burden): oxen, bulls, sheep, goats, pigs, horses, mules
REMES HAADAMAH (*reptilia*, reptiles)

---

into three categories: HHAIATH HAARETS (*bestiae terrae*, beasts of the earth), BEHEMAH[58] (*iumenta*, beasts of burden), and REMES HAADAMAH (*reptiles*, reptiles). The beasts of the earth were in turn divided into three principal groups and a fourth group of small wild beasts. Table 10.3 shows the classification of the animals of the earth suggested by this interpretation. The examples used in the table are those Arias Montano gave in his general discussion of the nature of animals of the earth.

As Solomon had done, Arias Montano organized the descriptions of individual land animals in order of their "dignity": from the most magnificent to the lowliest. Curiously, he concludes the section on animals in the *History of Nature* without ever discussing the genus of reptiles or explaining the omission. The lion occupies the first place and is followed by brief descriptions of other beasts. Most of descriptive material is taken from the Sacred Scriptures and supported with a number of biblical citations. In some cases Arias Montano was unable to find a Latin name or other contemporary referent for a particular beast mentioned in the Bible. This does not seem to have troubled him, for his intention in this section of the *History of Nature* was not to be exhaustive, but to provide biblically based classification schemes. The exegete in him, however, dictated that he find a place in his scheme for all the

---

58. Arias Montano's explanation of the use of the word BEHEMAH to denote cattle is curious. He argued that the phonetic elements of the word mimic the sound of bellowing cattle, particularly if one takes into account that the root of the word is HAMAH, "to make a sound.": *Corpus*, 327–31; *HistNat*, 439–41.

animals mentioned in the Bible. He did this by unraveling and synthesizing multiple biblical references for each beast—nineteen, for example, for dogs. On the other hand, to remain faithful to the biblical text, he chose to ignore a number of common plants and animals—house cats, for example—that are not mentioned in the Bible. He nonetheless devoted extensive sections to two animals: mules and mollusks. I discussed earlier his etymological analysis of the Hebrew word for mule and how the animal's inability to reproduce placed it among the *spuria* of the earth, an animal that exists separate from others as a misguided work of human artifice. While he viewed mules with total disdain, Arias Montano lavished praise on mollusks. He had to put all his exegetical skills to work to find a specific biblical reference to these creatures, however, and in doing so he discovered something remarkable.

He found the key in the use of the word for "act" or "deed," MAGASSE (מַעֲשֶׂה, *maʿăśe, opus*), in Psalms 107:23–24 VUL.[59] In this psalm and elsewhere in the Bible, the word signified something made using a specific technique and in a precise form. Furthermore, this psalm suggested that things of this kind could also be found in the deep. For Arias Montano, the wonder lay in that "the first author and master our Lord" had given amazing dexterity and skills to such simple animals with such delicate bodies, so that without bones or extremities and using only their spirit and ingenuity, they could fashion such admirable homes. This last aspect, that these animals construct their own shells, was something Arias Montano alleged that none of the ancient Greek or Latin writers had ever mentioned but that was indicated clearly in the Sacred Scriptures. He considered this another instance of a divine gift bestowed by the reading of the Sacred Scriptures—a gift of knowledge he held plainly and undoubtedly but confirmed only after years of observation of innumerable examples.[60] God, he explained, preferred to teach using clear experiments and examples (*experimenta et exempla*) rather than through reasoning: "that is, with induction rather than syllogism."[61]

In the earlier general description of aquatic animals, Arias Montano noted the multiple forms mollusks could attain, particularly the way they ranged widely in size within each genus. This was indicated by studying the great variety of aquatic animals, but it was confirmed by scripture: "This great and wide sea, wherein are things creeping innumerable, both small and great beasts"

59. "Some went off to sea in ships, plied their trade on the deep waters. They saw the works of the LORD, the wonders of God in the deep" (Ps. 107:23–24 NAB).

60. *Corpus*, 285–86; *HistNat*, 395–96.

61. *Corpus*, 286; *HistNat*, 396.

FIGURE 10.1. Mollusks (top) and essential shapes of all mollusks (bottom). Arias Montano, *Naturae historia*, 283, 289. Biblioteca Histórica de la Universidad Complutense de Madrid (BH FLL 14614).

(Ps. 104:25 KJV). The careful observation of shells in his collection confirmed this as well. He had shells in the shape of obelisks that he interpreted as demonstrating the range of sizes these animals of the waters could attain, from a wide base (*maximum*) to a small tip (*minimum*); from great to small, just as the psalmist had exclaimed (fig. 10.1, top).

His description of the *testacea*, or mollusks, is followed by descriptions of what he understood to be the animals' constituent parts: slugs (*limaces*) and shells (*conchylia*). The *limaces*, despite their tenuous and slimy bodies, can transform their nutrition into the viscous and gummy fluid that they excrete and leave behind as slimy trails. Those that do not form shells transform these liquids within their bodies, sometimes even changing color as they do so, but eventually they excrete the liquid as a friable matter akin to an eggshell or a brick. The *limaces* that form these shells got the name *conchylium* because of this ability. Arias Montano explained that the *limaces* have some limited sense perception, like other aquatic animals. *Limaces*, as far as he could ascertain, have the senses of taste, sight, and touch. He noted, however, that he had been unable to discover whether they can hear, either by consulting scripture or by conducting experiments (*experimenta*). This and the ability to speak or smell would be useless anyway, he added, since these faculties require properly disposed air in order to function.[62] The eyes of the *limaces* are on two protrusions on their heads. Another pair of longer protrusions or horns on their heads work like fingers or the trunk of an elephant and can manipulate the viscous fluid they excrete—not unlike what bees and spiders do.

In the following section, Arias Montano found that the double genera (*duplex genus*) of the *limaces* and *conchylia* in turn consisted of two forms established by the command IEHI. One is in the form of a helix, a twisted line, or a spiral, like the trajectory of the sun through the "circle of the signs of the sun," while the other resembles the changing size of the face of the moon as seen from the heavens and the earth.[63] He included two images illustrating these shapes (see fig. 10.1, bottom). Beyond these established shapes, he granted the genus the possibility of fabricating infinite variations of shapes and designs, some distinct to certain regions, but generally constant within each group. He noted as well that the mollusks could adapt to their environment, mentioning that those inhabiting rocky shores with restless seas tended to make their shells thicker and sturdier than similar shells of mollusks from quieter seas. Although likely aware of Aristotle's theory that *testacea* were the product of spontaneous generation, Arias Montano did not refer to the animal's production. With rhapsodic eloquence, he sang the praises of the constructions these little creatures were capable of building, unrivaled in their perfection and beauty by anything made by architect or artist. They were an example for the best human artisans, some of such perfection that they are very difficult to imitate. Arias Montano's fascination presaged the following century's baroque collectors' obsession with shells. And although his narrative shows concern with describing and classifying the variety of species, his objective here—as it was for plants—was to show how God was able to endow these tiny spineless,

---

62. *Corpus*, 289; *HistNat*, 398.

63. "Solis intra signorum circulum perpetuo observatam: Alterum verò ea praefert linea quam lunae facies e caelo e terris conspecta variat.": *Corpus*, 289; *HistNat*, 399. I thank Joshua M. Smith for clarifying the unconventional Greek terms and misprints of the 1601 edition.

nerveless bodies with such an ingenious spirit, made manifest in the wonderful examples of their efficiency.

*

Arias Montano saw his contribution in the *History of Nature* as the rediscovery of the divine plan that operated during Creation, a plan that organized the plant and animal kingdoms into different genera and that, once deciphered and understood, would reveal both the magnificence of its Creator and nature's utility for humanity. The way the plan unfolded gave Creation a third purpose. In Arias Montano's view, God had also created living things to serve as the ornament of this world. By arranging their order of creation from the simplest (aquatic animals) to the most complex (large terrestrial animals) and culminating in humanity (the most perfect), God had demonstrated that the teleology of the world was directed toward perfection. If the smallest and simplest creatures could attain such perfection, should not humanity follow the same path? His objective concerning natural history was to propose taxonomic systems that encompassed all of nature, whether discovered or yet to be discovered. It was a scheme he had found after a lifetime of carefully reading the Sacred Scriptures. In them he had identified a "compendium and method" that allowed him to achieve "exact knowledge of one and another group" of animals, plants, and trees.[64]

The final major section of the *History of Nature* brings Montanian natural philosophy to the creation of the human, from the formation out of the fertile earth ADAMAH to the particular constitution of the human soul and an exploration of human biology. It is the anthropological background to the first part of the *Magnum opus*, the *Anima*. Here again the exposition is chronological from the moment of creation of man, and later of woman, to what the happy existence in Eden reveals about the prelapsarian human condition, both carnal and spiritual, and finally to the damage caused by the Fall. It concludes with a reflection on the "animal life of man" (*De animali hominis vita*) and the human life cycle. A detailed discussion of this section beyond what I have already discussed in chapter 7 awaits further study. The *History of Nature* ends abruptly at this point; as Arias Montano explained in its preface, it was only the first part, meant to be the "head" of a grander project. His death in 1598 cut short the *Magnum opus*.

64. *Corpus*, 238; *HistNat*, 345–46.

CHAPTER 11

Disciples and Detractors

Arias Montano's natural philosophy was not without its devoted followers, ambivalent critics, and powerful detractors. Two disciples proved to be the most committed, largely by how steadfastly they defended his legacy and reputation as a devout Catholic but, as we will see in this chapter, despite not always completely understanding the implications of his natural philosophy. The Hieronymite friar José de Sigüenza (1544–1606), the most enthusiastic follower outside his most intimate circle of family and friends, endured an inquest by the Spanish Inquisition for positions he voiced at the monastery of El Escorial that were largely influenced by his devotion to Arias Montano's style of biblical exegesis. Sigüenza also labored for years on a Spanish version of sections of the *Magnum opus*. Pedro de Valencia (1555–1620), who served as one of Arias Montano's secretaries and whom Arias Montano considered a son, became his mentor's interlocutor and later protected his legacy when Spanish censors turned their attention to his oeuvre, following Rome's cue. Although neither Sigüenza nor Valencia published works based on the Montanian philosophy of nature, they both tangled with its main premises and natural philosophical postulates. The correspondence between them on the subject reveals how they struggled to understand and explain their mentor's complicated system, which they could do only by resorting to the tired language of Scholastic natural philosophy. Other biblical exegetes, including the Dominican Tomás Maluenda, found the Montanian natural philosophical proposals largely unintelligible, and their engagement with Arias Montano's corpus remained restricted to the rigid discourse of hexameral commentary.

THE DISCIPLES

Beyond his circle of friends in Seville, Arias Montano's influence as a natural philosopher was most apparent among a segment of the community of the Hieronymite monastery of El Escorial. Other friars besides José de Sigüenza, among them the next two librarians of the monastery, Juan de San Jerónimo and Lucas de Alaejos, also followed in Arias Montano's footsteps. They copied and amended some of the works that date from his stays at El Escorial, commenting on them and creating their own versions in the same style and with similar substance.[1]

Arias Montano exerted a significant influence among a select number of Hieronymites at El Escorial, yet the broader impact of his followers is difficult to gauge because they never embarked on a successful program to publish derivative works following his philosophy. Doing so would have acknowledged the merits of the Montanian exegetical approach and made a commitment to the natural philosophical postulates he derived from his exegesis. Furthermore, it would have implied that the authors believed Arias Montano had arrived at his approach by a revelation. To an attentive reader of the *History of Nature*, it would have come as no surprise to hear that the author believed the philosophical insights he derived from deciphering the arcane language of the Bible were divine revelations. He says as much in at least two works: in the votive elegy that opens the tome, throughout which he repeats it, with more or less modesty, and also in an ex voto hymn that closes his commentary on Zachariah.[2] During the gestation of the *Magnum opus*—which coincided with his last lengthy stay at El Escorial—he apparently also confided to his most intimate associates that he believed he had received divine revelation.

José de Sigüenza and the Monastery of El Escorial

Among all his disciples, the Hieronymite friar José de Sigüenza left the largest written testimony to how Arias Montano's theological, exegetical, and natural philosophical ideas influenced his followers. Sigüenza completed his studies at the seminary school of Párraces in 1575, afterward visiting El Escorial occa-

---

1. An example of this type of exercise is Arias Montano's commentary on Ecclesiastes, which survives in three versions at the Library of the monastery of El Escorial: Arias Montano, *Discursos sobre el Eclesiastés de Salomón*, 11–16.

2. Arias Montano, *Commentaria in duodecim Prophetas*, 821–22.

sionally to preach, though he did not join the community officially until 1590.[3] In 1591 he became the royal librarian, later he was rector of the school, and he twice served as prior of the monastery, in 1603-4 and again before his death in 1606. His duties at the library brought him close to Arias Montano, whom he admired and loved. It is most likely that the friendship blossomed during Arias Montano's last visit to El Escorial (late 1591 to April 1592), precisely the years he devoted to his *Magnum opus*. Sigüenza's most significant projects—his *Life of Saint Jerome*, the second and third parts of the history of the Hieronymite order, the unpublished paraphrase in Castilian of sections of the *Magnum opus*, and his sermons and poetry—are all clear testimony to the influence of his mentor's teachings.[4] Luis Villalba Muñoz, one of Sigüenza's biographers, proposed that the two discussed the ideas behind the *Magnum opus* as Arias Montano continued working on the *History of Nature* and reviewing the proofs of the *Anima* during those months at El Escorial. According to this historian, Sigüenza began his paraphrase of the *Magnum opus* during this time, its elaboration interrupted only by his inquisitorial trial. Recent scholarship, while not dismissing this scenario, situates most of Sigüenza's work during the early years of the seventeenth century, after the published edition of the *History of Nature* and later Montanian works became available.[5] Some of his surviving sermons also show how Sigüenza incorporated his mentor's exegetical approach into his preaching. His conversion to Arias Montano's beliefs had been complete.

Sigüenza listened carefully to Arias Montano during the winter and spring

3. Sigüenza's biography is known mostly through the monastery's necrology, the *Memorias sepulcrales*, reprinted with commentary in Sigüenza, *Historia del Rey de los reyes* (1916), 1:vii-xx, xxiv-xxxiii. Other biographical material can be found in the introduction to Sigüenza, *Historia de la orden de San Jerónimo* (2000 ed.), and in Andrés, "Nuevos datos sobre la genealogía del padre José de Sigüenza," and Ozaeta, "Arias Montano, maestro del p. José de Sigüenza." It was through his association with Arias Montano that Rekers also included Sigüenza as part of the "cell" of the *familia charitatis* at El Escorial. Current scholarship challenges Rekers's thesis, see chapter 1, note 5.

4. Sigüenza, *Vida S. Gerónimo doctor de la Santa Iglesia* (1595), and *Tercera parte de la Historia de la orden de San Geronimo, doctor de la iglesia* (1605). The paraphrase of the *History of Nature* was edited in 1961 and 2000. Over thirty-four of his sermons remain in manuscript, BME MS Ç-III-13, but his poetry in Spanish, along with that of Arias Montano, has been published: Arias Montano, Sigüenza, and García Aguilar, *Poesía castellana*. For Sigüenza's bibliography, see Campos y Fernández de Sevilla, "Bibliografía de y sobre el p. José de Sigüenza, OSH."

5. Sigüenza, *Historia del Rey de los reyes* (1916), 1:lxviii-lxxii; for a more nuanced account of the work's gestation, see the preliminary study in Sigüenza, *Historia del Rey de los reyes y Señor de los señores* (2016), 12-15.

of 1591–92. He was himself an accomplished and popular preacher, and his sermons were said to have been delivered forcefully and earnestly, a style that made him popular with the courtly audience at the monastery and the palace.[6] He also listened to Arias Montano preach, later translating four of his sermons into Castilian; these translations were collected along with thirty-four of his own sermons.[7] Given that Sigüenza most likely delivered these sermons after his mentor's last visit, they show how Arias Montano integrated his historical premises, hermeneutics of nature, and natural philosophy into a broader theological program. Yet, because the transcriptions were made from memory at a later date, they do not permit us to scrutinize the variations (and there are many) in biblical citations and compare them with the nuanced translations Arias Montano undertook in his Pagnino edition of the Hebrew Bible of the Antwerp Polyglot. What is certain is that they echo many of the themes of the *Magnum opus* and center on explaining scripture, including drawing on Hebrew terms.

Arias Montano's sermons use several examples from the natural world to drive home the moral lessons relevant to the liturgical calendar. In these instances he resorted to his own natural philosophy, or sometimes to examples from Pliny when building similes to explain the readings. For example, he cited Pliny's description in the *Natural History* (book 7, chapter 42) of the ability of an "admirable fish" to feed itself by opening its mouth and luring other fish inside to set up a parallel with the way Christ attracted people by his Word: "Now, by the fragrance it exhales, or the secret force it has over fish, it attracts to itself a great quantity of them and feeds on them."[8] Considering Luke 21:25 in his sermon for the first Sunday of Advent—"And there shall be signs in the sun, and in the moon, and in the stars; and upon the earth distress of nations, with perplexity; the sea and the waves roaring" (KJV)—Arias Montano chose to focus on the differing perceptions prophets have had of the nature of God: where some saw only mercy, other saw a terrible and punishing God. To ex-

6. Ozaeta, "Tres sermones inéditos del p. fr. José de Sigüenza," 154–55.

7. These sermons place Arias Montano at El Escorial two months earlier than previously stated by Gregorio de Andrés. Núñez Rivera dates the sermons to 1585–92, but they are most likely from the 1592 visit: Arias Montano, *Sermones castellanos*, 16. The sermons were delivered for the first Sunday of Advent, the feast of Saint Thomas the Apostle (21 December 1591), the feast of Saint Peter the Apostle, and *Dominica in Sexagesima*, or the eighth Sunday before Easter. If we take the feast of Saint Peter to refer to the feast of the Chair of Saint Peter (22 February), the dates of the sermons would have fallen roughly between the last Sunday of November 1591 and February 1592, since Easter fell on 29 March that year according to the Gregorian calendar.

8. Arias Montano, *Sermones castellanos*, 109. For his discussion of whales, see *Corpus*, 284; *HistNat*, 393.

plain the many facets of God and how those who have witnessed his majesty describe as wondrous whatever aspect they perceive, Arias Montano drew on Pliny's description of the opal, a multicolored stone that, depending on how you look at it, can seem like a diamond, a ruby, or an emerald.[9] The sun serves as a similar example; its light is so preeminent that it could seem that its only purpose is to illuminate the world; yet if one considers the heat of sun's rays, one finds that the same rays enfold everything and vivify all they reach.[10]

In these sermons Arias Montano stated unequivocally his thoughts about natural philosophical systems devised by humans. The sermon he delivered the eighth Sunday before Easter on the parable of the sower (Luke 8:4–8) gave him the opportunity to equate the heart of someone learned in philosophy, but with no room in it for the Word of God, to the infertile ground where the sower scattered his seed. The Word of God, he preached, holds no falsehood or deceit, whereas the utterance of the "wise of the world [*sabios del mundo*], who among a thousand lies hardly teach a single truth, is divided into a thousand opinions and personal fancies while still being mistaken, in that there is not a clear thing, nor a truth so certain, in which they have not erred."[11] They speak only to our reason, not to our will.[12]

There survive two sermons Sigüenza delivered for the first Sunday of Advent that comment on Luke 21:25 ("And there shall be signs in the sun, and in the moon"). Comparing them with the one by Arias Montano shows that Sigüenza did not readily interlace his sermons with reflections drawn from a type of Montanian biblical exegesis.[13] The earlier of his two sermons for that liturgical day[14]—if we are to trust the order of the manuscript—centers on the generation and regeneration of the world (*descruso y sucesión*), which Sigüenza drove home with three examples: the image of the ouroboros (a snake eating its own tail); the story of how ancient Egyptians, when the Nile failed to flood, fed

---

9. Arias Montano took some liberties with Pliny's description of opals (*Natural History*, book 37). Whereas Pliny compares the stone's qualities to "the gentler fire of the ruby," "the rich purple of the amethyst," and "the sea–green of the emerald," Arias Montano calls out "the light of the diamonds," "the cold red of rubies," and the "light green we see in emeralds.": Arias Montano, *Sermones castellanos*, 147.

10. "Da ser a las criaturas, cría debajo de la tierra las venas de oro y en las secretas conchas el precioso nácar, etc.": Arias Montano, *Sermones castellanos*, 147–48.

11. Ibid., 115.

12. Ibid., 116.

13. BME MS Ç-III–13, fols. 121r–22v. "Sermones de fray José de Sigüenza," n.d. The manuscript contains thirty–four sermons by Sigüenza and four by Arias Montano.

14. BME MS Ç-III–13, fols. 121r–24r.

their dead to the fish in the river so they in turn could feed on the fish; and finally, the myth of Saturn eating his own sons. He presented these as examples of mistaken ancient beliefs about the circularity of the material and spiritual world that were corrected with the message of the Apocalypse: when God wills it there will be signs in the heavens before the end times. Yet he used these same mythological examples to illustrate the continuity of the church and its liturgical cycle, a "revolution" that will persist until the end of the world. Although the first theme developed in the sermon dwells on circularity, Sigüenza makes no reference to the Montanian notion of GHOLAM, nor does he use biblical citations in Hebrew to support his arguments on this point, though the sermon topic would have been an ideal fit for such reflection. The later sermon for the same liturgical day, however, uses the properties of the sun and moon to describe the types of signs these celestial bodies could deliver to presage the Second Coming.[15] Although it reveals that Sigüenza was well aware of the technical aspects of astrology, he was more interested in the analogy between man's three states of being (*estados de ser*)—body/soul/intellect—and the earth/moon/sun.[16] This allows him to reflect on what an eclipsed moon signifies for the soul at the end times or how the moon's "stains and shadows" reflect the relation between the sun and the earth, or by analogy how stains on the soul are the consequence of the interaction of the body and the intellect. He used a large variety of sources, from astrological notions of aspects (sextiles and trines) and a blood moon, to Andrés de Acitores's *Theologia symbolica* and several references taken from Pliny. He punctuated his arguments with quotations from the gospels of Saint Augustine and Saint Clement, among others. At least in this small selection of Sigüenza's sermons, the influence of Arias Montano seems marginal—likely for good reason, as we will see below. Sigüenza did find an opportunity to fold his mentor's natural philosophy into one of his more ambitious works.

Sigüenza's Inquisitorial Inquest

By the late spring of 1592 the monastery and school at San Lorenzo de El Escorial were in an uproar. Soon after Arias Montano had left El Escorial, the head of the Hieronymite order called for an internal canonic visit (*visita canónica*) to examine the unusual opinions of some preachers and confessors of the monastery, as well as its school.[17] Within days after the visit got under

15. BME MS Ç-III-13, fols. 203v-6r.
16. BME MS Ç-III-13, fol. 204r.
17. The original documentation for the case (*proceso*) survives at the University of Halle, Yc, 20, 2, Dd 4, and was published in Andrés, *Proceso inquisitorial del padre de Sigüenza*, 79-294.

way, José de Sigüenza made his way to the supreme inquisitorial court in Toledo in an effort to preempt with his own arguments some twenty propositions
the visitor and subsequent evaluator (*calificador*) found suspect. He presented
a letter from the monastery's prior Diego de Yepes—the likely instigator of the
whole affair—containing a series of accusations against him that had come to
light during the visit, asking that they be examined. This first visit was followed
by an official inquest by a local Inquisition official in May, which found twelve
propositions suspect. Evaluators at the Inquisition's tribunal in Toledo found
three of these opinions "daring," another "thoughtless," and one "heretical."
Sigüenza was taken into custody on 20 July and housed in the Hieronymite
monastery of La Sisla in Toledo. He would remain there for three months as
the witnesses were deposed and he prepared his defense.

Historians agree that in this whole affair Sigüenza stood as a proxy for
Arias Montano—considered untouchable because he was favored by King
Philip II—their assessment based on the nature of the propositions denounced
by the Hieronymites and on personal jealousies that plagued a monastery
where royal favor, and that of other important persons, was so close at hand.[18]
Yepes in particular saw his authority as prior and theologian undermined by
Arias Montano and was irked by Sigüenza's criticism of his preaching style,
as he states in his deposition.[19] Several of the propositions certainly indicate
that Arias Montano's influence was a factor, but so were Sigüenza's irascibility
and his propensity for blunt speaking. The witnesses pointed out the excessive
praise Sigüenza gave Arias Montano, extolling his exegesis of Hebrew scriptures as surpassing that of the holy fathers of the church. Sigüenza, it seems,
privileged Montanian exegesis above all others. One witness testified that he
had heard Sigüenza say, "If they left me a Bible and Arias Montano's books, I
would forgo all the books in my cell—which are many and among them those
of the saintly doctors."[20] The inquisitorial examiners found all instances of
excessive praise for Arias Montano to be "daring." In his defense, Sigüenza
explained that he had found in Arias Montano's vast works a useful synthesis
of the doctrine of the church fathers: "He has drawn the substance from all of
them." Another proposition alleged that Sigüenza had said Scholastic theology was not necessary and was a waste of time. This one was deemed poten-

18. Bataillon, Rekers, and Gregorio de Andrés concur on this opinion. For a more recent
assessment, see Gómez Canseco, *Humanismo después de 1600*, 33–43.

19. Andrés, *Proceso inquisitorial del padre de Sigüenza*, 190.

20. "Si le dejasen la Biblia y los libros de Arias Montano, dejaria cuantos libros tenía en
su celda, que son muchos y entre ellos los santos doctores.": Andrés, *Proceso inquisitorial del
padre de Sigüenza*, 246.

tially heretical, since it implied that the works of church fathers such as Saint Thomas were not worthwhile. In his defense Sigüenza explained that, yes, too much time is wasted on many *cuestiones*, with new ones coming up every day and being subjected to a thousand opinions that only lead to competitions and fights. But he had studied and advocated the study of Scholastic philosophy, "as all those who read Hebrew with doctor Arias Montano can attest," and had made that remark in reference to an interminable discussion about the nature of angels.[21] In light of his explanation, the inquisitors later deemed the accusation without merit. The only proposition deemed heretical was an alleged claim that "barbarians and gentiles, Turks and Moors" could be saved as long as they believed in one God and lived according to natural law. To this Sigüenza responded that the sole priest making this claim, not having *letras* or university studies, had not understood the point he was trying to make, since he had structured his explanation following the line of Saint Thomas.

Throughout his defense, Sigüenza mostly qualified his alleged statements by situating them as having been made in casual conversation and not in any way meant as formal statements intended to undermine church doctrine. He stated his admiration for the work of Arias Montano, but he also acknowledged the importance of church doctors. Sigüenza's defense convinced the inquisitors, and he was absolved of all charges, with the definitive acquittal coming a few months later.[22] He quickly returned to El Escorial and resumed his duties at the monastery's library. In the meantime Arias Montano had been elected prior of his order's monastery, Santiago de la Espada in Seville, and never returned to El Escorial. We do not know to what extent he continued communicating with Sigüenza. The historical record resumes only after Arias Montano's death, when over the course of several years Pedro de Valencia replied to Sigüenza's questions about Arias Montano's works.

The *Historia del Rey de los reyes y Señor de los señores*

Far from being an original work, Sigüenza's *Historia del Rey de los reyes*[23] serves up generous portions of his mentor's works, often simply translating them into Spanish and on occasion clarifying Arias Montano's Latin prose. But to what extent then is the *Historia* a strictly derivative work? Villalba

21. Ibid., 120.

22. Ibid., 50–54.

23. Sigüenza draws from Revelation 19:16: "And he hath on his vesture and on his thigh a name written, KING OF KINGS, AND LORD OF LORDS."

Muñoz was dismayed by Sigüenza's unattributed borrowing and described Arias Montano's unacknowledged presence in Sigüenza's work as that of a "mystic in complete and profound obsession with an idea" to whom Siguenza turned for answers about the nature of the world.[24] García Aguilar meanwhile found Sigüenza's borrowings entirely in keeping with authorial customs of the time.[25] Yet in spite of its extent, most scholarship has considered it little more than a paraphrase. It is indeed difficult to disentangle Sigüenza's personal contributions from those of his mentor; the very process of translating into the vernacular and explaining the material for a particular audience can blur distinct contributions. But it is possible to identify certain passages where Sigüenza embarks on exegetical exercises not found in Arias Montano and through these to understand how the Montanian philosophy of nature was understood—at least by this one disciple.

Since Sigüenza's *Historia* remained unfinished and the surviving manuscripts might be truncated, its lack of a preface or dedication requires that we turn to the text itself to determine the author's plan and motivation for this project. For García Aguilar, Sigüenza conceived of it as falling within the historical genre, a project that, as the historian of the Hieronymite order, he was well prepared to undertake. The *Historia* was planned as a history of Christ, not just during his incarnation, but showing how his presence has been manifested through the ages in the Sacred Scriptures, as stated in Hebrews 13:8 (KJV): "Jesus Christ the same yesterday, to day, and for ever." Sigüenza had the broadest possible audience in mind, all the "vassals, faithful, brothers, sons, and disciples of this same king, father, and teacher."[26] He followed a more strictly chronological arrangement than the *Anima* or the *History of Nature*, yet he also began with a description of the attributes of God (*Historia*, part 1, book 1), which are made manifest in the Creation account (part 1, book 2) and culminate with the coming of Christ (part 2). While for Arias Montano it was essential to first explain the relationship between the Divinity and humanity—

24. "Un vidente, un místico en la plena y profunda obsesión de una idea, en la posesión de un sistema propio, de un principio supremo eje principal y resorte de todas las cuestiones.": Sigüenza, *Historia del Rey de los reyes* (1916 ed.), 1:xxxvi.

25. See García Aguilar's introductory study for an analysis of Sigüenza's amplifications of Arias Montano's texts, as well as the rhetorical and stylistic devices used in his paraphrase. He makes a convincing case that the second part of the *Rey de los reyes* contains more original contributions by Sigüenza than the first part, which relies more heavily on the *History of Nature*: Sigüenza, *Historia del Rey de los reyes* (2016 ed.), 20–21. On the manuscript's transmission, see ibid., 74–75.

26. Ibid., 99.

the main topic of the *Anima*—before explaining the acts of Creation, for the historian in Sigüenza it made more sense to rearrange the material testifying to Christ's presence throughout history to suit his penchant for chronology: "I offer only to compile it and give it some order so that we may see through it, if possible, the infinite treasure that this history encloses."[27] In this scheme Arias Montano's *History of Nature* serves as the principal source of the natural "treasures" that came about during the days of Creation. Sigüenza copied large portions of Arias Montano's text and also adopted his hermeneutic and methodological model for those few instances when he undertook his own exegesis.

In translating, Sigüenza often shed the more abstruse and technical of Arias Montano's etymological justifications and inserted everyday examples to make points. In the second part—particularly in his discussion of the Gospels—his contributions and amplifications become more prevalent. Yet ultimately his project simplified the Montanian exposition, though not so much that a general reader would doubt that the origin of its precepts and philosophy was the original Hebrew or Greek scriptures; anyone familiar with Montanian thought would recognize its true source. It is precisely the rearrangement and simplifications Sigüenza introduced in his translations that suggest his purpose was to bring Montanian theology and natural philosophy to a broader vernacular audience. With Arias Montano gone, Sigüenza was trying to take up his mentor's mantle, but instead of continuing the discourse at the same erudite level, he sought to broaden the appeal of the message by writing in Spanish, using a traditional genre, drawing on familiar examples, and constructing a straightforward chronological narrative. Suggesting that the project only touched on the treasures that can be derived from the Sacred Scriptures, he wrote, "We will not dwell on them, but rather run through then as if making a list of some treasures and jewels of this infinite archive."[28] He also directed some notes of caution to overzealous followers of Montanian-style exegetical exercises, as when he discusses the importance of proper names. These names, he complained, can easily be interpreted incorrectly by persons who ascribe to them some prophetic quality. Sigüenza took pains to explain that the proper names a man or a woman give to their offspring can refer only to past or present memories, whereas those names given by God or by someone God has enlightened, such as Adam, can be analyzed for how they foretold the individual's

27. Ibid.
28. "Y no nos detendremos en ellas, sino corriendo, como quien hace padrón o lista de algunos tesoros y alhajas de este archivo infinito.": ibid., 119.

future.[29] He seems to be reacting to an unnamed person or persons who had adopted Arias Montano's exegetical approach using Hebrew etymologies but had taken this mode of analysis too far, coming up with "ridiculous" interpretations of scripture.[30]

In the *Historia* Sigüenza rehearses the same historical premises and hermeneutics of nature found in the *Magnum opus*. In his more personal remarks he revels in calling out the errors of the gentile philosophers and the fruitlessness of defending or fighting over the ideas of the likes of Epicurus, Zeno, and Plato.[31] Even Aristotle—"to whom nature gave such a great ingenuity"—maintained no small number of "ignorant" notions, including a low opinion of God and his providence. He also thought the only angels were those needed to move the heavens about—heavens, he pointed out, that neither Aristotle nor others figured out correctly in number and composition. Aristotle also erred in his ideas about the beginning of the world, thinking it was *ab eterno* like God and produced like some sort of natural emanation, without needing a first cause. Sigüenza continued:

> [Aristotle] spoke in such a way about the principles of things that each one interprets them as they please. And when the nature of things is finally known, by one way or another, it bears such little fruit and is known so poorly that there is as little to it after it is known as before. And he is the best one [of them], at least the most methodical, coherent, and logical, the most ingenious of all that have come down to us from the days of the gentility (unless we also say along with Cicero, *Platonem semper excipio*).[32]

He, like Arias Montano, was not going to rely on "profane authorities or erudition" to explain the history of Creation. He would neither compete against nor dismiss the ideas or opinions of others—as the Peripatetics liked to do—although he had a thousand opportunities to do so. His disdain for Scholasticism was legendary—as his accusers pointed out to the Inquisition—yet a decade after that fracas he minced no words when he discussed the consequences of mixing profane philosophy with the sacred. It was akin to the

29. García Aguilar identifies the two relevant passages but explains Sigüenza's clarification as a deviation from Arias Montano's ideas as expressed in the *De arcano sermone*: ibid., 45–48, 319–20, 436.

30. Ibid., 436.

31. Ibid., 290.

32. Ibid., 100.

"monster" that results from the mixing that produces a mule, "as we see in many books and schools. Because what they call theology is the invention of Plato and Aristotle mixed with the Holy Word, and thus it is neither one nor the other, but rather a mule against the precept of God."[33] Studying their philosophy—the fancies of men—was futile. After expending oil, work, and even a life on it, the only thing you will know is that you know nothing (*es saber que no se sabe nada*):

> I could tell you briefly here what has been learned from all the philosophers who have dealt with the nature of things, for a young man can learn it in a school in a month and afterward knows as much of the matter as when he knew nothing at all; and by saying "prime matter," "form," and "privation" and other unfounded imaginations—about which anyone can make up things at will.[34]

Sigüenza's blunt style distilled some of the most abstruse Montanian arguments in ways that erased his mentor's intentional subtlety, as when he reinterpreted Arias Montano's view of astronomy. As I discussed earlier in this book, Arias Montano expressed his views on observational astronomy in very general terms, remarking that all the necessary aspects celestial bodies shared could be noted by experienced observers of the heavens without the need to imagine orbs for each of them. He, of course, maintained that there were no crystalline spheres in the heavens but rather a single region, SAMAIIM. Sigüenza was far more specific in his criticism of the astronomical arts. After following Arias Montano's text closely and explaining that the celestial movements are so consistent that expert observers can easily determine their "positions and stations," he added bluntly that this was the case "without having the need to put so many cycles, shells, eccentrics, concentrics, and epicycles, as well as another hundred imaginations born from the schools of the Chaldeans to save the appearances, which were not taught by the divine word or divine history."[35]

How did Sigüenza deploy Arias Montano's method and draw on his natural philosophy? Two examples of Sigüenza's works must suffice here. These are instances where I believe the analysis was mostly his own and not a translation of a specific passage from Arias Montano. (It is nonetheless possible that a devoted reader of Montaniana might be able to find an analogous passage in the

---

33. Ibid., 264. García Aguilar points out that Villalba Muñoz omitted the last sentence from his edition, since it seemed to imply that Scholastic theologians were mules.

34. Ibid., 179.

35. Ibid., 218.

hundreds of thousands of words our biblist wrote.) When Sigüenza tackled the acts of Creation, for example, he chose to structure them as a sequential commentary on the text of the first book of Genesis, devoting one chapter to each day (something Arias Montano never did).[36] He began with the Vulgate version of Genesis 1:2 ("Terra autem erat inanis et vacua et tenebrae super faciem abyssi et spiritus Dei ferebatur super aquas"). His attention, like that of countless hexameral writers before him, was drawn to three Hebrew words: *thohu* (תֹהוּ, *tōhû*, *inanis*), *bohu* (בֹהוּ, *bōhû*, *vacuus*), and *hhosech* (חֹשֶׁךְ, *ḥōšek*, *tenebrae*).[37] On the same page of the Antwerp Polyglot where the phrase first appears, Sigüenza could read three Latin translations of it: Saint Jerome's, that of the Greek LXX, and the Chaldean paraphrase. He had no quarrel with the different Latin translations, but he goes an interesting step further by defining the antonyms of the words, or rather the terms they are the opposites or contraries of. While *thohu* means "privation," its opposite is "that part of nature we call 'matter,'" so the world described as being *thohu* would possess none of the characteristics of matter such as weight, density, or extension. It is what "the Greeks called '*hyle*.'" *Bohu* also means "privation," but it stands opposite "form." When the world was described as *bohu*, it meant it was formless and without distinct features. He finds that *hhosech*, in addition to "darkness," can mean "confusion" and so is the opposite of "order." These opposites then allow him to explain the situation of the world at the beginning:

> Thus the first thing Moses teaches us of all this visible machine, heavens and earth, the state of that, coming out of the pure *nihil* and from the nothing, was a sort of body or mass or—saying it using known terms—a matter indisposed or undigested that, although he calls it "earth," did not have weight or gravity or lightness; nor was it soft or hard, nor did it have width or length or depth, but a privation, a *Thohu* of all this; nor did it have genus or species or definite nature; that is, it was *Bohu*, and therefore neither did it have any use or order or duties or correspondence with another thing: a pure confusion, a *Hhosech*, a fog.[38]

Where did Sigüenza get this interpretation? It is absent from the *History of Nature*, yet Arias Montano might have provided a hint in his commentary

36. Part 1, book 2, chapters 2–8, with chapters 9 and 10 devoted to the creation of the woman and the dignity of the human condition: ibid., 180–298.

37. Ibid., 180–81.

38. Ibid., 181.

on Isaiah, where he refers to *thohu* as the state of the world without form.[39] Yet it is clear that Sigüenza consulted a reference work on hexameral Hebrew to get a definition of these words and the vocabulary to describe them. He might have consulted the terse explanation in the thesaurus of the Antwerp Polyglot,[40] Fagius,[41] or alternatively the Pagnino/Kimhi thesaurus. There he would have found, under the entry for *bohu* a note taken from Rabbi David (Kimhi) explaining that scholars declared that *thohu* is unformed matter apt to receive all forms—called *hyle* in Greek or *materia* in Latin—and describes the earth before things were "designated, digested, and distinct," while *bohu* was defined as the form given to that matter.[42] Yet despite his willingness to resort to philosophical terms as descriptors, Sigüenza admitted that the biblical text did not really explain how God had acted to bring order to the world; Moses goes only so far as to say that ELOHIM agitated these waters, an action for which Sigüenza admitted there was no suitable philosophical term: "Humans lack a name to describe, in general terms, the manner of how and with what all nature was made. In particular, it was through the agitation of the water by the spirit Elohim."[43] At this point Sigüenza included a paraphrase of Arias Montano's discussion of the four principles: CAUSA, ELOHIM, IEHI, and MAIM. But not forgetting his own contribution, Sigüenza returned to a brief reflection on *hhosech*, *bohu*, and *thohu*. He found that these three "privations" coincided well with opposites of Solomon's description of how God brought order to Creation: "You have disposed all things by measure and number and weight" (Wisd. of Sol. 11:20 NAB). *Hhosech* stood for a confused state impossible to measure, *bohu* was a disorder that could not be counted, and *thohu* was a weightless matter; thus the act of Creation consisted in imparting measure, number, and weight to the world.

But whereas Arias Montano studiously avoided using philosophical terms in his explanation of Genesis, Sigüenza deliberately resorted to them when he felt he needed to explain one concept or another. Besides using "privation" and "form," he also drew on the Aristotelian notion of prime matter. Thus, after discussing the formless earth in the quotation above, he continued:

39. "Ut enim Thohu, quod reapse nullius formae aut generis fuerat, terram à Deo factam accepimus, id quod à Deo factum est.": Arias Montano, *Commentaria in Isaiae*, 1309.

40. Under *Bohu*: *est forma quae dat esse materiae*: Raphelengius, "Thesauri hebraicae linguae," 6:8.

41. Fagius, *Exegesis sive expositio dictionum hebraicarum literalis et simplex*, 9.

42. Pagnino, *Thesaurus linguae sanctae* (1529), 2715, 179.

43. Sigüenza, *Historia del Rey de los reyes* (2016 ed.), 181.

. . . which, if you look closely, are the conditions proper to what we all call prime matter, namely, that it does not have a body, or qualities, or being, or distinction, or use, or order. We are going along using known and common terms, and thus we call this "matter" or "like matter," warning beforehand that this name "matter" was not used by the prophets or the apostles.[44]

He clearly was aware that by using philosophical terms he was not remaining faithful to his mentor's program or to the biblical narrative. Later, when he returned to the notion of matter, words again failed him and he resorted to calling MAIM, the dual waters, "true matter":

> The matter from which everything was made was an earth that did not have nature or body, although it was already called thus; it was only privation of all form and of qualities and accidents and order and office. The true matter— which was close to receiving its particular perfection—was the two liquors or waters.[45]

The relation between natural philosophical terminology and the events of Creation was clearly on his mind to a degree not apparent in Arias Montano. For whereas the master had simply banished them from his explanation by relying almost exclusively on the power of the Hebrew lexicon, Sigüenza still felt he had to explain the events in the language of Scholastic natural philosophy. In the following important, if somewhat lengthy, passage, he explains the history behind the terms used by ancient philosophers, the limitations imposed by the nature of the Creation narrative, and how a reader should understand the philosophical principles there in relation to contemporary understanding of philosophy:

> As I said, what the philosophers call *prime matter* belongs more to metaphysics or mathematics than to the nature of bodies; what Moses is teaching us here concerns what the philosophers call *elements* and the way they compose themselves and what they resolve themselves into after all their transmutations. Nor do the Holy Scriptures know this word [elements] with this meaning; it uses the more appropriate one given by the meaning of the letters and the first principles of science. The most ancient philosophers among the Greeks

44. Ibid., 181.
45. Ibid.

were satisfied with knowing the four qualities that can be experienced, without searching for simple bodies into which they affixed themselves first or in a greater degree. Later they invented this business of fire, air, earth, and water and called them *elements*, considering that this world was like a poem consisting of many letters, enunciations, or words that stood for these bodies. And the simplest parts were elemental letters, the final thing they resolve themselves into, and indivisible, such that they are not the most simple, nor like letters, nor yet like syllables, but rather compounds and divisible (Plato, *Timaeus*, and Galen, *De natura hominis*). And because of this they searched for other indivisible principles, which are *matter* and *form*, and we do not challenge them with what we have been discussing. Yet the Holy Scriptures do not discuss this, nor do they [Plato and Galen] consider this, or provide much benefit or clarity. It deals only with visible and palpable things that are the earth and these two liquids, the oily and salty, in which we see the two passive qualities of humidity and dryness. And the other two qualities, cold and heat, are found in other tenuous, permeable bodies. And the Holy Scriptures refer to these bodies as *spiritus*, and they are the most notable and sensible among the spirits of Elohim that serve nature.[46]

Sigüenza chose to use a historical framework to explain the relation between the language of contemporary natural philosophy inherited from ancient Greece and the language used in Genesis to explain natural principles. He highlighted the origins of natural philosophical terms such as "prime matter," "elements," and "fire" in Greek analogies or metaphors rather than giving explanations derived from observation of the origins of these principles. For Sigüenza the natural philosophical terms of Greek philosophy existed in the realm of language and were not necessarily ontologically real. Even "matter" and "form" were simply terms the Greeks found when the earlier language of elements did not suffice. Consider the continuation of the passage quoted above, in which he explains the many embodiments of the primordial liquids:

The many different forms are derived from the trajectory and joining of these two liquors, varying and multiplying them in proportion, because the forms and species of things, although our philosophers call them "formal causes," could better be called by them the "effects" of the work of the efficient on mat-

46. Ibid., 186.

ter, which is in scripture the word *Iehi*, and the spirit of *Elohim*, who carries out the dominion of the world.[47]

The philosophers, Sigüenza tells us, have chosen to use the term "formal causes" to describe the composition of complex bodies, but what *really* lies behind it is an efficient cause that acts in the commands issued by IEHI and the action of the spirit ELOHIM.

Like Arias Montano, Sigüenza understood that in contrast to what Greek philosophers had done by constructing an ex post facto natural philosophy, the Sacred Scriptures did in fact describe what truly happened at Creation, although in a language that was somewhat vague, even maddeningly obscure. What Sigüenza did, which Arias Montano avoided, was to map the language of natural philosophy onto that of Genesis. Perhaps aware of the consequences of tampering with the finely tuned apparatus of natural philosophy, he clarified that he was not challenging the notions but presenting an alternative biblical way of arriving at their equivalents. Arias Montano would not have approved.

Pedro de Valencia

Pedro de Valencia was one of Spain's foremost late Renaissance humanists, a Hellenist, theologian, and commentator on economic and political matters whose career spanned the scholarly and civic spheres. He lived during an era his biographer Goméz Canseco incisively described as the "no-man's-land" between the flourishing years of Renaissance humanism and the Enlightenment.[48] An arts licentiate from Salamanca, sometime in 1578 he was introduced to Arias Montano, with whom he continued biblical studies and learned Hebrew. Although Valencia lived in a separate household after his marriage in 1587, he served as Arias Montano's copyist and secretary until his teacher's

---

47. "De la marcha y junta de estos dos licores, variándola y multiplicándola a proporción, se sacan muchas diferencias de formas, porque las formas y las especies de las cosas, aunque nuestros filósofos las llaman *causas formáles*, mejor las llamarían *efectos* de la obra del eficiente en la materia, que es en la Escritura el verbo *Iehi*, y el espíritu de *Elohim*, que ejecuta el imperio de la palabra.": ibid., 186–87.

48. Gómez Canseco, *Humanismo después de 1600*, 13. See Valencia's *Obras completas* in the Colección humanistas españoles edited by Morocho Gayo and published by la Universidad de Léon, 1993–2010. In addition to the preliminary studies in that collection, this section follows Gómez Canseco's book, as well as Suárez Sánchez de León, *El pensamiento de Pedro de Valencia*, and García Gutiérrez, "Arias Montano y Pedro de Valencia."

death in 1598. (His cousin and brother-in-law, Juan Ramírez [1574–1625], also worked as Arias Montano's secretary.) Valencia entered the seventeenth century under the protection of the powerful Conde de Lemos, joining the royal service in 1607 as chronicler of the Indies and Castile, a post he accepted reluctantly, since the duties kept him away from his biblical studies. During the post-Montanian years his profile in court increased, and his political treatises (*arbitrios*) could count an attentive readership among those close to the king and his favorite *valido*, the Duke of Lerma. He also maintained a cordial and consultative relationship with the archbishop of Toledo and inquisitor general of Spain, Bernardo de Sandoval y Rojas—the Duke of Lerma's uncle. Over the years Pedro de Valencia became more than Arias Montano's literary executor; he proudly assumed the role of protector of his intellectual legacy. And it was in this capacity that José de Sigüenza wrote to him while working through the *History of Nature*.

Pedro de Valencia published only one work in his own right, a treatise on the variants of ancient skepticism, *Academica, sive De iudicio erga verum* (1596).[49] Arias Montano surely helped shepherd the book through publication at the Plantin Press, although Valencia demurred that his Sevillian friends, unknown to him, had gotten the book published. In the *Academica*, Valencia examined the skeptical positions of various ancient schools. Working from Greek sources, he sought to clarify certain sections of Cicero's *Academica*, contrasting the probabilistic approach of the Academics and Carneades with the more radical dogmatism of Pyrrhonian skeptics. He was most concerned with ancient positions regarding the criteria for truth; though he leaned heavily on Cicero in collecting ancient opinions on the topic, he also engaged with other skeptical philosophies. Throughout the book it becomes apparent that Valencia shared Arias Montano's opinion on the futility of trying to find true knowledge through any philosophical school, since only God could grant access to true wisdom. ·

The book came in handy when José de Sigüenza described the iconographic program of the frescoes of the Royal Library of El Escorial as part of his grander history of the Order of Saint Jerome. One fresco, titled *The School*

49. There are two Spanish translations in 1987 and in volume 3 of his *Obras completas*. For more on Valencia's skepticism, see Suárez Dobarrio, "Filosofía y humanismo crítico en Pedro de Valencia"; Gómez Canseco, *Humanismo después de 1600*, 96–102; Suárez Sánchez de León, *El pensamiento de Pedro de Valencia*; Valencia, *Obras completas: Academica*, 3:52–63; and Laursen, "Pedro de Valencia's *Academica* and Scepticism in Late Renaissance Spain," 118–23.

of Athens, was tellingly placed beneath the allegorical figure of philosophy. The fresco depicts a debate between the Stoics, led by Zeno of Citium, and Socrates and the Academics, whom Sigüenza describes as skeptics (*dudosos*). They are debating, according to Sigüenza, whether man—chained as he is to the senses—has the means to determine what is true so he can understand the nature of things. Stoics sought definitions and precepts based on sense experience, while Academics held that knowledge gained by the senses was false and "could fool you a thousand times." Thus the Academics inferred "with much evidence" that humanity could never affirm what such false witnesses attest.[50] If readers want to further explore the philosophical postures depicted in the image, Sigüenza refers them to Valencia's book. And in fact, on the very first page we read that ancient philosophers can be divided into dogmatists and skeptics.

Sigüenza might have been aware of Valencia's work as early as 1590, but it was not until after its publication in 1594 that the two men corresponded on the subject.[51] In a letter dated August 1603—just as Sigüenza was writing the description of the library—Valencia responded to the friar's request to clarify Saint Augustine's position concerning the skepticism debate and referred him to the section in his book where he discusses the subject.[52] The section describes in detail the debate on the notion of cognitive impression between Zeno of Citium, who maintained the position of *non opinaturus sapiens*—that the wise should not opine about what is not perceived and that only the corporeal is known—and Arcesilaus, who countered with *nihil approbaturus sapiens*—because nothing can be perceived, man can assert nothing. In his book, Valencia stages the debate as a "disputation between a pair of students," finally resolving the issue by citing Saint Augustine's sentence on the debate. Augustine sided with the Academicians, preferring not to assert anything rather than to agree with Zeno that only the corporeal exists in nature.[53]

For Sigüenza, however, the fresco was there to remind viewers of two things: the difficulty of learning about nature using only sense perception, and the futile and seemingly endless speculations of the ancient philosophers, who had

50. Sigüenza, *Historia de la orden de San Jerónimo* (2000 ed.), 2:616–17.

51. Valencia might have completed the *Academica* as early as 1590: Valencia, *Obras completas: Academica*, 3:93–100.

52. Letter from Pedro de Valencia to José de Sigüenza, 9 August 1603, Zafra, in Antolín, "Cartas inéditas de Pedro de Valencia al p. J. de Sigüenza," 42:294.

53. Valencia, *Obras completas: Academica*, 3:207–13.

never resolved the matter, "but instead spend their intellect in finding arguments for both sides" (sino que gastan el ingenio en hallar razón por entrambas partes).[54] His explanation betrayed his allegiance to the Montanian program. Sigüenza's position should not suggest a negative, dogmatic attitude about the possibility of asserting anything as true, as Academic and Pyrrhonian skeptics postulated. Quite the contrary, the Montanian program never abandoned hope of finding intractable truths. Its prescriptive methodology sought to attain certain knowledge by anchoring it in physical and metaphysical principles derived directly from the Sacred Scriptures.[55]

This was not the only matter on which José de Sigüenza consulted Pedro de Valencia. Seventeen letters from Valencia to Sigüenza survive, written from 1593 to 1605.[56] In them the humanist and the friar discussed, among other things, theological and natural philosophical issues raised by Arias Montano's exegesis. The first recorded discussion finds Valencia hard at work copying the *History of Nature* and the *Commentary on Isaiah*. In September 1594 he informed Sigüenza that the manuscript of the first part of the *Corpus*, the *History of Nature*, had arrived in Antwerp. He had great hopes for its publication; it would reveal to all the greatness of the *History of Nature*, not just because of its quality but also because of its size.[57] Valencia was clearly in awe of his master's work, particularly his discussion of the human condition at the time of the Fall (*de homine et eius casu*), which in Valencia's view surpassed anything Arias Montano had previously published. Difficulties ensued with its publication, however. Plantin's successor, Jan Moretus, had grudgingly continued honoring Plantin's commitments to the author, after Arias Montano's

54. Sigüenza, *Historia de la orden de San Jerónimo* (2000 ed.), 616.

55. Sigüenza's skeptical critique has not gone unnoticed by historians. Colin Thompson notes Sigüenza's skepticism about the ability of human reason to ascertain truth purely by means of philosophy: Thompson, "Reading the Escorial Library: Fray José de Sigüenza and the Culture of Golden-Age Spain," 91. Ramiro Flórez notes the affinity with Pedro de Valencia and traces its roots to the philosophical perspective of Arias Montano: Flórez and Balsinde, *El Escorial y Arias Montano*, 223, 351–64.

56. BME MS L–I–18, fols. 7r–40v, letters from Pedro de Valencia to José de Sigüenza. These were published by Antolín in four volumes (41–44) of the journal *La ciudad de Dios* (1896–97) and in Valencia, "Cartas de Pedro de Valencia al p. Sigüenza," in *Epistolario español*, 62: 43–45.

57. "Por aquel primer miembro se echará bien de ver cuán grande será el cuerpo, y que con razón no solo por la calidad, sino por la cantidad se llama *Opus magnum*.": Letter from Pedro de Valencia a José de Sigüenza, 5 September 1594, Zafra, in Antolín, "Cartas inéditas de Pedro de Valencia al p. J. de Sigüenza," 41:349.

death—and that of Luis Pérez, his agent in Antwerp—prompting Valencia to frequently remark to Sigüenza on the unwillingness of those in Antwerp to publish unprofitable books, likely referring to Moretus's more sanguine approach to publishing.

In 1596 it appears Sigüenza sent Valencia a draft of a project he was working on, perhaps an early portion of the *Rey de los reyes*, asking him to clarify some natural philosophical concepts. Among other things, Valencia responded to a query concerning Arias Montano's interpretation of *fiat lux* from Genesis 1:3. Valencia tried to convey what he thought Arias Montano understood to be the nature of this first light:

> In this place in Genesis 1, *fiat lux*, I gather from Arias Montano's writings that he believes it to be not material light, but rather the kind that suited the eyes that existed then, those of God and spiritual creatures, which is taking things from the darkness of nothing and giving them being; so that in being they can later be known and be conspicuous. It is as if saying, "let there be created being." About this we could discuss presently: naturally this light is dark to our eyes: "Until that day dawns, for God, who commanded the light to shine out of darkness, hath shined in our hearts," as another apostle said. Your paternity will consider it.[58]

In the *Rey de los reyes*, Sigüenza incorporated Valencia's clarification as follows:

> With His imperial word he produced and created the heavens and the earth out of nothing, as the first examples of his light. And this creation through which it was possible to distinguish what was from what was not and no being

---

58. "El lugar del l.º del Genesis, *Fiat lux*, colijo de los escritos de Arias Montano que allí no entiende luz material, sino la que lo es para los ojos que entónces habia, de Dios y de las criaturas espirituales, que es sacar á las cosas de las tinieblas del nada, y darles sér; que en siendo, luégo pueden ser conocidas, y de suyo son conspicuas. Así que será como decir: haya sér criado. De esto podriamos tratar presentes: de suyo es oscura para nuestros ojos aquella luz: *donec dies illucescat, et Deus, qui dixit ex tenebris lumen splendescere, ipse illuxit in cordibus nostris*, dice otro apóstol. Vuestra paternidad lo considerará.": letter from Pedro de Valencia to José de Sigüenza, 8 September 1596, Zafra; published in Valencia, "Cartas de Pedro de Valencia al p. Sigüenza," 44. Valencia seems to have conflated quotations from 2 Corinthians 4:6 and 2 Peter 1:19.

became being, from privation, abyss of nothing, and emptiness to something, fullness, and truth, He called light, saying: *Fiat lux, et facta est lux.*[59]

Both Valencia and Sigüenza resorted to the philosophical language of ontological categories when having to explain Arias Montano's thoughts, and Sigüenza went a step further and used the Scholastic term "privation" to explain the state of nonbeing. What is evident from this interpretation is that in the Montanian program this first light was not light per se but a word used to represent a metaphysical event: coming into being.

After an apparent five-year gap in the letters, their surviving exchanges resumed in 1603. These letters suggest that Sigüenza had been reading the *History of Nature* very carefully and often writing to Valencia seeking clarifications. In a letter dated 5 May 1603, Valencia responded to several queries, some concerning the meaning of a Greek word (ἅρπαγμα, the act of robbery or seizing) used in Philippians 2:6; Valencia not only explained the word but suggested the theological consequences of his interpretation.[60] For our purposes, however, the explanations that follow concerning light and celestial movements are particularly illustrative of how Arias Montano's closest disciples wrestled with the natural philosophical ideas embedded in the *History of Nature*. Sigüenza had apparently said he considered certain passages from the *History of Nature* particularly difficult. Valencia conceded as much but reassured him that some of the problems came from this branch of philosophy, which was often condemned without being understood. He had discussed (*controvertimos i disputamos*) this philosophy with his discoverer (*el descubridor della*)—meaning Arias Montano—and the doctors Tovar, Oropesa, and Aguilar.[61] First of all, he explained, to defend this philosophy it is necessary to work from within its own principles and not mix other doctrines with it or judge it according to propositions from other *viae*, or paths, simply presumed

59. "Con su palabra imperial produjo y crio de nada los cielos y la tierra, como las primeras muestras de su luz. Y esta creación con que se vio y se distinguió lo que era de lo que no era y salió de no ser a ser, de privación y abismo de nada y de vació al algo y al lleno y la verdad, llamó *luz*, diciendo: *Fiat lux, et facta est lux*.": Sigüenza, *Historia del Rey de los reyes* (2016), 183.

60. Letter from Pedro de Valencia to José de Sigüenza, 7 May 1603, Zafra, in Antolín, "Cartas inéditas de Pedro de Valencia al p. J. de Sigüenza," 42:130.

61. See chapter 6 for Arias Montano's relationship with these Sevillian doctors. Valencia added, in a self-effacing manner concerning his interlocutors: "Loquebar in conspectu regum et non confu[n]debar" (I will speak of thy testimonies also before kings, and will not be ashamed [Ps. 119:46 KJV])—a curious choice, since the phrase was also the epigraph of the Augsburg Confession!

to be true. It is not entirely clear what Valencia means by the *via* of philosophy, since the context suggests he could be referring to philosophical schools or to disciplines such as astrology, cosmology, or physics. What is evident from his remarks, however, is that Sigüenza was not the first to be puzzled by the Montanian program.

The first topic Valencia touched on concerned Sigüenza's inquiry about "prime matter." This concept, he wrote, was one of those dogmas that were presupposed and had more in common with metaphysics than physics.[62] Even Tertullian had noted that neither the prophets nor the apostles knew the word "matter," suggesting that it was tangential to Montanian biblical natural philosophy. The subject Arias Montano was discussing in the unspecified passage was not "prime matter" but concerned *cuerpos criados*—not those without form, but very simple forms that enter into composition with the bodies we see. They were something akin to the "elements," not as what they later came to signify—fire, air, and such—but in the way the ancient Greek *physicos* had intended. They had identified four qualities with four letters so they could describe the world's coming together as letters and syllables forming words. Yet as Plato explains in the *Timaeus*, these four elements (στοιχεῖο) are themselves compounded and divisible, which is the reason the ancient Greeks, "whom nobody contradicts, and neither do we," in an effort to find indivisible principles "sought and found" matter and form.[63] The Sacred Scriptures, Valencia continued, do not mention matter and form but speak instead of other "second principles" (*principios segundos*) that are visible, are palpable, and contain the two passive qualities dryness and wetness: *arida* (dry earth) and the two liquors. These two become affixed to other tenuous bodies that are penetrable, referred to in the Sacred Scriptures as *spiritus*; among them are the qualities of heat and cold, "the two most notable ones among the *Spiritus Elohim*." In his explanation, Valencia had managed to attach the four qualities of ancient Greek philosophy—dry, wet, hot, and cold—to the Montanian interpretation, something Arias Montano had gone to great lengths not to do.

At this point Valencia offered his interpretation of what was apparently another of Sigüenza's concerns: Arias Montano's attribution of all action in the world to Spirit. Sigüenza wondered why Arias Montano attributed this agency to ELOHIM rather than to IEHI. To illustrate the relation between the

62. "Aqui no se tratta de *Materia prima*, que esse dogma se presupone i casi pertenesce a trattacion metaphysica más que physica.": Letter from Pedro de Valencia to José de Sigüenza, 7 May 1603, Zafra, in Antolín, "Cartas inéditas de Pedro de Valencia al p. J. de Sigüenza," 42:131.

63. Letter from Pedro de Valencia to José de Sigüenza, 7 May 1603, Zafra, ibid., 42:132.

two, Valencia drew an analogy between the microcosm of the human body and the macrocosm of the universe. If we consider man as a microcosm, then intellect—*ánimo* (ὁ νοῦς, *mens*)—acts as his principal and universal cause. As such, it represents the Father, or *Pater Spirituum*, that has its seat in the heart. This sends the animal spirits to the head, which seems to be where the principle of voluntary motion resides; from there the spirits cause the motion of the various members of the body. A similar relation exists in the universe, Valencia proposed. For whereas the Father attributed to himself creation and government by means of the word (*in verbo*), IEHI, the execution of everything and each thing is through the Spirit (*per Spiritum*), which encompasses all the earth and fills the world. It is through this Spirit pervading everything that we live, move, and exist.[64] Furthermore, he continued, a similar relationship exists between the Father, Son, and Holy Spirit. In the Trinitarian version of the explanation that followed, Valencia equated the function of the Holy Spirit to that of ELOHIM, citing Psalms 104:30 (KJV), "Thou sendest forth thy spirit, they are created: and thou renewest the face of the earth." So the Spirit of God acts like a soul of the world of the natural spirits created for the agitation and fermentation of the liquids, just as in humans the rational soul through the agency of those corporeal spirits—"as doctors refer to them"—agitated, conserved the humors, and moved the body.

What did Sigüenza do with this explanation? If we turn to the *Rey de los reyes* we find that on the vexed issue of whether the regeneration of natural things on earth was due to the animated nature of an ensouled earth, Sigüenza seemed unconcerned about (or oblivious to) any controversy this notion might elicit and repeated Valencia's words almost verbatim: "It suffices to say that this natural creation uses the spirit of God—which is like a soul of the world—the natural spirits born of the agitation and fermentation of the [dual] liquors, just as the rational soul agitates and conserves the humors and moves the body by the mystery of those corporeal spirits."[65]

Finally, Valencia turned to the concerns stemming from Sigüenza's interpretation of Arias Montano's notion of the heaven.[66] This page belongs to a broader section where Arias Montano reappraises the parts of the world based on the principles defined earlier, but now using observations to qualify earlier definitions about the composition and arrangement of the heavens. We can intuit from Valencia's response that Sigüenza was puzzled by the events

64. Ibid.
65. Sigüenza, *Historia del Rey de los reyes* (2016 ed.), 179.
66. *Corpus*, 180; *HistNat*, 285–86.

of the first day and the beginning of the reckoning of time. The first confusion stemmed, as Valencia understood it, from the way the author, not being entirely divorced from other doctrines, had inadvertently spoken *ex opinione aliorum* and seemed not to be following his own principles. Valencia never identified the language that caused the confusion, but in this very section Arias Montano refers to the movement of heavens as marking the day and night of the first day. (This movement would not have been possible until day two, when the waters were separated and the world divided into a heaven and an earth.) The Bible indeed refers to "day" and "night," but, Valencia notes, before the separation of the waters the universe was a confused globe beyond which there was nothing, "no material body that gave out light, nor diaphanous [body] to capture it, nor color that it could make manifest, nor eyes to see, and lacking any one of these four, the other three are useless."[67] Only SAMAIIM existed, composed of the attenuated and rarefied dual liquids, but there were no heavens or their movements to mark time.

Sigüenza was also puzzled by the delay between the creation of light on the first day and the creation of the luminaries on the fourth. An old question to be sure, as Valencia pointed out, and one that added to the obscurity and mystery of the first chapters of Genesis. However, counting days and nights against this first light should not be a problem because "corporeal things were already created, and they have duration, which is measured with time, although there might not be a clock to measure it." So the words *factumque est vespere et mane, dies unus* means a period of twenty-four hours, although they were all without light, at least the light we attribute to the sun. Valencia submitted all his explanations to Sigüenza's censure, begging off and explaining that these were matters for greater intellects than his. In subsequent letters Valencia clarified some parts of the *Anima* concerning the precepts given to Adam, but their correspondence on natural philosophical matters falls off at this point.

Their letters show the difficulty even those very close to Arias Montano had in understanding the natural philosophical precepts he had derived from the Bible. Words failed them, and as Valencia pointed out, words had on occasion failed Arias Montano himself. The difficulty derived from the effort to explain metaphysical and natural philosophical concepts in a language other than that of Scholastic Aristotelianism or Platonism. Even a disciple like Sigüenza, who had learned the precepts directly from Arias Montano, found it impossible to convey Montanian philosophy without resorting to the accepted lexicon.

67. Letter from Pedro de Valencia to José de Sigüenza, 7 May 1603, Zafra, in Antolín, "Cartas inéditas de Pedro de Valencia al p. J. de Sigüenza," 42:133.

Valencia was marginally more successful, but he was also under the spell of the natural philosophical ideas ingrained in the Western philosophical tradition. His responses show that his first instinct was to resort to qualified analogies in his explanations, but when these did not suffice he turned to the familiar concepts and language of Scholastic Aristotelianism.

THE DOUBTING THOMISTS

Within the broader community of biblical exegetes, Arias Montano's natural philosophical proposals seem to have caused barely a ripple. Some readers, even attentive ones, were largely unaware of or ignored the implications of his *Magnum opus*. One such reader was fray Andrés de Acitores (d. 1599), a Cistercian theology professor at Salamanca trained in the trilingual exegetical approach of Cipriano de la Huerga (as Arias Montano had been). Acitores's *Theologia symbolica* (1597) makes frequent references to the *De arcano sermone*; in fact, he wrote that he wanted to supplement its brief literal definitions with his symbolic and allegorical meanings. Yet he ignored the admittedly vague references to Arias Montano's hermeneutics that appear in the introduction to the *De arcano sermone*.[68]

The Unknown Inquirer

Others seeking clarification tried to engage Arias Montano's intellectual executor, Pedro de Valencia. There is an unpublished manuscript at the National Library of Spain in Madrid that forms part of a volume of works that are attributed to or once belonged to Arias Montano and Pedro de Valencia. One document in particular shows to what extent Valencia acted as Arias Montano's posthumous prolocutor.[69] The unsigned tract consists of a series

---

68. Acitores, *Theologia symbolica, sive Hieroglyphica*, Prolegomena, 3, 26, 55, 79.

69. The manuscript has been attributed to Pedro de Valencia, largely because it was placed in the amended edition of Nicolás Antonio's influential *Bibliotheca Hispana Nova* (Madrid, 1788), among the works of other Genesis commentators. Yet the unsigned manuscript has some internal references that suggest a different but as yet unidentified author who is *addressing* Valencia. In addition to those mentioned here, in folio 193 there is a curious reference to meeting with a *Licenciado Moreno, cura de Llerena*, who had a unicorn horn that he had shown to Don John Vicentello, followed by a discussion of whether unicorns existed or were confused with the rhinoceros. This same priest had also shown him an image (*figura o rostro*) of a thirty-one–year-old Christ and the handwritten letter Christ had sent to King Abgar of Edessa, a story then being promoted by Cardinal Cesar Baronius in his *Annals*: BNE MS 149: Pedro de Valencia, "Exposición del Génesis," n.d., fols. 184r–95r.

of queries concerning the first chapter of Genesis. The writer was clearly well schooled in Thomist theology and Scholastic natural philosophy, often switching from Spanish to Latin when he addressed theological or natural philosophical ideas. The queries seem to have arisen after the writer had corresponded with Valencia or had personal conversations with him. On that occasion he had received a curt response concerning the nature of the light mentioned in Genesis 1:3. This time he begged Valencia to indulge him in this new series of questions. They were the product of a recent careful rereading of Genesis 1, he wrote, adding apologetically that the queries were surely a consequence of the "tenuousness" of his understanding, the chapter's "profundity," the brevity of Valencia's earlier response, and the "newness of the doctrine." Over the next twenty-one folios the writer engages with many of the traditional topics of hexameral commentaries, but also with several new ones, the product of what he understood of the "new doctrine" (which I understand to be the Montanian explanation of Genesis). The writer urged Valencia to declare his own opinions and views on these questions and also those of Doctor Arias Montano, "in glory be."[70]

The writer's doubts are arranged around nine hexameral phrases. Each of these in turn provoked a number of questions that—as a good Scholastic— Valencia's correspondent parsed into objections and responses. What is clear from the exposition is that the writer was seeking an explanation of the first chapter of Genesis along the lines of a traditional hexameral commentary and found the redefinitions and conceptualization of the Montanian program largely incomprehensible, not only because Arias Montano had not addressed the usual sticking points of the hexameral genre, but also because the Montanian program took pains to avoid Scholastic terminology in presenting its explanations. It seems that the writer did not have the *History of Nature* at his disposal but based his understanding of the Montanian interpretation of some salient points of Genesis on the *De arcano sermone*, as is most evident in his discussion of light.[71] In what follows I will examine just a few of these sets of queries, though they all merit careful study for what they tell us about the intersection of hexameral and natural philosophical treatises during this time.[72]

---

70. "Declare su opinion y parecer, y del S[eñ]or Dottor Arias Montano que S[an]ta Gloria aya.": Valencia, "Exposición del Génesis," fol. 184r.

71. Arias Montano, *Libro de José*, 126–28, 421–22, [13–14].

72. The nine phrases are: "And God said, Let there be light" (Gen. 1:3), "And God called the light Day" (Gen. 1:5), "And the evening and the morning were the first day" (Gen. 1:5), "Let there be a firmament in the midst of the waters" (Gen. 1:6), "Let the waters under the heaven be gathered together unto one place" (Gen. 1:9), "And God called the dry land Earth" (Gen. 1:10),

The first set concerned the nature of light when it is first mentioned in Genesis. The writer began by copying Valencia's original short response:

BRIEF RESPONSE BY SEÑOR LICENTIATE PEDRO DE VALENCIA

By the "light" of the first chapter of Genesis, Arias Montano understood the distinction of things that intrinsic light makes visible, although there are no eyes to see them, as there were then no corporeal ones.[73]

Earlier we saw Valencia's response to a similar question by Sigüenza. This time, however, Valencia distilled the explanation into two concepts: light as representing the act of the distinction of things, and the "intrinsic-ness" of light acting to make visible certain things that are invisible to human eyes. The former was in line with the explanation Valencia had given Sigüenza, but the notion of "intrinsic light" was new. Intrinsic means essential or, as Covarrubias defined it in his dictionary of 1611, "something that is within another thing but does not manifest itself."[74] If we take into consideration his earlier explanation, Valencia might have had this intrinsic light stand in for the idea that with *fiat lux* things had come into being in such a manner that they could later be made conspicuous to corporeal eyes.

Before proceeding with the analysis of the writer's queries that had arisen from Valencia's brief response, let me point out a few key ideas about this first light as Arias Montano described it in the *De arcano sermone*. Arias Montano subscribed to the notion that the first line of Genesis indicated that God had created the whole world in the beginning and that the actions of the subsequent days assigned each thing its particular place and office. Light was the first of these, so that "all the other things, as is written, were created before

---

"Let there be lights in the firmament" (Gen. 1:14), "Let the waters bring forth abundantly the moving creature that hath life, and fowl that may fly above the earth in the open firmament of heaven" (Gen. 1:20), "God created man in his *own* image, in the image of God created he him; male and female created he them" (Gen. 1:27). All citations KJV.

73. "*Respuesta breve del S[eñ]or L[icencia]do Pedro de Valencia* | Por la luz del Cap. 1° del genesis, entendia Arias Montano la distinccio[n] del las cosas que las [*sic*] da luz intrinseca ser visibles, aunque no aya ojos que las vea como no los avia entonces corporales.": Valencia, "Exposición del Génesis," fol. 184r.

74. Sebastián Covarrubias, *Tesoro de la lengua castellana*, 506v (1611 edition): http://fondosdigitales.us.es/fondos/libros/765/16/tesoro-de-la-lengua-castellana-o-espanola/.

light, but light existed through the Word of God."[75] Theologians, Arias Montano pointed out, have not been able to discern anything beyond this about its creation, nor do the Sacred Scriptures say more than this. Light is therefore the first thing that conveys the beginning and opportunity (*initium et opportunitas*) for the distinction and separation of things (*distinctio et separatio rerum*). Note that he does not add the qualifier "intrinsic," nor does he suggest that this light carried out the act of distinguishing or differentiating among things. In typical Montanian fashion, he was happy to leave the matter unsettled. In the *History of Nature* he did not go much further, suggesting that light could belong to something or be obtained by some things.[76] What Valencia had done was draw together Arias Montano's notion of the Creation as an action granting distinction to things and the notion that light can reside in things. As we will see, Valencia's conflation of these two notions proved most unsatisfying to his anonymous correspondent.

Valencia's correspondent began by discussing the opinion of Saint Thomas from the first part of *Summa theologica* (q. 67), where he says that light can be interpreted in two ways: as visible light, and as knowledge made manifest and explained.[77] The writer commented that he liked Valencia's, and by extension

75. This and other pertinent quotations: "Por lo cual, en el principio mismo, cuando fueron creados cielo y tierra, antes incluso de que existiera el día o algún otro tiempo, surgió la luz, la primera de todas las cosas que se dice que fueron creadas por la palabra de Dios. En efecto, todas las demás cosas, como está escrito, fueron creadas antes que la luz, pero la luz existió por la palabra de Dios." "Ahora bien, es cierto que luego los cielos obtuvieron su lugar, sitio y oficio por obra de la palabra divina; pero tuvieron su naturaleza por obra de la creación." "En efecto, acerca de la luz, excepto que la materia precedente queda excluida, nada mas claro han resuelto los teólogos." "[La luz es] la primera que trae principio y ocasión para la distinción y separación de las cosas.": Arias Montano, *Libro de José*, 126, 421, [13].

76. *Corpus*, 151; *HistNat*, 253.

77. S[an]to. Tho[mas], [ST] 1.p[arte]. q[uestio]. 67 ar[ticulo]. 1. in cor. dize, que este nombre, lux, fue primero instituido a significar aquello que haze manifestacion, y declara algo en el sentido de la vista, y despues dice S[an]to. Thomas se extendio a significar todo aquello que es causa y razon de manifestar y declarar algo por qualquiera conocimiento, ora sea de sentido ora sea de entendimineto, y segun esto, bien le quadra el nombre de luz, a la distinccion de las cosas, pues fue causa que se manifestassen, y estuviessen patentes, claras, y manifiestas para de qualquiera potencia ser conocidas **per lucem namque universae creaturae, factae fuerunt conspicuae et visibiles etc.** / viene bien con esto, que suelen llamar a la disticcion y division del equinoco [*sic*] y oblicuxo [*sic*], lux y claridad, y al Maestro que procura distinguir y huir equivocaciones y obscuridades, llama[n]lo distintto y claro.": Valencia, "Exposición del Génesis," fols. 184r–v. (The Latin text in the original manuscript is always in a heavier ink suggesting boldface.)

Arias Montano's, doctrine on this. In particular, he thought it resolved a number of issues that often came up with respect to whether this first light was the light of the sun, which is not mentioned until the fourth day. These concerned the impossibility of having a light (which is an accident) occur without its subject (the sun), the existence of day and night without a sun, the possibility that this was some other type of light or whether it later became corrupted, and the nature of its relation to the darkness that hovered over the abyss. He asked Valencia to concede that his brief explanation presupposed what Saint Thomas in question 68 said had been the opinion of Chrysostom: that the first line of Genesis is a summary of the act of creation, with the exposition that follows being a description of gradually unfolding events, thereby supporting the notion that the first light was not that of the sun, since the sun was not created until the fourth day. Given this, the writer synthesized his doubts about Valencia's views as follows.

First, he noted that before this first light or "distinction" (as Valencia referred to it), distinctions already existed, such as between the earth and water and between the elements according to place (*terra erat sub aquis*), so there was no chaos. Furthermore, he argued, "distinction does not intrinsically give things visibility, because to be [visible] comes from real intrinsic light, which is in things, although there might not be this 'distinction' you mention."[78] The difficulty in the interpretation Valencia proposed was that light does not give distinction to things *ex se*. In this correspondent's estimation, the actual situation described in Genesis was that things were visible *in potentia remota*, "as the philosophers say." They were visible *in actu*, not by this type of light Valencia called "distinction," but rather because there was real light—that of the sun—which was nearby and diaphanous.[79] Therefore, in order for Valencia to uphold his argument about distinction, he would have to accept that the first line of Genesis describes a complete Creation and that this light was in fact that of the sun.

---

78. "Porque de serlo proviene de lux intrinseca real, que esta en las cosas, aunque no esta distinccion de que V[uestra] M[erce]d habla.": Valencia, "Exposición del Génesis," fol. 185v.

79. "De manera que todas las cosas corporales que Dios crio, antes que uviesse esta luz o distinccion **ex se, ratione lucis intrinsecae, erant visibiles, in potentia remota (et dicunt Philosophi)** aunque no uviesse ojos corporales quales viessen, **quaeque quidem fiunt visibiles in actu, non per lucem quae vocatur distinctio, sed per lucem realem seu per lumen, cuique subiectum est diaphanum, quae requiritur ex parte medii, et obiecti**, y esta luz ya la avia, antes desta distinccion, pues avia Sol el qual fue criado quando Dios crio el cielo, como hemos dicho y me paresce que V. md. a de dezir y assi nada les faltaria a las cosas para ser visibles.": Valencia, "Exposición del Génesis," fol. 185v.

Second, the writer pointed out, Valencia seemed to suggest that this "distinction and manifestation of things" happened only on the first day. This implies that the work of distinction was twofold, which is against what theologians maintain: that the works of distinction are threefold, and all three are for the sake of ornament.[80] Perhaps this light should just be called "day," rather than "distinction," he suggested, if not with all propriety, at least metaphorically.

Third, he questioned Valencia's explanation because it suggested two different scenarios, neither of which seemed reasonable to him. One scenario held that this light of distinction extended over the six days of Creation and was involved in new productive and creative actions; in the other scenario, this distinction manifested and distinguished only what was created or produced without its involvement. Valencia could not maintain either opinion, the writer challenged. The first contradicted the interpretation of the first line of Genesis as suggesting a complete creation. Neither could he affirm the second interpretation, because the text of the six days includes many verbs demonstrating productive and creative actions, so these other days contain activities of new things' being made, thus going beyond distinguishing and separating what has been made. He was concerned that Valencia opened the door to an interpretation that saw the six days as devoid of subsequent creations, which would be contrary to what was suggested by phrases such as *producat terra, producant aquae*.

After setting up all these objections, the writer put forth his own interpretation and submitted it for Valencia's opinion. He again found himself unable to express his ideas wholly in Spanish. His view was that the light of the first day "was nothing other than beauty, decoration, and perfection" (*nihil aliud sit quam quaedam pulchritudo, decor, ornatus, et perfectio*) because although the earth was created in the beginning, it lacked "beauty and decoration, which is indicated by the words "and the earth was empty and void" (*pulchritudo, vel ornatus, qui significatur illis verbis, terra autem erat inanis et vacua*).[81] So, he continued, although on the first day there was distinction in terms of substantial forms and other proper and common qualities, there was also distinction

80. "Opera distinctionis sunt tria, et alia tria opera ornatus.": Valencia, "Exposición del Génesis," fol. 185v. Here the writer was drawing from Aquinas's Q. 65 and Q. 70, where he maintained that the work of Creation could be thought of as unfolding in three phases: creation, formation or distinction, and adornment. The writer is essentially paraphrasing a commonly held Scholastic interpretation, such as the one in Francisco Suárez's disputation on Aquinas's commentary on the six days of Creation, *De opere sex dierum ac tertius de Anima*, 63.

81. Valencia, "Exposición del Génesis," fol. 186v.

concerning location and place (*sitio y lugar*), since each element had its natural place. If place is also considered in terms of conservation and necessity for the generation of things, without doubt place may also provide beauty, decoration, and ornamentation to adorn the act of generation. This is signified by the words "Let the waters be gathered into one place and let the dry land appear." So if we say that the light of the first day is "this beauty and ornament" (*haec pulchritudo, seu ornatus*), it would encompass all that is said in the first chapter of Genesis, since it is not incompatible with this statement to suppose that some action or production is taking place, such as those *de accidente* that took place on the second day, or *de substantia* as in the production of mixed things.

The next query I will examine here belies the anonymous writer's frustration with the never-ending debate about the word "firmament."[82] After summarizing several positions typical of those discussed in hexameral commentaries, the writer asked Valencia to please respond to the following questions at length (*no sea corto*): (1) What is understood by the word "firmament"? (2) Is it *de natura* of the fifth essence or *de natura elementaria*? (3) What did God do on the second day, and what did He produce? (4) How can the words *fiat firmamentum* be verified, since the Hebrew word יְהִי (*yĕhî*, becoming) suggests continued production and new existence? (5) Are the waters above the firmament real waters, or of the same nature as the sky? (6) Why did the interpreter refer to it as "firmament"? And finally, (7) Why did God not say "it was good" as he did on the other days?

Genesis 1:9—the congregation of the water in one place—posed no small number of concerns either, including those of "a modern" (*un moderno*) who maintained that in the beginning the earth had been covered with water much as it was now.[83] Again he asked Valencia to clarify the arguments of this unnamed *moderno*. He was also confused by the arguments Aristotle put forth in the second chapter of the *Meteorology* concerning the origin of springs as the result of the sun's heating vapors inside the earth. At this point the writer makes a telling remark that reveals his acquaintance with Arias Montano's work: "They tell me [*dizen me*] that doctor Montano believed that the springs originated in the sea and had explained away arguments alleging that the water could not be so pushed upward to the top of mountains by resorting to an explanation of water pumps."[84] He begged Valencia to please confirm this last point and tell him how he understood the matter.

82. "Que sea firmamento no acaba[n] los authores de dezirlo.": ibid., fol. 188r.
83. Ibid., fol. 189v.
84. Ibid.

Then followed more queries concerning the nature of the heavens. Did *fiant luminaria in firmamento caeli* mean, as he thought Valencia maintained, that in the beginning God had created "heaven and the heavens" (*cielo y cielos*) and the other orbs and planets, the moon, the sun with its light, and stars, all of which are encompassed by the word *caelum*? He doubted that all these things had been prepared and created according to essence and substance in the beginning and then had not had their perfecting qualities communicated to them until that day.[85] His final query lay at the very junction of the hexameral tradition and natural philosophy and concerned the celebrated question of the aqueous nature of the heavens. Here the writer seems to be referring to an earlier conversation where Valencia had suggested their watery nature. The writer now wanted Valencia to clarify whether this implied the corruptibility of the heavens and that the stars are not affixed to orbs but travel through them "as fish in the sea," which was the opinion of Jerónimo Muñoz, professor of astronomy in Salamanca. Furthermore, Valencia had also suggested that the heavens are made of the same species throughout and that the matter the heavens were made from is shared with the inferior world. Finally, he wanted to know whether Valencia agreed with other philosophers that each heaven has its own "intelligence" or an angel moving it.[86]

Unfortunately, we do not have Valencia's responses or even know whether he responded at all. Yet the queries tell us that Arias Montano's works were read *and explained* against the tradition of late hexameral commentaries in which the interpretation of the works of the seven days was intertwined within the natural philosophical parameters of Scholastic Thomism. Although this is easy to see in the type of arguments and philosophical terminology the anonymous writer put forth, the approach is also present in Valencia's effort to explain Arias Montano's ideas. Earlier we saw that for all his devotion to his mentor, José de Sigüenza was unable to break away from the concepts and

85. "Omnia esse condita et creata secundum essentiam et substantiam, et tunc non habuisse qualitates perfectivas, virtutes, influentias, motus etc., sed haec omnia illis hac die fuisse comunicata.": ibid., fols. 190r–v.

86. "Pareceme si no me engaño que explicandome V. Md. algo de este capitulo tomando ocasion de los principios de las cosas naturales, dixo V. Md. ser los cielos de naturaleza de agua, y si esto es assi, dira V. Md. que son corruptibles de suyo, y que las estrellas no estan fijas en ellos, sino que discurren por ellos haciendo sus circulaciones, como peces en el agua, lo cual era opinion del Maestro Muñoz, cathedratico de astrologia en Salamanca. Y tambien que los cielos no difieren en especie, y que la materia de los cielos, y la de estos inferiores sera de la misma especie. Tambien no pondra V. Md. a cada cielo una inteligencia o angel que la mueva, y que le paresce a V. Md. de esta opinion de los Philosofos.": ibid., fol. 190v.

linguistic convention this tradition also implied. For them and for many of Arias Montano's readers, the issues raised by the Creation story could not be discussed on purely cosmological grounds; they also had to be in concert with accepted parameters of scriptural interpretation. By citing Jerónimo Muñoz, the anonymous writer showed familiarity with alternative cosmological ideas and with treatises not intended to serve as hexameral commentaries. What was at stake for this doubting Thomist, however, was not so much whether Arias Montano's ideas proposed a radically different view of the natural world, but whether they could be rationalized in a way that was in concert with the accepted philosophical and theological frameworks.

Tomás Maluenda, OP

Like the unknown author of the queries addressed to Pedro de Valencia, the Valencian Tomás Maluenda, OP (1566–1628), engaged with Arias Montano's proposal within the strict confines of the hexameral commentary.[87] The formidable Dominican was also trained in the trilingual tradition within his order at the convent of Llombay, but though familiar with the *Apparatus*, as Acitores had been, he also read the *Magnum opus* very carefully. Among Maluenda's works are commentaries on the Antichrist (1604) and on paradise (1605). Both show that he was well aware of Arias Montano's work as a Hebraist. In the *De paradiso*—largely a compilation of views on the true nature and location of paradise—he made frequent references to several of Arias Montano's treatises in the *Apparatus*, particularly the *Phaleg*. In Rome from 1601 to 1608, the Dominican has been associated with Arias Montano's expurgation in the *Index* of 1607. As we will see in the next chapter, however, although he was well aware of the expurgations of Arias Montano's works, his involvement in the Index did not pertain to the Montanian corpus but was largely limited to inserting into the Index his expurgation of the *Sacrae bibliothecae sanctorum patrum* (Paris, 1589), in whose pages he is himself cited seventeen times, as his editor indicated in the posthumously written *Testimonia* that prefaces Maluenda's *Opera*.[88]

---

87. Most of the material we have on Maluenda is derived from a brief biography written by Nicolás Figueres and inserted as a preface to his *Opera*. See also Ximeno, *Escritores que han florecido hasta el año M.DC.L.*, 1:312–17; Robles, "Documentación para un estudio sobre Tomás Maluenda, OP (1565–1628)."

88. Maluenda, *Commentaria in Sacram Scripturam*, 1:xiv.

It is in Maluenda's literal exposition of the Old Testament, based on his *verbo ad verbum*—word for word—interpretation of the Hebrew Bible, that we find evidence of his engagement with the natural philosophical consequences of Montanian exegesis. He labored on the translation and commentary from 1615 until his death but got only as far as the book of Ezekiel. The work was published posthumously in 1650. The Hebrew text was not included in the published edition, but the Vulgate and Maluenda's new interpretation based on the Hebrew appear in side by side (fig. 11.1), with each section followed by a mostly lexical commentary explaining and justifying the alterations to the biblical text. He clearly engages with Arias Montano in his commentary, though he never mentions him by name except in the list of authors cited.

In his exposition on the primordial waters MAIM, Maluenda referred directly to the liquor with a dual nature and described it in the same terms Arias Montano had used in the *History of Nature*. He goes so far as to cite—and slightly alter—the section where Arias Montano describes the action of the spirit ELOHIM hovering over the waters, agitating and warming, and explains how this agitation became the agent for all subsequent creation.[89] (Earlier in the exposition of the word *ferebatur*, "was hovering," he had described the action of ELOHIM over the chaos in Montanian terms as an agitating force that brings order and creates.)[90] He did not challenge either of these interpretations, yet he qualified them as "new and curious" (*novè et curiosè*) and considered them "subtle and strange" (*subtilia et aliena*) (fig. 11.2). The RAKIAGH or "firmament" in Maluenda's interpretation described the whole of the celestial body, which was once fluid but formed so that it stood firm and expanded in the heavens. His definition included the notion that this firmament might refer to orbs that had originally had a fluid nature but had been turned into a very hard, stable material, like the hardening of molten metal.[91] He did mention, however, that "others" thought it stood as something separating the watery mass into two regions, as Arias Montano (and Saint Basil) had proposed. As for the separation of the waters, he also cited Arias Montano's description of the division of the unctuous and salty liquids and the meaning

89. "Et spiritus ELOHIM MERAHHEPHETH, id est, efficaciter motitans, confovens, ac agitans super facies gemini liquoris.": *Corpus*, 150; *HistNat*, 253. Maluenda left out the word MERAHHEPHETH (מְרַחֶפֶת, *mĕraḥepet*), "was moving or hovering," when he quoted Arias Montano. Maluenda, *Commentaria in Sacram Scripturam*, 1:5.

90. Ibid., 1:5–6.

91. Ibid., 1:6–7.

5 Februa-
rij 1621.

LIBER
GENESIS,
ID EST GENERATIO,
Seu Procreatio mundi:
HEBRAICE BERESCIT,
HOC EST,
IN PRINCIPIO
QVOD ITA ORDIATVR.

CAPVT PRIMVM·

De mundi creatione, rerum creatarum distinctione & ornatu, déque hominis for-
matione ; cui subiecit Deus omnia quæ creauerat.

Vulg.

Act.14.c.15.
17.f 24.
Psalm.32.a.
6. 135.a.5.
Eccle.18.a.1.

 1 N principio creauit Deus cælum & terram.

2 Terra autem erat inanis & vacua, & tenebræ erant super faciem abyssi: & Spiritus Dei ferebatur super aquas.

Hebr.11.2.3. 3 Dixítque Deus: Fiat lux. Et facta est lux.

4 Et vidit Deus lucem quòd esset bona: & diuisit lucem à tenebris.

5 Appellauítque lucem Diem, & tene-bras Noctem : factúmque est vespere & mane, dies vnus.

6 Dixit quoque Deus: Fiat firmamen-tum in medio aquarum: & diuidat aquas ab aquis.

Maluenda, In libr. Genes.

Malu.

 1 N principio creauit ELOHIM ·cælos & terram.

2 Et terra fuit inordina-tio & vacuitas: & [a] tenebrositas super facies molis-aqueæ : & spiritus ELO-HIM contremiscere-faciens super fa-cies aquarum. [a obscuri-tas, tenebræ]

3 Et dixit ELOHIM Erit lux. Et fuit lux.

4 Et vidit ELOHIM lucem quia bona: & separauit ELOHIM inter lucem, & inter tenebrositatem.

5 Et vocauit ELOHIM luci, dies : & tenebrositati vocauit nox. Et fuit ves-pera, & fuit mane dies vnus.

6 Et dixit ELOHIM : Erit expansio in medio aquarum , [b] & fuit separans inter aquas, [c] aquis. [b v: sit c dariuisit vel ad aquas]

A   7 Et

FIGURE 11.1. Tomás Maluenda, *Commentaria in Sacram Scripturam una cum nova de verbo ad verbum ex Hebraico translatione* (Lyon, 1650), 1. Biblioteca Histórica de la Universidad Complutense de Madrid (BH DER 18036).

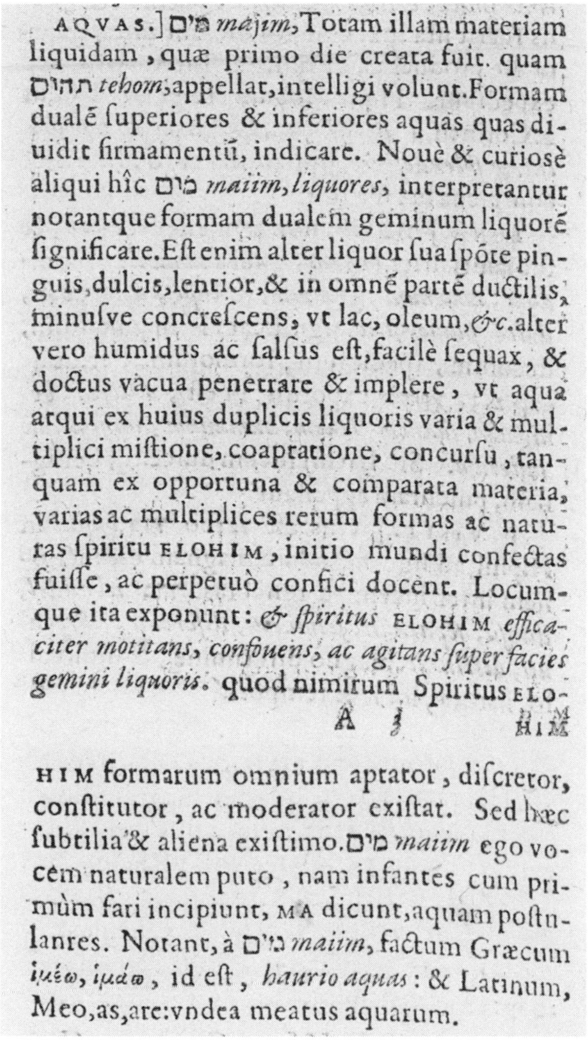

AQVAS.] םימ *májim,* Totam illam materiam liquidam , quæ primo die creata fuit. quam םוהת *tehom,* appellat, intelligi volunt. Formam dualē fuperiores & inferiores aquas quas diuidit firmamentū, indicare. Nouè & curiosè aliqui hîc םימ *maiim, liquores,* interpretantur notantque formam dualem geminum liquorē fignificare. Eſt enim alter liquor fua ſpóte pinguis, dulcis, lentior, & in omnē partē duſtilis, minuſve concreſcens, vt lac, oleum, *&c.* alter vero humidus ác falfus eſt, facilè ſequax, & doſtus vácua penetrare & implere , vt aqua atqui ex huius duplicis liquoris varia & multiplici miſtione, coaptatione, concurſu, tanquam ex opportuna & comparata materia, varias ac multiplices rerum formas ác naturas fpiritu ELOHIM, initio mundi confeſtas fuiſſe, ac perpetuò confici docent. Locumque ita exponunt: *& ſpiritus* ELOHIM *efficaciter motitans, confouens, ac agitans ſuper facies gemini liquoris.* quod nimirum Spiritus ELO-

A ჳ HIM

HIM formarum omnium aptator , diſcretor, conſtitutor , ac moderator exiſtat. Sed hæc fubtilia & aliena exiſtimo. םימ *maiim* ego vocem naturalem puto , nam infantes cum primùm fari incipiunt, MA dicunt, aquam poſtulantes. Notant, à םימ *maiim,* faſtum Græcum ἰμέω, ἱμάω , id eſt , *haurio aquas* : & Latinum, Meo, as, are: vndea meatus aquarum.

FIGURE 11.2. "Aquas." Tomás Maluenda, *Commentaria in Sacram Scripturam,* 5–6. Biblioteca Histórica de la Universidad Complutense de Madrid (BH DER 18036).

of SAMAIIM,[92] and he again qualified the Montanian explanation as *subtilia et curiosa.* Ultimately, although he acknowledged that many of the most excellent scholars affirmed the aqueous nature of the heavens, he declared himself unable to adjudicate the matter.[93]

92. *Corpus,* 152; *HistNat,* 255.
93. Maluenda, *Commentaria in Sacram Scripturam,* 1:7.

*

What defined the engagement of both the anonymous doubting Thomist and Maluenda with the Montanian exegesis was the way they read the *History of Nature*. They clearly did not read it as a cohesive exposition of a new metaphysics and natural philosophical system based on an exegesis of Sacred Scriptures. They read it as another gloss on the meaning of certain key Hebrew terms that Arias Montano then culled for his own translation or as a hexameral commentary, where the author simply presented a plausible interpretation without necessarily engaging with the finer points of philosophy. Indeed, they both identified some of Arias Montano's innovations, but they failed to acknowledge, or maybe even to notice, the grand plan of knowledge reform behind the *Magnum opus*. Meanwhile his disciples José de Sigüenza and Pedro de Valencia understood the philosophical implications of their preceptor's work, but in their exchange we see two committed if somewhat puzzled students who had to fall back on natural philosophical terminology and concepts to explain Montanian natural philosophy.

CHAPTER 12

Expurgated

During his lifetime Arias Montano fought a number of critics suspicious of his orthodoxy, most notably León de Castro in his campaign against the Antwerp Polyglot, yet for the most part his relationship with church censors seems to have been smoot.[1] This all changed in 1607 when five of his major works appeared in the Vatican's first expurgatory index, *Index librorum expurgandorum*, prepared under the direction of Giovanni Maria Guanzelli, OP (b. Brisighella, 1557–1619), Master of the Sacred Palace. The focus of the expurgations, as historians have pointed out, were some theological positions concerning grace and original sin that Arias Montano had articulated a bit too ambiguously for Roman censors in the midst of the heated *de Auxiliis* debate between Jesuits and Dominicans, but that had little to do with natural philosophy. Spanish censors could not avoid addressing the Roman censorships once they undertook their own expurgatory *index* in 1612; this issue would be addressed again in subsequent Spanish indexes. In both Rome and Spain, Arias Montano's reputation and merit as a scholar and a man of faith made him worthy of expurgation rather than prohibition. Whereas prohibition was reserved for books expressing views of known heretics, challenging accepted church doctrine, or published anonymously, expurgation or the stipulation of *suspensus donec corrigatur* (banned until corrected) was reserved for books

---

1. Arias Montano's translation of the *Itinerary of Benjamin de Tudela* was placed on the Spanish *Index of Prohibited Books* in 1583 because it violated the index's fourth rule, an interdiction against books on Jewish customs: Quiroga, *Index et catalogus librorum prohibitorum*, 2, 12; Martínez de Bujanda, Higman, and Farge, *Index des livres interdits*, 6:189.

considered meritorious, written by pious authors but containing some errors of doctrine.[2] It was thought then that expurgation was a benevolent service undertaken by the church on behalf of an author whose work was deemed deserving. That Arias Montano's works would grace the first Roman expurgatory index is not without irony, however. This new censorial genre was devised at the Council of Trent, but it was Arias Montano who first rehearsed it in 1570 while in Antwerp to "save" the works of Catholic authors from prohibition.[3]

This chapter studies the original Montanian censorship documents that have survived at the Archive of the Congregation of the Doctrine of the Faith (ACDF) in the Vatican. They present a significant correction to the historiography of the censorship history of Arias Montano's works, since they reveal that the expurgations resulted from a denunciation by a former prior of the Hieronymite monastery of San Miguel de los Reyes, fray Juan San Esteban y Falces (Ioanne S. Stephanus et Falces),[4] and were coordinated and compiled for publication by Cardinal Roberto Bellarmino, SJ.[5] After a reasoned and vigorous defense by Pedro de Valencia, slightly different expurgations followed in the Spanish indexes throughout the seventeenth century.

Arias Montano's posthumous expurgations have been of much interest and were studied attentively by John A. Jones and Melquiades Andrés Martín.[6] Their invaluable research, however, relied on the expurgations as they appeared in the published Roman and Spanish indexes and on archival material from the Spanish Inquisition deposited in Spanish archives, mostly the Archivo

2. For a synthetic explanation of the process of expurgation, see Baldini and Spruit, *Catholic Church and Modern Science Documents*, 1:53–59.

3. Arias Montano, *Index expurgatorius librorum* (1571).

4. Giovanni Falces, OSH (Juan Suárez or Juan San Esteban y Falces, b. Azanuy, date unknown, d. Brindisi, 1637) was appointed archbishop of Brindisi in the Kingdom of Naples by Philip III on 4 July 1605. A native of Aragon, he professed at the monastery of San Miguel de los Reyes in Valencia and became its prior in 1603. He studied at the monastery of El Escorial beginning in 1582, returning to San Miguel de los Reyes at an unknown date: Sebastián Castellanos de Losada, *Biografía eclesiástica completa*, 26:210–11. For his career after he became archbishop, see Leo and Guerriero, *Dell'antichissima città di Brindisi*, 109–11.

5. Morocho Gayo first called attention to the documents at the ACDF and to Bellarmino's intervention in Morocho Gayo, "Estudio introductorio," in *Obras completas de Pedro de Valencia*, 4.2: 291n483. Godman later confirmed his participation: Godman, *Saint as Censor*, 182–84.

6. Jones, "Pedro de Valencia's Defence of Arias Montano: The Expurgatory Indexes"; Jones, "Advertencias de Pedro de Valencia y Juan Ramírez"; Jones, "Pedro de Valencia en su correspondencia"; Jones, "Pedro de Valencia's Defence of Arias Montano: A Note on the Spanish Indexes"; Andrés Martín, "Teología de Arias Montano"; Andrés Martín, "Declaración de Pedro de Valencia sobre algunos lugares teológicos de Arias Montano."

Histórico Nacional (AHN). With the opening of the Vatican's Archive of the Congregation of the Doctrine of the Faith to researchers in 1998, historians can now consult the original Roman documents from the Congregation of the Index and study the process of Arias Montano's expurgation. Among the documents are the minutes of the meetings of the Congregation of the Index, the original reports from the various examiners assigned to study Arias Montano's works, and the compilation and emendation of the proposed expurgations at the hands of Cardinal Bellarmino. This rich trail of documents opens a new perspective into why and how the church undertook the Montanian expurgations.[7]

For our purposes, the most striking feature of the expurgations— particularly those of the *Anima*, since the *Apparatus* of the Antwerp Polyglot and the *History of Nature* remained untouched—was that they ignored Arias Montano's natural philosophical postulates and left his methodology largely unquestioned. To explain this, the final section of the chapter returns to the cosmological theories of Roberto Bellarmino, in particular those he shared with Arias Montano about the fluidity of the celestial region. He articulated these theories late in life in correspondence with Federico Cesi of the Accademia del Lincei in 1618; Arias Montano is cited in this correspondence as an authority on the subject. In Bellarmino we have a figure who, like Arias Montano, valued exegesis of the Hebrew Bible, was open to non-Aristotelian natural philosophical ideas, and felt deeply the natural theological reflections of many of his contemporaries. In sum, like his Spanish brethren, he shared in the disquiet that had overtaken natural philosophy in the late sixteenth century.

THE ROMAN *EXPURGATORY INDEX* OF 1607

After the tumultuous years that followed the publication of the Antwerp Polyglot, Arias Montano seems to have escaped having to answer formal accusations before the Inquisition or the church's other censorial bodies. Rumors persisted about what lay behind his Hebraism, as did unease about his disdain for Scholastic theology. Enemies often have long memories, however, and may wait for an opportunity to strike; when the Roman Congregation of the Index announced its intention to prepare its first expurgatory index, it seemed to some a propitious time to revisit the Montanian oeuvre.[8] Philip II, Arias Mon-

7. Godman copied in the appendix only the annotations in Bellarmino's hand: Godman, *Saint as Censor*, 182–84, 291–97.

8. This expurgatory index followed Clement VIII's vast prohibitory *Index Expurgatorius* of 1599. Published in 1607, it was the first of its kind prepared in Rome. It included fifty works,

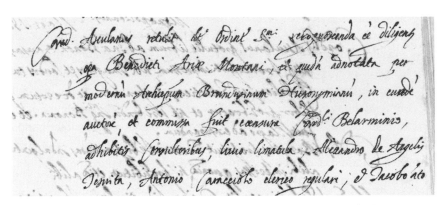

FIGURE 12.1. Entry for 30 September 1605 calling for the examination of Arias Montano's works. ACDF, Index, *Diarii*, 1, fol. 178r. By permission of the Archive of the Congregation of the Doctrine of the Faith, all rights reserved.

tano's great protector, was dead, and the new king, Philip III, although favoring Pedro de Valencia, was not likely to intervene forcefully with Rome. The first hint of Roman censors' interest in Arias Montano's works was recorded in the daily journal of the Congregation of the Index, or *Diarii*. On 30 September 1605, Cardinal Asculanus (Girolamo Bernerio, OP), then serving as the secretary of the Commission of the Roman and Universal Inquisition, relayed to the Congregation of the Index an order by Pope Paul V to have the works of Arias Montano examined with diligence[9] (fig. 12.1). The matter had been brought to the pope's attention by the newly appointed archbishop of Brindisi, the Hieronymite Juan San Esteban y Falces, though the entry recording the event does

some of which had been declared in an earlier index or decreed as prohibited until corrected, but also others, like those of Arias Montano, that had never been publicly singled out as suspicious. Only the first volume of a promised series was published, since the project drew significant criticism, and the lone published volume was recalled. A proposed revision was eventually suspended by Pope Paul V. On this index and Guanzelli's participation, see Martínez de Bujanda, "Indici dei libri prohibiti, Roma," 2:780–83, and Palumbo, "Guanzelli, G. M.," in ibid., 2:740. For a general history of the Roman indexes and the fraught relationship between the Roman Holy Office, the Congregation of the Index, and the Master of the Sacred Palace, see Fragnito, "Central and Peripheral Organization of Censorship," 13–49.

9. ACDF, Index, *Diarii*, 1, fol. 178r. "Card. Asculanus retulit ab ordine S^m, recognoscenda esse diligentia opera Benedicti Ariae Montani, et quaedam adnotata per modernum Archiepiscopum Brundusinum Hieronymianum, in eundem auctorem, et commissa fuit censura Cardinli Belarminio, adhibitis Consultoribus, Livio Limatola, Alessandro de Angelis, Jesuita, Antonio Caracciolo clerico regulari et Jacobo Antonio clerico St. Pauli decollati."

not mention the nature of his allegations. The investigation was assigned during the same meeting to Cardinal Roberto Bellarmino, and the names of some potential examiners were put forth: Bishop Livio Limatola, the Jesuit Alessandro de Angelis, the regular (Theatine) cleric Antonio Caracciolo, and Jacobo Antonio of the Congregation of Regular Clerks of Saint Paul (Barnabites).[10]

All the examinations (*censurae*) and recommended expurgations were completed by 7 August 1606, and Bellarmino presented them to the commissioner and to the Congregation of the Index.[11] A little over a month later Bellarmino reported that he had compiled all the reports into one censure.[12] The Congregation recommended they be sent to the Master of the Sacred Palace to be included in the index he was preparing for publication. The original censorships survive, and they show that Limatola and de Angelis submitted reports, but apparently neither Caracciolo nor Jacobo Antonio did so. (Father Teófilo, a Theatine priest, had complied with an expurgation in place of Caracciolo.) The examinations were assigned as follows: Limatola expurgated only the *Commentaria in duodecim Prophetas*;[13] de Angelis expurgated three works, the *Elucidationes in quatuor Evangelia*,[14] the *Elucidationes in omnia Sanctorum Apostolorum scripta: Eiusdem in S. Ioannis Apostoli et Evangelistae Apocalypsin significationes*,[15] and the *Liber generationis et regenerationis Adam*;[16] and Father Teófilo expurgated the *Commentaria in Isaiae Prophe-*

---

10. Livio Limatola (b. Maddaloni, 1563–1611) became bishop of Bitetto in 1606. Antonio Caracciolo (1562–1642) was a Theatine regular cleric and ecclesiastical historian. He chronicled the lives of the founders of his order in *Synopsis veterum religiosorum rituum atque legum notis ad constitutiones clericorum regularium* (1612). Alessandro de Angelis, SJ, attended the mathematics courses at the Collegio Romano and was later theologian of the cardinal legate in Ferrara. He wrote *In astrologos coniectores libri quinque* (Rome, 1615). On de Angelis, see Baldini, "Academy of Mathematics of the Collegio Romano from 1553 to 1612," 92n89. I have been unable to identify Jacobo Antonio.

11. ACDF, Index, *Diarii*, 1, fol. 185r, entry dated 7 August 1606. The event was also recorded among the *Decreta* of the Holy Office (1606–7) on 9 August 1606, fol. 138v.

12. ACDF, Index, *Diarii*, 1, fol. 185v, entry dated 18 September 1606. In addition, a certain Pro. Gutierrez, perhaps a *procurador*, whom I have been unable to identify, was also notified.

13. ACDF, Index, Protocolli Z, fols. 47v–48r: Livio Limatola, "Qua animadversione digna sunt in *Commentarii . . . duodecim Prophetas*."

14. ACDF, Index, Protocolli Z, fols. 45r–v, Alessandro de Angelis, "Que corrigenda videntur in Elucidaciones in 4up Evangelia et acta Apostolorum."

15. ACDF, Index, Protocolli Z, fols. 46r–47r, Alessandro de Angelis, "Quae animadversione digna sunt in Elucidacionibus Benedicti Ariae Montani in omnia SS. Apostolorum scripta."

16. ACDF, Index, Protocolli Z, fols. 53r–54r: Alessandro de Angelis, "In libro generationis . . . notatu digna videntur."

tae sermones.[17] Notably, the second part of the *Magnum opus*, the *Corpus* or *History of Nature*, was not expurgated. Despite the book's having been published in 1601, and though it never achieved the broader circulation of Arias Montano's earlier works, it was certainly known. The ACDF records do not show whether it was examined or even considered for examination. Finally, Bellarmino compiled the expurgations of the five books—not without first making some significant adjustments to them—added the notorious general admonition, and made clean copies for those preparing the *Index* for publication.[18]

Since the documentary evidence discovered so far does not indicate the specific issues fray Juan San Esteban y Falces raised about Arias Montano, let me speculate on his motivation for alerting the pope about his works. That the new archbishop would devote time to the subject during a papal audience surely speaks to the importance he gave it. It is likely—but still undocumented—that he coincided with Arias Montano at some point during his frequent visits to El Escorial and might have met him or listened to his sermons or classes. Regardless, he surely would have known about Sigüenza's *proceso* before the Inquisition and understood the nature of the allegations. (Falces does not appear on the witness roll for the *proceso*.) What we know for certain is that he was a committed Scholastic theologian and advocated an approach to theology diametrically opposed to the biblism of Arias Montano.

Among San Esteban y Falces's published works was one on how to identify heretics and convince them of the error of their ways: *Ars ad solvenda omnia argumenta haereticorum* (1623). In the book, he proposed an art against which, he alleged, heretics were defenseless. The art consisted of challenging heretical positions using Aristotelian Scholastic logic deployed in imitation of Saint Thomas Aquinas. The defender of the faith would establish clear terms as the basis of truthful propositions. These were found, in order of authority, in the literal sense of the Sacred Scriptures as explained by church fathers, decrees by church councils, church tradition, consensus of the sacred fathers, and finally, consensus of the Scholastic doctors according to Saint Thomas Aquinas—all these presented as succinctly as possible without adding any-

17. ACDF, Index, Protocolli Z, fols. 52r–v, 55r: Teófilo Theatino, "Censura del P. bt Teofilo Theatino sopra il comentario di Beneditti Aria Montano in Isaiam."

18. Bellarmino's working notes survive in his hand (according to an archivist's note on the document and confirmed by Godman) with the title "Corrigenda in Arias Montano," in ACDF, Index, Protocolli AA, fols. 308r–9r, 326r–v. Copies of this document are also found in ACDF, Index, Protocolli Z, fols. 35r–38, 43r, 44r, and 51r–v. These clean copies contain the same text that appeared in print in the *Expurgatory Index* of 1607.

thing superfluous or citing the authority of anyone outside the church.[19] His art then proceeded with identifying truthful and fallacious propositions, arguing against the false ones using syllogisms, or undermining those used by heretics. It is reasonable to suppose that such a committed Scholastic theologian would have been at odds with Arias Montano and that he took advantage of having the pope's ear to raise the matter in Rome, especially after Sigüenza's inquest came to naught.

THE ROMAN EXAMINERS' AND
BELLARMINO'S INTERVENTIONS

While a line-by-line study of each examiner's suggested expurgations is beyond the thematic scope of this book—although they merit close study—here I will briefly examine the justifications for the expurgations some examiners included in their reports and the general tenor of the expurgations Bellarmino ignored as well as others he added. Let us first consider the expurgations Father Teófilo suggested for Arias Montano's commentary on Isaiah.[20] This expurgation has no prefatory comments or explanations for why Father Teófilo deemed particular passages questionable. If he did make some remarks to this effect, say, in a letter accompanying the expurgations, they do not survive or have become separated from the rest of the folios. Of the five expurgations he suggested—the Theatine priest indicated the precise text to be excised—Bellarmino disregarded one and altered the other four.[21] The first expurgation was from chapter 1, page 43, where Arias Montano discussed Isaiah 1:11 (NAB), "What do I care for the multitude of your sacrifices? says the LORD." Father Teófilo wanted nearly six lines deleted, but Bellarmino reduced the expurgation to three lines.[22] Later in the chapter (page 53) Bellarmino reversed course; instead of Teófilo's eleven lines, he selected over thirty for deletion. Bellarmino disregarded the expurgation Teófilo suggested in chapter 10, page 252, where

19. San Esteban y Falces, *Ars ad solvenda omnia argumenta hæreticorum*, 1:218.

20. They are cited, but not copied, in Godman, *Saint as Censor*, 182n109.

21. Guanzelli, *Indicis librorum expurgandorum* (1607), 39.

22. Teófilo singled out the following text for expurgation: "Duo sacrificiorum genera sunt antiqua, vel naturae, vel legis aetatibus factitata, et nostro, id est, Novi Testam[enti] tempore a Christi mortem inchoatum, et absolutum, atque ad nos usque deductum sacrificum." Bellarmino added to the censorship the sentence that followed: "Neque enim nos illud sacrificium offerimus, sed illud ipsum Christi repraesentamus, et referimus; incruento (ut dicitur) modo repetimus id, quod Christus cruentum obtulit; de quo alias plura, et cognitu dignissima explicanda sunt."

Arias Montano commented on Isaiah 10:22 and had put forth an interpretation based on Pagnino's translation of the Bible instead of the Vulgate.[23]

In the case of Limatola's report on the *Commentaria in duodecim prophetas*, the bishop pointed out twenty-one instances where he detected questionable ideas but did not identify the specific passages to be expunged. Bellarmino accepted only six of these, apparently identifying for himself the specific segments of text to be excised, as his working notes suggest.[24] Among Limatola's propositions that Bellarmino ignored were three instances (pages 175, 179, and 272) where Arias Montano had based his commentary on a version of the Bible other than the Vulgate, in these cases the Hebrew or Chaldean version. Most of the other expurgations Bellarmino disregarded concerned concupiscence. But whereas Limatola had not found anything objectionable in the commentary on Malachi, Bellarmino certainly had. He added an extensive censorship of almost one hundred lines and explained the expurgation with a note in his own hand that he added at the end of Limatola's report—but that did not appear in the published index:

> Pag. [ ] In the first chapter of Malachi Arias Montano explains the words: "I have no pleasure in you, saith the LORD of hosts, neither will I accept an offering at your hand. For from the rising of the sun even unto the going down of the same my name shall be great among the Gentiles." This is to be understood as referring to the sacrifices of the Gentiles, which were offered to idols. This explanation seems to be contrary to the apostle, who in the first [letter to the] Corinthians [chapter] 10 says, "that the things the Gentiles sacrifice, they sacrifice to devils, and not to God." Also contrary to almost all the Latin and Greek fathers, who explain this place of Malachi as the sacrifice of the Mass.[25]

23. Arias Montano preferred Pagnino's use of the singular third person future passive indicative for "to convert."

24. ACDF, Index, Protocolli AA, fol. 308v.

25. "pag. [ ] in primum caput Malachiae exponit Benedictus Arias illa verba: 'Non est mihi voluntas in vobis, et munus non accipiam de manu vestra. Nam ab ortu solis usque ad occasum magnum est nomen meum in gentibus.' [. . .] [H]aec esse intelligenda de sacrificiis gentilium, quae offerebant idolis. Quae expositio videtur contraria Apostolo, qui prima Corinth. 10 dicit: 'Quae immolant gentes daemonibus immolant et non Deo.' Item contraria est omnibus fere patribus Latinis et Graecis, qui locum hunc Malachiae exponunt de sacrificio missae." ACDF, Index, Protocolli Z, fol. 48r: Limatola, Livio. "*Qua animadversione digna sunt in commentarii . . . Duadecim Prophetas.*" As transcribed in Godman, *Saint as Censor*, 291. Bible verses in quotation are from Mal. 1:10–11 (KJV).

It seemed to Bellarmino that Arias Montano had found too much merit in the offerings of the Jews and Gentiles and had not condemned them in the strong terms used by Saint Paul. This and other instances show that Bellarmino carefully read all of Arias Montano's works, evaluated the material himself, and, where he deemed it necessary, intervened in the expurgation.

Bellarmino's influence is most evident in the expurgations recommended by Alessandro de Angelis. Of the eighteen expurgations de Angelis suggested for the *Elucidationes in quatuor Evangelia*, Bellarmino chose twelve and added five of his own. In the *Elucidationes in omnia Sanctorum Apostolorum* only five expurgations remained of de Angelis's original nineteen, while Bellarmino added seven of his own. De Angelis, however, included a preamble to the expurgations in which he highlighted six points where Arias Montano needed reprobation, particularly on the roles of grace and concupiscence in Christian perfection. He also found the treatises *De Christi Iesu veritate* and the section on the Apocalypse particularly questionable. In response to de Angelis's remarks concerning these treatises, Bellarmino excised three sections, but he added a note in the *Index* to the effect that *De Christi Iesu veritate* could be completely deleted, since it was intended to demonstrate grace in the New Testament but seemed to suggest that the *fomes peccati* ("tinder of sin") was extinguished in the elect—a proposition that was theologically untenable.[26] The section on the Apocalypse suffered a similar fate. De Angelis had begun expurgating it but stopped after only three passages, noting that Arias Montano interpreted the Apocalypse using a "dubious" notion of the triple essence of man. Bellarmino translated de Angelis's observation into another recommendation that the entire treatise on the Apocalypse could be eliminated because it discussed a "new and dangerous" notion of a triple essence of humans consisting of a divine part, an uncorrupted part, and a corrupted part.[27]

De Angelis also censored the *Liber generationis et regenerationis Adam*, or the *Anima* of Arias Montano's *Magnum opus*. Here again he rarely pointed out specific passages to be deleted but referred to sections he found questionable. In this case Bellarmino concurred with eight of de Angelis's suggested twenty-five expurgations and added a series of general propositions to them. De Angelis began by noting that in the section *Summa ac brevis de veteris Testamenti*

---

26. "Fortasse etiam tota haec disputatio tolli posset, quia magna ex parte consumitur in demonstranda gratia novi Testamenti, qua fingitur fomes peccati in electis plane extingui.": ACDF, Index, Protocolli AA, fol. 309v.

27. Guanzelli, *Indicis librorum expurgandorum* (1607), 44.

authoritate responsio of the book's preface, Arias Montano taught that Moses's miracles did not convince men of his authority. He opined that this contradicted Exodus 4. (Perhaps de Angelis was irked that Arias Montano had cited Tibullus as evidence that wizardry in ancient times did not necessarily testify to God's presence.) Bellarmino sidestepped this critique and the next one consisting of a general admonition because of the profusion of references to ancient idolatries. He also ignored instances where de Angelis misunderstood Arias Montano, such as in his interpretation of the depravity of humanity before the time of Noah and a reference to an *arcanum negotium* during the times before Moses that de Angelis believed referred to concupiscence but was really an allusion to human salvation. De Angelis's next complaint concerned Arias Montano's treatment of concupiscence. His interpretation of the text led him to believe that Arias Montano maintained that with the aid of grace concupiscence could be dominated and even vanquished. This time Bellarmino agreed, deleting passages on pages 91 and 92. The same reasons operated in the deletions on pages 202 to 208. After these expurgations, de Angelis seems to have found Arias Montano's erroneous interpretation of the extinction of concupiscence all over the book, and only Bellarmino's careful reading saved the author from over a dozen expurgations de Angelis would have liked to see. In the extant report by de Angelis it seems that the further he advanced in examining the book, the more suspicious he became that Arias Montano had argued that under the Old Law holy persons could have escaped concupiscence, suspecting most instances where Arias Montano drew parallels between the sacrifices of the ancient Hebrews and those in the New Testament. Bellarmino navigated these passages and exonerated most of them. Yet while doing so he must have realized that Arias Montano's language could derail even someone as learned as his Jesuit confrere. At the bottom of his working notes compiling all the expurgations to the Montanian oeuvre, Bellarmino added a general admonition, which was interpreted as covering all of Arias Montano's works and condemning seven propositions he supposedly held.

The general admonition stated that, given that it had not been possible to expunge from Arias Montano's works all the passages that might deserve it, the following propositions were hereby noted as "new and erroneous" and might be expunged wherever they were found in any of his books:

> First proposition: grace is twofold, the one acquired, the other received.
> Second: grace that was acquired was common to fathers of the Old and New
>   Testament, while the received grace is proper to the Gospels.
> Third: grace that was acquired was not inherent but imputative.

Fourth: received grace extinguished all inclination to sin.

Fifth: everyone who makes good use of acquired grace will arrive at being worthy.

Sixth, received grace is inherent justice and what will make the sons of God.

Seventh, every stirring of concupiscence is sin, yet a human being might be able, if he wished, to extinguish concupiscence.[28]

The theological problem stemmed from Arias Montano's having– in the eyes of the censors– established different states and types of grace for Old and New Testament believers—*comparata* and *accepta*—states that were not recognized by Scholastic theology but that he had derived from his scriptural interpretations.[29] The ongoing controversy about grace had been framed in Thomist terms, such as *gratia efficax* and *gratia sufficiens*. During the early years of the seventeenth century it had been shaped by the debate between Jesuits and Dominicans that resulted in Pope Clement VIII's convening the Congregatio de Auxiliis Divinae Gratiae. Largely in recognition of the futility of the debate, Pope Paul V closed the congregation in 1607 and forbade further discussion by the factions. As some contemporaries stated—and as Jones and Andrés pointed out—many of the problems arose from Arias Montano's penchant for developing these subtle ideas using his own terminology and not embracing the well-worn Scholastic theological language the church used.[30] The result was the ambiguity that had tripped up de Angelis and that even Bellarmino found alarming. In his defense of the expurgations Valencia argued that the whole matter had been mostly a problem of semantics: "It would have been very easy for the Roman censor to reduce all the expurgated parts to Catholic doctrine and not deduce from them erroneous propositions."[31]

The seven condemned propositions Bellarmino detected in Arias Montano

28. "Prima propositio, Gratia duplex est, altera comparata, altera accepta. / Secunda, gratia comparata communis fuit Patribus Testamenti veteris et novi, accepta autem propria est Evangelii. / Tertia, gratia comparata non erat iustitia inhaeren<s> sed imputativa. / Quarta, gratia accepta extinguit omnino fomitem peccati. / Quinta, omnis qui bene utitur gratia comparata, perveniet ad acceptam. / Sexta, gratia accepta est iustitia inhaerens, et quae faciet filios Dei. / Septima, omnis motus concupiscentia est peccatum, quia posset homo, si vellet, extinguere concupiscentiam.": Guanzelli, *Index librorum expurgandorum* (1607), 45–46; No, 'os' also in ACDF, Index, Protocolli AA, fol. 326v.

29. For Valencia's defense of these terms, see Domenichini, *Analecta montaniana: Il Tractatus de perfectione Christiana*, 141–42.

30. Jones, "Pedro de Valencia's Defence," 130.

31. Domenichini, *Analecta montaniana*, 142.

fell within the broader theological question of how far every human remained chained to original sin as a consequence of Adam's fall—a sinner's condition that smoldered in the soul like tinder on a grate (*fomes peccati*)—and the role of grace in releasing humanity from this state to attain salvation. The Lutheran position argued for a perpetual state of sin, despite baptism, that could be redeemed only by the righteousness of Christ, whom the sinner had to willingly embrace. This made justification an external and imputed state. The post-Tridentine Catholic position reiterated the redemptive function of baptism and the inherent justification that resulted from Christ's sacrifice. Concupiscence, however, continued to smolder, not as original sin but as a continuing test of human virtue and commitment to a Christian life. The censors had wondered, Was Arias Montano questioning God's ability to grant inherent grace to believers in the Old Testament? Could he be claiming that the apostles had achieved such a state of grace that they were free from concupiscence?

What is evident from the original *censurae* and Bellarmino's intervention, particularly when we consider what was left untouched in the *Anima*, is that Montanian natural philosophy, or at the very least the postulates of it he included, were not deemed troublesome. Neither was his exegetical methodology or his Hebraism. In Bellarmino Arias Montano's works had found a tough but sympathetic censor. Bellarmino was trained in Hebrew and was very much inclined to biblical interpretation using the Hebrew Bible.[32] From 1578 to 1583 Bellarmino headed a commission in charge of expurgating Hebrew works, principally commentaries on the Pentateuch. He was also tangentially involved in Cardinal Sirleto's efforts to censor Hebrew works and might have known of his effort to recruit Arias Montano for this project. Yet by 1593 this earlier, more open climate toward Hebrew works had hardened and resulted in Clement VIII's two bulls, one of which, *Quum Hebraeorum malitia* (28 February 1593), banned the Talmud.[33] Bellarmino followed suit, preparing the expurgations to the Lyon edition by Bartholomaeus Vincentius of Pagnino's *Thesaurus linguae sanctae* (1577), singling out the contributions of Jean Mercier and Antonine Chevalier, both of whom were prohibited authors.[34] Bellarmino was inclined toward censoring books he thought were "infected" not just by

---

32. Bellarmino, *Autobiografia* (1613), 50.

33. Parente, "Index, the Holy Office, the Condemnation of the Talmud," 174.

34. ACDF, Index, Protocolli N, fols. 211r–14r:, Bellarmino, "Expurgatio *thesauri linguae sanctae*." The prohibition to this and the Stephani edition appeared in the Clementine Index of 1596 as *donec expurgentur*. The 1577 edition was later expurgated in 1607. Martínez de Bujanda, Higman, and Farge, *Index des livres interdits*, 9:725–26. On the expurgation of Christian Hebraists see Burnett, *Christian Hebraists*, 226-47.

Protestant notions but also by rabbinical ones. Among the books he examined were the Psalms of the Chaldean paraphrase of the Antwerp Polyglot, deeming some translations "rabinismus" or "redolent fabulas Talmud."[35] He also prepared guidelines for a future *emendatio* to the Greek translation of the New Testament of the Antwerp Polyglot.[36]

In censoring Arias Montano, had Bellarmino been searching for *rabinismus* to eradicate, he would not have had particular difficulty. Yet the expurgations generally avoid this aspect of his exegesis. Had Bellarmino chosen examiners friendly to this type of exegesis? More research is needed to confirm this. Yet in Bellarmino Arias Montano's works found someone who appreciated the value of the *hebraica veritas* in Christian exegesis, although not the use of rabbinical authorities to elucidate the Bible. He read Arias Montano's treatises in the *Apparatus* of the Antwerp Polyglot carefully and cited them in his own work.[37] What neither the examiners nor Bellarmino forgave were newfangled theological formulations, let alone some that might feed the flames of the raging *de Auxiliis* controversy. The controversy was complicated enough, and the thought of having to insert new formulations into the debate must have sparked apprehension.

THE SPANISH *EXPURGATORY INDEX* OF 1612 AND SUBSEQUENT INDEXES

The issue of Arias Montano's expurgation was repeatedly revisited in Spain during the seventeenth century, each time one of the four expurgatory indexes published during that century was prepared.[38] Jones framed his studies of the Montanian expurgation in the Roman and Spanish indexes within the growing rivalry between the two bodies in charge of the expurgations, particularly the desire of the Spanish to act independently of the Roman body and of the Romans to show they had the upper hand.[39] The rivalry is evident in the Spanish *Expurgatory Index* of 1612 prepared under the direction of Inquisitor General Bernardo Sandoval y Rojas. Once again the Montanian oeuvre was examined, this time by Spanish censors who took pains to act independently of the Roman

---

35. Bellarmino and Le Bachelet, *Auctarium Bellarminianum*, 660–61.

36. Ibid., 457.

37. Bellarmino, *Disputationes de controversiis christianae fidei*, 1:293, 1:982.

38. For a synthesis of the processes involved in the elaboration of the Spanish indexes, see Pinto Crespo, "*Índices* de libros prohibidos," and Pinto Crespo, "Aparato de control censorial y las corrientes doctrinales."

39. Jones, "Pedro de Valencia's Defence," 122–23; Godman, *Saint as Censor*, 183.

body, and they found only three books worthy of expurgation: the *Commentaria in duodecim prophetas*, the *Elucidationes in omnia sanctorum apostolorum*, and the *Commentaria in Isaiae*.[40] Again the *History of Nature* escaped examination. The Spanish censors did not order that passages be deleted from any of the books, but instead added the milder, marginal *caute lege* (reader beware!) note to nine passages in all, only five of which coincided with the Roman expurgations.[41] This more lenient treatment of Arias Montano by the Spanish in 1612 was due in great part to respect and admiration for his memory,[42] but more concretely, to the reasoned and persuasive defense mounted on his behalf by Pedro de Valencia and Juan Ramírez.[43] Just four years later, Valencia and Ramírez came to Arias Montano's defense again when in 1615 Andrés de León of the Friars Minor proposed a "corrected" version of the Chaldean paraphrase that sought to amend the Chaldean text so it coincided better with the Vulgate.[44]

Arias Montano did not fare as well in subsequent Spanish expurgatory indexes, however. When a new index was being prepared in 1628, the directors showed greater deference to the 1612 Roman index. Even despite the Roman index's having been criticized so much that it was rescinded soon after publication, the new Spanish index added the 1607 Roman expurgations of Arias Montano to those identified in 1612. Furthermore, the 1632 index prefaced the Montanian expurgations with a warning about the ambiguous language Arias Montano employed in discussing matters of dogma.[45] The cautionary preface captured the concerns of the Roman censors and foregrounded the issues that arose from his unconventional use of theological language. After praising Arias Montano as a scholar of Sacred Scriptures and ancient languages, it pointed out that, as such, he was not accustomed to Scholastic disputations and exercises. Therefore when discussing dogma and sacred things

40. Sandoval y Rojas, *Index librorum prohibitorum et expurgatorum*, 55–56.

41. Jones, "Pedro de Valencia's Defence," 134.

42. As Pineda recalled, the bishop of Canarias, fray Francisco de Sosa, OFM (1551–1618), was the most vocal defender among them: AHN Inquisición L. 291, fol. 293v. Cited in Andrés Martín, "La teología de Arias Montano," 37.

43. AHN Inquisición 4467, no. 38. Pedro de Valencia's defense dated 24 August 1611 has been studied in the articles cited above by Jones and Andrés Martín and published in Domenichini, *Analecta montaniana*, 116–44.

44. Jones, "Advertencias de Pedro de Valencia"; Gómez Canseco, *Humanismo después de 1600*, 85–87; Morocho Gayo, "Transmisión histórica y valoración actual del biblismo de Arias Montano," 148–66; Magnier, *Pedro de Valencia and the Catholic Apologists*, 221–28.

45. Zapata, *Novus index librorum prohibitorum et expurgatorum*, 86–89; Jones, "Pedro de Valencia's Defence of Arias Montano: A Note on the Spanish Indexes of 1632, 1640 and 1667," 84; Andrés Martín, "Declaración de Pedro de Valencia," 211.

he tended to use a somewhat lax and unaccustomed (*laxioris et insolentioris*) terminology that lacked the precision of the Scholastic language. Sometimes, it continued, as a most Latin person (*homo Latinissimus*) he tended to wander too far from what was accepted and used new words that only further obscured difficult matters. The prefatory note also highlighted the issues of grace and justification that had troubled the Roman censors.

Historians have attributed this note and the stricter treatment of the Montanian works to Juan de Pineda, SJ (1558–1637);[46] not having to contend this time with Pedro de Valencia, who had died, he could approach the Montanian expurgations anew. The Sevillian Pineda was then serving as examiner general of books and libraries (*visitador general de libros y librerias*), by appointment of the inquisitor general. He had had a part in preparing the 1612 *Expurgatory Index*, but the 1632 version was under his direction. The Jesuit had taught philosophy and Sacred Scriptures at various schools, including the Colegio Imperial de Madrid and the University of Évora. He was a biblical scholar and Hebraist in his own right, publishing, among other works, a massive *Commentary on Job* (Madrid 1597–1602) that was reprinted at least nine times.

Arias Montano's use of theological terms came up yet again during the preparatory meetings for the 1640 *Index* overseen by Inquisitor General Antonio Sotomayor.[47] This time the prefatory statement concerning language was significantly shortened in response to the outcry it had caused; Arias Montano still had defenders, particularly among members of his religious order. Among the documents solicited before this *Index* was prepared is an undated memorandum, signed only with initials, that recounts the controversies that followed the publication of the 1632 *Index*. The letter was written in response to a royal decree that sought to account for the many complaints elicited by the *Expurgatory Index* of 1632. The author writes that the outcries about Arias Montano's inclusion were the loudest, explaining, "the weightiest and noisiest were from the clerics and knights of the Order of Santiago."[48] They had published pamphlets declaring that Arias Montano had been wronged by the expurgations and by the asperity of the words used in the prefatory comments. His defenders insisted that everything that had been censored was instead worthy of universal acclaim (*aplauso universal*). They based their defense on the way Arias Montano approached translating the Greek and Hebrew, explaining that when he found no single

46. Caro, *Varones insignes en letras naturales*, 109n136; Peña Diaz, "Pineda, Juan de," 3:1210.

47. Jones, "Pedro de Valencia's Defence of Arias Montano: A Note on the Spanish Indexes of 1632, 1640 and 1667," 86–87.

48. AHN Inquisición L. 291, fols. 30r–31v, "Sobre el [índice] espurgatorio: Arias Montano," n.d.

word in Latin that fit the meaning of the original, he never attempted to say more than was in the text ("no significaba mas, sino menos, para declarar el texto"). Instead, he would use one, two, or even three Latin words to approximate the true meaning of the Hebrew or Greek text, "increasing the style because the Hebrew increased it or lowering it because the [Hebrew] lowered it. And if this notable doctor found, as he wrestled with the Latin language to adjust it to the original Hebrew, that there were still no suitable words to finish translating the Hebrew, in this case he used new words in the translation."[49] His defenders said this required great effort and science. The anonymous writer of the memorandum summarized the opposing view about Arias Montano's style, stating that "it was noted as a novelty and then some" (*se calificó por novedad y algo mas*).

This same memorandum also singled out Juan de Pineda as a steadfast detractor of Arias Montano. His displeasure went back to the Antwerp Polyglot, its use of the Chaldean Paraphrase and its Latin translation, and the use of Hebrew "Talmudic" sources in the *Apparatus*. Pineda also faulted Arias Montano for not being more forthcoming with his criticisms of heretics, whom he had lived and written among in Flanders. But his worst faults were on matters of faith. On this point Pineda had argued that in some cases Arias Montano had been treated with too much indulgence in 1612, with the censors not erasing a single word and only calling for marginal notes. One such case was the exposition of Malachi 1:11. Pineda claimed Arias Montano had interpreted it as solely a sacrifice of the Gentiles, "forgetting about the sacrifice of the Mass and of the Catholic church." The writer of the memorandum noted that Pineda cited a number of authorities who expressly rejected Arias Montano's interpretation, including Bellarmino, Father Christobal de Castro, Cornelius à Lapide, Gregorio de Valencia, and Francisco Suárez. In sum, Pineda maintained, since the Council of Trent all commentators had considered Arias Montano's interpretation of this passage "Jewish and heretical." Pineda was amazed that this author, especially on matters that had been discussed at Trent and taught in schools, would still use "extraordinary words and ways of talking not known or used by the church . . . making up new words and new ways that when they do not scandalize are truly dangerous."[50] All this was discussed during

---

49. "Ya subiendo fuese el estilo, porque subia el Hebreo, ya fuese baxando, porque baxaba. Y si forcejando el insigne doctor, con la lengua Latina, para ajustarla al original Hebreo, veia que aun no avia en toda ella vocablos con que acabar de declarar la Hebrea, en este caso, usaba de voces nuebas que la declarasen.": AHN Inquisición L. 291, fol. 30r.

50. "Palabras y modos the ablar extraordinarios y no conocidos, ni usados en la iglesia . . . en todo lo qual inventa nuebas palabras y nuebos modos que quando no escandalisen es cierto que son peligrosos.": AHN Inquisición L. 291, fols. 31r-v.

the preparatory meetings, and it was decided that prefatory words would be deleted but the expurgations would remain the same in 1632.

Juan de Pineda had come well prepared to the conferences for what would become the 1632 *Expurgatory Index* and deployed an impressive arsenal of opinions against Arias Montano, including those of his fellow Jesuits Bellarmino and Francisco Suárez. Pineda knew Arias Montano's works very well and cited not only the *Apparatus*, but also the *History of Nature* in his *Commentary on Job*.[51] Yet from his recorded testimonies about the genesis of the earlier Spanish index, it does not seem he was aware of how much Bellarmino had participated in the original Roman expurgations. Or did he remain silent on the point in his testimony? Had Pineda become aware of the personal involvement of Cardinal Bellarmino, an acknowledged authority on doctrinal matters, it surely would have emboldened him to adopt the Roman censures. In the end Pineda, and through him Bellarmino, had the final word. The expurgations of 1632 continued to appear in Spanish indexes until the nineteenth century.

The original documents at the ACDF also clarify an error introduced into the historiography by a memorandum written by Juan de Pineda in 1628 recalling how the original Roman expurgations came to be. He inaccurately described the involvement of Tomás Maluenda and Jerónimo de Tiedra in the Roman expurgation of Arias Montano.[52] Perhaps Pineda's memory of the conversation was hazy or Maluenda had wanted to take credit for the expurgations—although they were not initiated until 1605—but the documents at the ACDF show no trace that either Maluenda or Tiedra specifically

51. Pineda cited Arias Montano but also often disagreed with him. For example, see his comments on wild goats (יָעֵל, *yā'ēl*), onagers, and cedar trees, among many others, in Pineda, *Commentariorum in Job libri tredecim*, 2:601, 2:605, 2:635. Pineda was not always convinced by Arias Montano's interpretations, judging, for example, the one for wild goats as *cabalístico* because Arias Montano referred to the gematria value of the word.

52. "En realidad de verdad aquel índice no lo hizo ni trabajó él maestro Joan María Brasichelense sino el maestro fray Tomás de Maluenda, que era uno del expurgatorio (como el mismo Maluenda me lo dixo a mí en Roma en 1603). El cual con su presencia y con la ayuda del maestro Tiedra, que después fue Arzobispo de Charcas y era de su orden y doctrina que siempre apoyaba, daba vida y aliento a aquellas sus notas y expurgaciones, que tenían nombre del Maestro del Sacro Palacio.": AHN, Inquisición, libro 291, fols. 293r–v; Juan de Pineda, "Informe autógrafo del P. Pineda al Consejo de la Inquisición de 24 de Julio de 1628," Madrid. Andrés Martín, following Pérez Goyena, sees in this account evidence that Maluenda and Tiedra were responsible for Arias Montano's expurgation in the 1607 Roman *Index*, while Morocho Gayo sees it as all Pineda's doing, possibly following Bellarmino's lead: Pérez Goyena, "Arias Montano y los Jesuitas," 314; Andrés Martín, "Teología de Arias Montano," 32, 43; Morocho Gayo, "Transmisión histórica," 147–48; and Morocho Gayo, "Estudio introductorio," 347–48.

intervened in them. Maluenda did contribute to expurgating other works in the Roman index, as I noted in the previous chapter, but not the Montanian ones, as the documents in the ACDF clearly show. He did, however, form part of the junta that oversaw the Spanish *Expurgatory Index* of 1612 and thus seems to have acquiesced to the gentler treatment Arias Montano received then.

BELLARMINO AND ARIAS MONTANO'S NATURAL PHILOSOPHY

To a historian of science, the name Roberto Bellarmino immediately brings to mind his role in the first phase of the Galileo affair (1615–16). On 26 February 1616 Bellarmino—acting as prefect of the Congregation of the Index— admonished Galileo for teaching Copernicanism, particularly the claim that the sun stood still at the center of the world and the earth moved around it.[53] A few days later the Congregation of the Index issued a decree censuring the notion of the earth's motion and indicating prohibition until expurgated for Copernicus's *De revolutionibus* and Diego de Zuñiga's *Commentary on Job.* It did not mention Galileo by name, but it prohibited contemporary works on the subject by the Carmelite Paolo Antonio Foscarini and five Protestant authors.[54] The admonishment and decree were the result of a *censura* prepared by eleven *consultores* (examiners) of the Roman Holy Office, the body to whom Galileo had been denounced. Paul V accepted the *censura*,[55] judging the Copernican thesis "foolish and absurd in philosophy" (*stulta et absurda in philosophia*) and heretical, since it explicitly contradicted the Sacred Scriptures. He instructed the less authoritative body of the Congregation of the Index through Bellarmino to admonish Galileo orally. The oral admonition, as Godman points out, "meant discipline in its mildest form."[56] At Galileo's

53. Here I follow the interpretations by Fantoli, *Galileo*, 175–88, and Finocchiaro, *Defending Copernicus and Galileo*, 138–43.

54. Pagano, *Documenti vaticani del processo di Galileo Galilei (1611–1741)*, 46–47; Finocchiaro, *Trial of Galileo*, 103–4. On Zuñiga, see Navarro Brotóns, "Reception of Copernicus in Sixteenth-Century Spain," 67–70.

55. Pagano, *Documenti vaticani*, 42–43.

56. Godman maintains that the fact that the decree was issued by the then largely ineffectual Roman Congregation of the Index, but after having been evaluated by *consultores* at the Holy Office, shows that the matter was not considered particularly troublesome and did not carry the weight some of the historiography surrounding the event has proposed: Godman, *Saint as Censor*, 218.

request, the cardinal even took the unusual step of describing the nature of the admonition in writing, clarifying that it had not involved abjuration or penance, let alone a trial.[57]

The historiography of the Galileo affair has been revised over the past several decades, and we now have a far more nuanced interpretation of the actions of those involved in the first phase of the Galileo affair. The mildness of the admonition to Galileo is now evident compared with other actions taken by the Holy Office and the Congregation of the Index. Furthermore, contextualizing Bellarmino's response against his approach to scriptural authority and natural philosophy makes it evident that at issue were not only the heliocentric and geokinetic theories per se, but also the novelty of a cosmological proposition that contradicted the Sacred Scriptures and had not yet been substantiated by philosophy or by astronomical observations. For Bellarmino, anyone positing a heliocentric cosmology was articulating a hypothesis, since there was no physical proof such a configuration operated in the heavens and on earth. He made his position clear in a letter to Foscarini a little less than a year before the Holy Office became involved in the Copernican question.[58] Bellarmino conceded that the Copernican scheme served astronomers well by saving the appearances. Yet holding the physical propositions that Copernicus has set forth as *real* was another matter entirely. He explained that there was no danger in speaking about the Copernican model as a supposition—as he believed had been Copernicus's intent—but believing in the reality of the proposition presented serious conflicts with scripture. He was unconvinced by Foscarini's scriptural interpretations in support of heliocentrism. To settle this last point, he reminded the cleric that the Council of Trent had prohibited the personal interpretation of scripture if it challenged the consensus of the holy fathers or the authority of the biblical authors. Finally, Bellarmino explained his position on how a biblical exegete should reconcile observation, demonstrations, and scripture:

> If there were a true demonstration . . . then one would have to proceed with great care in explaining the scriptures that appear contrary, and to say that we do not understand them rather than that what is demonstrated is false. But I will not believe there is such a demonstration until it is shown to me.[59]

57. Finocchiaro, *Trial of Galileo*, 105.

58. Letter of R. Bellarmino to P. Foscarini, 12 April 1615, in Finocchiaro, *Trial of Galileo*, 78–80.

59. Finocchiaro, *Trial of Galileo*, 79.

More than as a doubting Thomist, Bellarmino responded to Foscarini and to Galileo's belief in Copernicus's ideas and the epistemological problem it presented as a skeptical empiricist, but also as a theologian who, after serving for decades as Rome's principal censor, understood the limits of biblical interpretation and enforced church doctrine on the matter. Some historians of science have cast Bellarmino in the role of a policeman guarding the border between science and religion in his dealings with Galileo. If we instead situate him within the broader context of his activities as one of the era's premier Counter-Reformation polemicists, it is possible to explain far more comprehensively the postures he took relative to that affair and, we can add, to the Montanian corpus. We can conclude, then, that Bellarmino's commitment to the literal interpretation as the primary meaning of scripture was unwavering and his role as protector of the church as the interpretive authority of scripture was steadfast, as was his position relative to the authority of traditional scriptural interpretations by the church fathers. He was also a skeptic, wary of taking as real empirical observations with dubious epistemic merits.[60]

As an author of natural theological works and a hexameral commentator, Bellarmino was part of the tradition of Catholic exegetes I discussed earlier in this book. In one of his devotional books written late in life, *De ascensione mentis in Deum per scalas rerum creatorum opusculum* (The mind's ascent to God by the ladder of created things) (Plantin-Moretus, 1615), he presented the search for God as an exercise consisting of fifteen steps that led the soul from the confines of the microcosm of the human body to the macrocosm of the four terrestrial elements and the celestial ethereal realm, and finally to the spiritual entities and the presence of the Divinity. In the first seven steps Bellarmino uses the awe inspired by the marvels of the creation as testimony to God's omnipotence and benevolence. His meditations remind us of the natural theology of fray Luis de Granada:

> What various powers lie hidden in plants! What strange powers are found in stones, especially magnetic stones and amber. What strength do we see in animals such as lions, bears, bulls, and elephants. How clever, although tiny, are ants, spiders, bees, and flies. I pass over the power of the angels and the might of the sun and stars, which are far away from us. Finally, how great is human genius, which has invented skills such that we wonder whether nature surpasses art or art surpasses nature! Now lift up your eyes to God,

60. Blackwell, *Galileo, Bellarmine, and the Bible*, 29–40.

my soul, and reflect on how great is the strength, efficacy, and power of the Lord your God.[61]

In his *Explanatio in psalmos* (1611) Bellarmino also wished to inspire in readers a similar sense of awe. His commitment to the literal interpretation allowed him to call attention to the celestial waters as another sign of divine power. As he explains in his exposition of Psalm 103:3 (VUL),

> God's wonderful skill in placing the waters above the heavens, as if he put them on so much fire without the fire being quenched by the waters, or the waters being dried up by the fire. Without entering into the various theories propounded to explain this passage, let it suffice to say that the general opinion of the holy fathers is that there is water above the ethereal sky called the firmament, and they are the waters alluded to, and not the water in the clouds.[62]

Yet this same literalism made him believe in the immobility of the earth.

Some forty years earlier, as professor of theology at Louvain, Bellarmino had delivered some provocative lectures positing a non-Aristotelian cosmology. Those lectures were written in the hexameral tradition and followed Aquinas's discussion of the six days of Creation in the *Summa theologiae*.[63] As Robert Westman points out, the lectures did not adhere to the conventions of an astronomical lecture, let alone a treatise, but instead followed the logical structure of Aquinas's *Summa theologiae*. Thus Bellarmino was only making "a series of existence claims concerning the constitution of the heavens" to address Aquinas's questions.[64] These were existence claims, I would add, that nonetheless introduced alternative textual interpretations of Moses's account of the creation of the world and were also in concert with observed phenomena and profoundly anti-Aristotelian. So, to account for retrograde motion he was willing to admit the possibility that planets accelerated, decelerated, and maybe even moved in spirals—impossibilities in an Aristotelian cosmos—as long as it preserved appearances and could be reconciled with scripture. For Bellarmino, however, the one appearance that had to be saved was the immobility of the earth. If we return to his commentary on Psalms 103:5 (VUL)

61. Bellarmino, "Mind's Ascent to God." 73.
62. Bellarmino, *Commentary on the Book of Psalms*, 263.
63. Baldini and Coyne, "Louvaine Lectures," 5.
64. Westman, *Copernican Question*, 217–19.

("Who laid the foundations of the earth, that it should not be removed for ever," Ps. 104:5 KJV), within his epistemic horizon it was possible to maintain just as firmly on the authority of scripture that the earth did not move, and to use this surety to write in praise of the Divinity, "Your command is surer than any foundation, and such being your orders, the earth, dependent on its own gravity, will remain undisturbed forever."[65] Bellarmino's description of celestial motions in the lectures would not have satisfied an astronomer, but it was perfectly in concert with the tradition of biblical exegesis. It was, in fact, a Mosaic cosmology that stood on the strength of the biblical text, given the lack of observational evidence to the contrary.

Bellarmino essentially held these views all his life, though he grew unsatisfied with lingering questions concerning planetary motion.[66] This is evident in a 1618 letter exchange with Prince Federico Cesi (1585–1630), founder of the Accademia dei Lincei.[67] Astronomy was among Cesi's seemingly insatiable natural philosophical interests, and he worked for a long time on a cosmological treatise where he hoped to demonstrate the unitary nature of the heavens and their fluid nature. In an earlier exchange with Cesi it seems Bellarmino's initial questions concerned the motion of the fixed stars and planets, and in particular "how diverse motions in one and the same star are to be explained, if there is only one heaven, which is immobile." By these diverse motions he meant retrograde planetary motion and the apparently different rates of motion between equatorial and circumpolar stars.[68] In his response Cesi had not addressed these astronomical issues to the cardinal's satisfaction, instead focusing on explaining the physical properties of heaven based on citations from the Hebrew Bible. Cesi described heaven as spherical, still, easily penetrable, and of a tenuous nature like air—all aspects that Bellarmino found unproblematic and knew perfectly well, telling the young prince, "these things are already

65. See also his expositions on Psalms 92, 103, and 148: Bellarmino, *Commentary on the Book of Psalms*, 237, 263, and 376.

66. On the evolution of Bellarmino's astronomical thought, see the essential Baldini, "Astronomia del cardinale Bellarmino," 300–303, and Lattis, *Between Copernicus and Galileo*, 214–16.

67. Translated in Blackwell, *Galileo, Bellarmine, and the Bible*, 43–44; originally published in the *Rosa Ursina* of Christopher Scheiner, SJ, it also has been published in Bellarmino and Cesi, "Federico Cesi a Roberto Bellarmino [Acquaparta/Roma, 1618]," and with an Italian translation in Cesi, "De caeli unitate" (1618). On Cesi and Bellarmino, see Freedberg, *Eye of the Lynx*, 134–37. On Cesi's letter, see Ricci, "Federico Cesi e la 'nova' del 1604."

68. Blackwell, *Galileo, Bellarmine, and the Bible*, 43–44.

known" (*queste cose già le sapevo*).[69] Bellarmino had wanted an astronomical explanation that preserved the phenomena—that is, the observable motions of celestial bodies within this fluid heaven—but without recourse to Ptolemaic epicycles and equants, let alone Aristotelian crystalline spheres. Likely unaware of the cosmological content of Bellarmino's lectures at Louvain—they had remained in manuscript—Cesi had attempted to argue in the exegetical arena he thought the cardinal would prefer, only to discover the question had moved beyond that field. The cardinal now wanted empirical evidence that would convince him of the reality of a fluid heaven that was nevertheless a cosmos in which the earth stood still.

In Cesi's letter Bellarmino found arguments about the plurality of celestial waters as implied by the Hebrew word for "water," the interpretation of RAKIAGH not as "firmament" but instead as "extension" and "diffusion," and of SAHHAKIM as a fluid and possibly airy substance. In support of his arguments, Cesi enlisted the works of many church fathers who held some or all of these views, but he also referred to the recent works of three Spanish scholars who maintained the position he subscribed to: Benito Pereira, SJ, Juan de Pineda, SJ, and Benito Arias Montano. Cesi cited Arias Montano's paraphrase of Job 37:18 in the *History of Nature* to support the notion that the heavens were airy and pure: "Extendisti ne cum illo SAHHAKIM perseverantes ut specula fusionis."[70] From Pineda's *Commentary on Job* he took the explanation of the reference to a "molten looking glass," in the biblical verse to suggest a fluid celestial region that was as resistant as a molten solid.[71] And from Pereira he took the notion that the heavens are composed of the same—or a very similar—airy substance as the sky.[72] Cesi considered Pineda and Arias Montano the leading

69. Bellarmino and Cesi, "Federico Cesi a Bellarmino," 662.

70. Job 37:18: "Hast thou with him spread out the sky, which is strong, and as a molten looking glass?" (KJV) and "tu forsitan cum eo fabricatus es caelos qui solidissimi quasi aere fusi sunt" (VUL): *Corpus*, 179; *HistNat*, 285.

71. Pineda, in fact, liked the Pagnino translation in the Antwerp Polyglot: Pineda, *Commentariorum in Job libri tredecim*, 2:547–48. In his paraphrase of Job 37:18 Pineda explained the solid yet airy nature of heaven thus: "Quae quamvis mollis, fluxa, et labilis sit, per Creatoris tamen potentiam constans, perpetuaque consistit, ac si fuisset durissimo, ex aere conflata. Et quidem aeris istius quamvis mollissimi, atque tenuissimi, impetus vehementissimus, instar rei solidissimae est.": Pineda, *Commentariorum in Job libri tredecim*, 2:554.

72. Cesi did not cite a particular book by Pereira, but this idea can be found in Pereira, *Commentariorum et disputationum in Genesim*, 105. See also Randles, *Unmaking of the Medieval Christian Cosmos*, 47–48, and Remmert, "'Whether the Stars Are Innumerable for Us?,'" 164-66.

contemporary authorities on cosmological arguments derived from scripture. When it came time to describe the motion of the stars, however, he resorted to assuming angels as the source of the perceived stellar and planetary motions in what only seem to be celestial circles. While staying away from the heliocentric or geokinetic notions, he cited Galileo's telescopic observations as empirical proof of "new planets, new fixed stars, new aspects of the stars."[73] Cesi was clearly aware that Galileo had looked through his telescope with Copernican glasses, and this interpretation included the geokinetic and heliocentric aspects, which remained very difficult to reconcile with the Sacred Scriptures. Cesi simply skipped the subject, but Bellarmino had noticed this and emphasized in his reply that he was after evidence for precession and diurnal motion that Cesi, and Galileo for that matter, could not provide in 1618 without resorting to the earth's motion. Bellarmino's biblically inspired cosmological theory, firmly anchored in the hexameral narrative but also in a realist perspective, became an established tradition within the Society of Jesus, and along with the mathematical approach to astronomy of Christopher Clavius, converged to form the Jesuit cosmological views articulated by Giovanni Battista Riccioli and Gabriele Beati during the mid-seventeenth century.[74]

*

After his death Arias Montano's exegetical exercises came under scrutiny by inquisitorial authorities in a way he never had to contend with during his lifetime. The history of his oeuvre's tribulations is in many ways a chronicle of the way his work was read. It is likely that the problems fray Juan San Esteban y Falces, archbishop of Brindisi, discussed with the pope were not Arias Montano's pronouncements on grace and original sin, but rather the biblism and anti-Scholastic rhetoric he had brought to El Escorial. The Roman censors, however, read Arias Montano almost exclusively in light of the *de Auxiliis* controversy then raging in Rome. Spanish examiners in 1612, some of whom might have known Arias Montano and for whom his reputation loomed large, were eager to show their independence from Rome and dealt with their Spanish brother with what they perceived as leniency. Yet some twenty years later, with Arias Montano's defenders gone, the consequences of San Esteban y Falces's whispers proved impossible to erase.

---

73. Bellarmino and Cesi, "Cesi a Bellarmino," 651.
74. Magruder, "Jesuit Science after Galileo," 192.

Epilogue

He later abandoned the philosophy of Aristotle he had studied, judging that there was no better philosophy than scripture, whose author was the Holy Spirit. Maybe because of this and for having commented on the Holy Books without citing authors, his works have not been well received by some.[1]

Why the *Magnum opus*? Why devote half a lifetime to fashioning a new natural philosophy derived from first principles identified in the Bible? In searching for the answers to these questions I have situated Arias Montano's work within the Spanish disquiet, a growing skepticism about the ability of ancient philosophical systems to explain the natural world, characterized by a profound concern about the consequences of studying nature from a purely empiricist perspective. Some aspects of this dissatisfaction have been interpreted within the broader narrative of European early modern science as setting the stage for the rise of experimental natural philosophy during the seventeenth century. As we have seen, some of Spain's sixteenth-century theologians and natural philosophers, particularly those trained in the trilingual scriptural tradition of Spanish universities, wrestled with fundamentally the same philosophical concerns that drove the development of science elsewhere in Europe. But, by having tested the bounds of empiricism early on during the century of discovery, they had identified the shortcomings of the approach, echoed in the natural theologies popular at the time. These theologians cannot be neatly grouped into camps: Franciscans versus Dominicans, Thomists, or Platonists, heterodox versus orthodox. In fact, trying to fit someone like Arias Montano

---

1. "La filosofia de Aristoteles que estudio, dexo despues con mejor conocimiento, juzgando que no avia mas acertada filosofia que la de la Escritura, cuyo autor es el Espiritu Santo. Por esto (por ventura) i por aver comentado los Libros Sagrados sin citar Autores, no an sido bien recibidas sus obras de algunos.": "Benito Arias Montano," in Pacheco, *Libro de descripción de verdaderos retratos*, no. 46, fol. 91v.

into such taxonomies would do violence to the historical analysis of his works. While we may find some traces of certain perspectives on the study of nature associated with these groups, they remind us that Arias Montano avoided fitting into any category other than being a good Christian and a loyal Catholic and a believer that the Sacred Scriptures were the foundation of all truth.

During his lifetime—and decades before the die was cast in the direction of the new experimental science—Arias Montano set forth a biblical or Mosaic natural philosophy he hoped to use as a viable theoretical framework for the study of nature, one that corrected the vagaries that had infected Adamic knowledge and resulted in the misguided philosophies of antiquity. He was well aware that a cohesive natural philosophical system had to reconcile experiential knowledge with philosophy; to this end he developed epistemic approaches and methodological tools that allowed him to formulate a new metaphysics that would undergird a new cosmology, terrestrial physics, and biological sciences. His commitment to a *hebraica veritas* and his belief in prelapsarian Adamic knowledge were as much based on profound faith as buttressed by the humanist ethos that considered it possible to regain the knowledge of antiquity using philological tools. Hebrew, Greek, Aramaic, and Latin philology, along with an impressive command of the Bible, allowed him to identify a new set of metaphysical principles in the book of Genesis. Similar exercises in exegesis found the rudiments of a complete natural philosophical system in the rest of the Bible. The Sacred Scriptures served as an unassailable source within his epistemological framework, while empirical knowledge about the natural world as humans perceived it enriched his sometimes esoteric semantic exercises. Working within this framework allowed him to address the central issues of the Spanish disquiet: disenchantment with ancient natural philosophy and giving a suitable spiritual purpose to the quest for empirical knowledge. In the *Magnum opus* he set forth a comprehensive natural philosophical system where the only arbiters of truth were experience and the Word of God.

Montanian natural philosophy was predicated on a handful of premises derived from a historical narrative of the Bible. The biblical story provided Arias Montano with clear directives about the purpose of nature and its study, both of them determined by the ideal relationship that should exist between humans and God. It also supplied the key episodes that supported his interpretation of the postlapsarian corruption of knowledge about nature, which had led to the mistakes of natural philosophical systems inherited from antiquity and that had persisted, if somewhat changed, to his day. The biblical history also highlighted the epistemic limitations of sense perception and the pitfalls that awaited if empirical knowledge was the sole objective of studying nature. Finally, the divinely inspired Creation narrative, once correctly interpreted from the original Hebrew, set the parameters for a wholly determined world, where each thing possessed an unchangeable nature and was instructed by a divine command to exist. These premises, in addition to the exegetical exercises that followed, led Arias Montano to some unconventional postulates. His biblical metaphysics based on CAUSA, IEHI, ELOHIM, and MAIM—after some inspection—reveal parallels with other philosophical systems, but they were nonetheless wholly novel formulations. Most novel was his conceptualizing the spirit ELOHIM as a secondary causal agent operating throughout the world and acting as a constant manifestation—and reminder—of God's presence and creative power. And while the three-part, fluid-filled cosmos of self-moving celestial bodies he articulated drew from the long tradition of Christian and Jewish exegesis of the Hexameron, his exploration of the consequences of the cosmological structure he described far surpassed those typical of the commentary genre.

A comprehensive history of the reception of the *Magnum opus* remains to be done. Despite being published by one of Europe's most reputable and widely distributed presses, of the one thousand copies each of the *Anima* and *Corpus* that were printed, four hundred copies of the *Corpus* remained unsold as of 1604.[2] Montanian natural philosophy proved challenging for disciples and critics alike, in large part because its novelty lay in his having constructed a natural philosophy that skirted the Scholastic terms that had pervaded natural philosophy since medieval reformulations of Aristotle's work in the hands of Ibn Rushd (Averroes) and Saint Thomas Aquinas. To Juan de Mariana, for example, his division of things in the world was simply unintelligible. Others such as Federico Cesi and Cardinal Bellarmino engaged with Arias Montano as they would have with a biblical exegete, whose line-by-line analysis and expositions they compared with other such pronouncements in the atomized manner of innumerable hexameral commentaries. A quick survey of seventeenth-century English authors shows that Arias Montano was cited with some regularity, but almost exclusively as a historian of the biblical Hebrews through his treatises of the Antwerp Polyglot and for his exegetical writings,

2. The *Anima* had received wider distribution; as of 1604 only 120 copies remained unsold: Dávila Pérez, "Correspondencia Latina inédita de Pedro de Valencia," 264.

rarely as a natural philosopher.[3] We find his discussion of the origin of man as made from a special type of dust in the *Anthroposophia theomagica* (1650) of Thomas Vaughan (1622–66) and on the immobility of the earth in *The New Planet No* Planet (1646) by Alexander Ross (1591–1654), while John Webster (1610–82) in his *Metallographia* (1671) commented on his interpretation of the technological skill of Tubal-cain. Arias Montano was particularly admired by Thomas James, the librarian of the Bodleian and a strident antipapist, who had his likeness painted on the library's frieze and lamented the attacks Arias Montano had endured in the aftermath of the publication of the Antwerp Polyglot.[4] On the Continent, Marin Mersenne engaged with Arias Montano's definition of *raquia* from the *De arcano sermone* in his *Obervationes et emendationes ad Francisci Georgi Veneto* (1623). Yet these readers seem to have missed the grander plan Arias Montano had almost succeeded in completing. An essential component of this plan was to break definitively not just with the inherited natural philosophical ideas, but with the Scholastic language, logical argumentation, and tenets in which it was framed. But he never fully articulated this aim in a systematic way that got his readers' attention. Even his most loyal disciples, Pedro de Valencia and José de Sigüenza, struggled to fully understand, let alone internalize, their preceptor's conceptual revolution. As a result the *Magnum opus* continued to be read as another exegetical exercise or as a history of the ancient Hebrews.

3. Among English authors citing him as a biblical scholar, we find Henry Ainsworth (1571–1622?), Zachary Bogan (1625–59), Edward Brerewood (1565?–1613), John Edwards (1637–1716), George Gillespie (1613–48), Brian Walton (1600–1661), and Andrew Willet (1562–1621). Citing him as a historian of Jewish antiquities were Hugh Broughton (1549–1612), Richard Cumberland (1631–1718), and Pierre-Daniel Huet (1630–1721), and his work was cited in the collections of Samuel Purchas (1577?–1626). John Gerard (1545–1612) mentions his exchange of natural historical specimens with Ortelius.

4. Thomas James cites an "Apologie" by Arias Montano—which he treasured and kept at the Bodleian—in which the biblist defended his contributions to the Antwerp Polyglot and gave "a full satisfaction to all his Adversaries Objections, and the whole histories of his troubles; beginning, successe, & progresse of that costly work." According to James, it was written in Spanish and had been seized by the Earl of Essex in the English attack in 1596 on "Cales." The reference to Cales might suggest a misspelling of Cádiz. Another alternative is that the document came from the library of the grand inquisitor of Portugal, Bishop Ferdinand Martins Mascarenhas when the English attacked Faro later the same year. Inquiries to the Bodleian did not yield the precious document; but I speculate that this might have been the original or similar to MS B1351 at the HSA: Thomas James, *Treatise of the Corruption of Scripture*, part 3, 43. On the stollen material, see K. M. P., "Grand Inquisitor and His Library," 240, and Dávila Pérez, "Apología de la *Biblia Regia*," 285.

Arias Montano was the first European to articulate a complete Mosaic natural philosophy that went far beyond what Steuco, Lemnius, or Francisco Valles had attempted. The seventeenth century would be the golden age of Mosaic philosophies. While the extent of Arias Montano's direct influence on these is as yet unclear, what is certain is that the Creation narrative continued to drive natural philosophical discourses well into the eighteenth century.[5] The lure of recapturing Adamic knowledge persisted during the seventeenth century, although it became increasingly challenged, particularly as the Mosaic authorship of Genesis was questioned.[6] As new methodologies began to supersede the textual and philological approach Arias Montano employed, the project of recapturing Adamic knowledge persisted as a model for the proper purpose of the scientific enterprise. For Thomas Sprat the very purpose of the Royal Society lay in emulating Adam when he faced Creation:

> Hence he will be led to admire the wonderful contrivance of the Creation; and so to apply, and direct his praises aright: which no doubt, when they are offer'd up to Heven, from the mouth of one, who has well studied what he commends, will be more sutable to the Divine Nature, than the blind applauses of the ignorant. This was the first service, that Adam perform'd to his Creator, when he obey'd him in mustring, and naming, and looking into the Nature of all the Creatures.[7]

Arias Montano's rejection of Scholastic semantics when articulating theological notions exasperated some members of his audience. Because it touched on theological matters, however, his critics could demand that his works be examined by church authorities of the highest stature; there was little tolerance for theological ambiguity during the opening decade of the seventeenth century, particularly from such a well-known figure. As a result, the Montanian oeuvre was subjected (again and again) to the post-Tridentine censorial apparatus of the Roman and Spanish Inquisitions.

There are various reasons why Arias Montano's unorthodox natural philosophical pronouncements were ignored by the censors. Some of them had to do with the competency of the Congregation of the Index and the Inquisition, which only in rare instances ventured into the realm of purely natural philo-

5. Poole, "Francis Lodwick's Creation."

6. Levitin, *Ancient Wisdom in the Age of the New Science*, 138–80; Coudert, *Religion, Magic, and Science*, 140–44.

7. Sprat, *History of the Royal-Society of London*, 349–50.

sophical disputes. Censors such as Juan de Mariana were reluctant to accept natural philosophical novelties, as he demonstrated in his comments about the preface to the *De arcano sermone*. Yet, because this type of argumentation did not contain anything that impinged on dogma, it did not fall under the purview of the Inquisition or the index. Historian José Pardo Tomás reminds us of the limits of censorship in Spain and the contingent nature—a truly *maquinaria arbitraria*—of the ideology behind the preparation of lists of prohibited and expurgated books.[8] Within this arbitrary machine there was no systematic attack on natural philosophy. He points to the period 1583 to 1640 as having the most vigorous censorial activity, but he concludes that the enterprise was ultimately futile and ineffective as a means of control. Yet its presence, Pardo Tomás explains, shaped habits of mind of the intellectual elite who came of age during those years, giving them a fear of novelty and repressed curiosity committed to a Scholastic pedagogy and its deeply ingrained intellectual habits, an adherence to safe doctrines, little incentive to engage with foreign ideas, either through travel or through reading, and a propensity to exclude certain readers—mostly the uneducated masses—from the discourse of ideas.[9] It was not a propitious time or place for new natural philosophies.

In closing, let me relate Arias Montano's project to another project that also hoped to overturn the natural philosophy of the ancients, but that materialized a few decades later: Francis Bacon's proposed new science of *The New Organon* (1620). Arias Montano, by focusing his analysis of the nature of things on those aspects most useful to humanity, was—like Bacon—following the utilitarian agenda that would partly define the new experimental science. Yet a radical aspect of Bacon's proposal was the idea that laws of nature could be deduced from an accumulation of empirical facts (learned by means of his—albeit loose—definition of experiment), without the guide of a philosophical system based on first principles. Arias Montano's project, as we have seen, also embedded empirical knowledge into his natural philosophical system. Yet for him and other natural philosophers of his era who shared in the Spanish disquiet, knowledge claims based on experience *had to be* embedded in a philosophical system that stood on firm metaphysical and logical ground, did not contradict Christian doctrine, and was universal. Adopting Bacon's approach would have seemed to them negligent, or at the very least epistemologically sloppy.

8. Pardo Tomás, "Censura inquisitorial y lectura de libros científicos," 4.
9. Ibid., 14.

Yet the break with the philosophical structures of the past could not be definitive. Indeed, we find strewn throughout his *History of Nature* a perhaps unconscious reliance on Aristotelian and Neoplatonic patterns of thought. Like some of his contemporaries disenchanted with the philosophy of the ancients, Arias Montano wanted to cast anew the whole of the natural philosophical apparatus. Unlike them, however, he turned to the Bible in search of divine guidance for his study of the natural world and, discarding the old systems, sought to start fresh from the very beginning as "In the beginning God created the heavens and the earth."

A C K N O W L E D G M E N T S

There are many persons and institutions whose support made this book possible. First I must thank my colleagues and students at the Johns Hopkins University's History of Science, Medicine, and Technology Colloquium series for listening attentively on several occasions to various portions of this project. Thank you for your helpful advice. I also owe special thanks to the many Spanish librarians who have digitized important rare books in their collections and made them easily accessible. The work of the Biblioteca Histórica de la Universidad Complutense de Madrid and its willingness to make these volumes available through the Biblioteca Digital Dioscórides and Google Books must stand as an example for the digital humanities. I am grateful to Ann Blair for her insightful comments on a paper I delivered at the History of Science Society conference in 2010; they helped me situate Arias Montano within the broader tradition of Mosaic philosophies of nature. My thanks extend as well to Anthony Grafton and Adam Beaver for organizing an Arias Montano Colloquium at Princeton University (13–14 May 2011) that brought together a number of Montañistas and fostered important scholarly exchanges between us. In Spain, Francisco José Campos y Fernández de Sevilla shared with me his important work on José de Sigüenza. I am grateful to Antonella Romano for inviting me to discuss portions of this book is a series of conferences at the École des hautes études en sciences sociales in Paris during summer 2016. My work benefited greatly from her thoughtful comments. A small army graciously stepped in to help with the text: Jean-Olivier Richard initially transcribed the lengthy Latin quotations; Anita Dam helped correlate Arias Montano's censorships from the ACDF with the published version of his works; Maren Kristina Mueller proofread the Latin portions and translations; and Shai Alleson-Gerberg checked the Hebrew. My copyeditor Alice Bennett greatly improved the manuscript with her considerable expertise. Thank you all. The research on Francisco Valles de Covarrubias was supported by a 2010 NEH summer grant (FT57484-10). Finally, the manuscript's two anonymous reviewers pointed out mistakes and omissions in elegant and constructive ways. The errors that remain, whether of bibliography, of interpretation, or in translations or transcriptions, are all mine.

BIBLIOGRAPHY

Abellán, José Luis. *Historia crítica del pensamiento español*. 5 vols. Madrid: Espasa-Calpe, 1979.

Acitores, Andrés de. *Theologia symbolica sive hieroglyphica pro totius Scripturae Sacrae, iuxta primarium et genuinum sensum commentariis, aliisque sensibus facilè hauriendis*. Salmanticae: Sumptibus Ioannis Pulmani: Excudebat Didacus à Cussio, 1597.

Alcalá Galve, Ángel. "Arias Montano y el familismo flamenco: Una nueva revisión." In *Anatomía del humanismo: Benito Arias Montano, 1598–1998. Homenaje al profesor Melquiades Andrés Martín: Actas del simposio internacional celebrado en la Universidad de Huelva del 4 al 6 del noviembre de 1998*, edited by Luis Gómez Canseco, 85–111. Huelva: Universidad de Huelva, 1998.

———. "Peculiaridad de las acusaciones de fray Luis en el marco del proceso a sus colegas salamantinos." In *Fray Luis de León: Historia, humanismo y letras*, edited by Víctor G. de la Concha and Javier San José Lera, 65–80. Salamanca: Ediciones Universidad de Salamanca, 1996.

———. *Proceso inquisitorial de fray Luis de León: Edición paleográfica, anotada y crítica*. Salamanca: Junta de Castilla y León, Consejería de cultura y turismo, 1991.

Alcocer, Mariano. *Felipe II y la Biblia de Amberes*. Valladolid: Emiliano Zapata, 1927.

Aldana, Francisco de. "Carta para Arias Montano." In *Poesías castellanas completas*, edited by José Lara Garrido, 437–58. Madrid: Cátedra, 1985.

Almási, Gábor. *The Uses of Humanism: Johannes Sambucus (1531–1584), Andreas Dudith (1533–1589), and the Republic of Letters in East Central Europe*. Leiden: Brill, 2009.

Alpert, Michael. *Crypto-Judaism and the Spanish Inquisition*. Houndmills, UK: Palgrave, 2001.

Alvar Ezquerra, Alfredo. "Benito Arias Montano en Portugal." In *Arias Montano y su tiempo*, edited by Catalina Pulido Corrales, 189–214. Badajoz: Editora Regional de Extremadura, 1998.

———. "El Colegio Trilingüe de la Universidad de Alcalá de Henares (notas para su estudio)."

In *Congreso internacional contemporaneidad de los clásicos: La tradición grecolatina ante el siglo XXI*, edited by Rosa María Iglesias Montiel, 515–23. Murcia: Universidad de Murcia, 1999.

Alvar Ezquerra, Alfredo, María Elena García Guerra, and María de los Angeles Vicioso Rodríguez, eds. *Relaciones topográficas de Felipe II*. 4 vols. Madrid: CSIC, 1993.

Andrés, Gregorio de. "Nuevos datos sobre la genealogía del padre José de Sigüenza (aclaraciones a su proceso inquisitorial)." In *La Inquisición española: Nueva visión, nuevos horizontes*, edited by J. Pérez Villanueva, 821–30. Madrid: Siglo XXI de España Editores, S.A. 1980.

———. *Proceso inquisitorial del padre Sigüenza*. Madrid: Fundación Universitaria Española, 1975.

Andrés Martín, Melquiades. "Declaración de Pedro de Valencia sobre algunos lugares teológicos de Arias Montano." In *Humanismo y tradición clásica en España y América*, edited by Jesús María Nieto Ibáñez, 191–216. León: Universidad de León, 2002.

———. "Una espiritualidad ecuménica." *La ciudad de Dios: Revista Agustiniana* 211 (1998): 7–32.

———. "La teología de Arias Montano y los índices expurgatorios de la Inquisición." In *El humanismo extremeño: Estudios presentados a las 3as jornadas organizadas por la Real Academia de Extremadura en Fregenal de la Sierra, Aracena y Alájar en 1998*, edited by Mariano Fernández-Daza, Carmelo Solís, Francisco Tejanda, Manuel Terrón, and Antonio Viudas, 27–45. Trujillo: Real Academia de Extremadura de las Letras y las Artes, 1999.

Antolín, Guillermo. "Cartas inéditas de Pedro de Valencia al p. J. de Sigüenza." *La ciudad de Dios: Revista Agustiniana*, vols. 41–44 (1896–97).

Aquinas, Saint Thomas. *Summa theologiae: Latin Text and English Translation, Introductions, Notes, Appendices, and Glossaries*. Translated by William A. Wallace. 60 vols. Cambridge: Blackfriars, 1967.

Arceo de Fregenal, Francisco de. *Método verdadero de curar las heridas y otros preceptos de este arte: Método de curar las fiebres*. Edited by Andrés Oyola Fabián and José M. Cobos Bueno. Bibliotheca Montaniana 21. Huelva: Universidad de Huelva, Servicio de Publicaciones, 2009.

Arias Montano, Benito. *Antigüedades hebráicas: Antiquitatum iudaicarum libri IX. Tratados exegéticos de la Biblia Regia*. Edited by Luis Gómez Canseco and Sergio Fernández López. Bibliotheca Montaniana 25. Huelva: Universidad de Huelva, 2013.

———. *Antiquitatum judaicarum libri IX*. Antwerp: Franciscus Raphelengius, 1593.

———. *Benedicti Ariae Montani Hispalensis Commentaria in duodecim prophetas*. Antwerp: Christophe Plantin, 1583.

———. *Commentaria in Isaiae Prophetae sermones*. Antwerp: Jan Moretus, 1599.

———. *Comentarios a los treinta y un primeros salmos de David: Estudio introductorio, edición crítica, versión española y notas María Asunción Sánchez Manzano, vocabulario hebreo Emilia Fernández Tejero*. 2 vols. León: Secretariado de Publicaciones de la Universidad de León, 1999.

———. "De varia hebraicorum librorum scriptione et lectione commentatio." In *Biblia He-*

braica: Eorundem Latina interpretatio Xantis Pagnini Lucensis, recenter Benedicti Ariae Montani Hispal. et quorundam aliorum collato studio, ad Hebraicam dictionem diligentissime expensa, edited by Benito Arias Montano and Santes Pagnini. Antwerp: Christophe Plantin, 1584.

———. *Discursos sobre el Eclesiastés de Salomón declarado según la verdad del sentido literal.* Edited by Valentín Núñez Rivera, Luis Gómez Canseco, Eloy Navarro Domínguez, and Sergio Fernández López. Huelva: Universidad de Huelva, 2012.

———. *Historia de la naturaleza: Primera parte del Cuerpo de la Obra magna.* Edited by Fernando Navarro Antolín. Bibliotheca Montaniana. Huelva: Universidad de Huelva, 2002.

———. *Index expurgatorius librorum qui hoc seculo prodierunt, vel doctrinae non sanae erroribus inspersis, vel inutilis et offensivae maledicentiae fellibus permixtis, iuxta Sacri Concilii Tridentini Decretum.* Antwerp: Christophe Plantin, 1571.

———. *Liber generationis et regenerationis Adam, sive De historia generis humani: Operis magni pars prima, id est Anima.* Antwerp: Jan Moretus, 1593.

———. *Liber Ioseph*, sive *De arcano sermone ad Sacri apparatus instructionem* a Benedicto Aria Montano hispalensi concinnatus." In *Biblia Sacra Hebraice, Chaldaice, Graece, et Latine: Philippi II. Reg. Cathol. pietate, et studio ad Sacrosanctae Ecclesiae usum*, edited by Benito Arias Montano, Guido Fabricius Boderianus, F. Raphelengius, Andreas Masius, Lucas of Bruges, et al., vol. 8. Antwerp: Christophe Plantin, 1571.

———. *Libro de José o sobre el lenguaje arcano.* Translated by Baldomero Macías Rosendo and Fernando Navarro Antolín. Huelva: Universidad de Huelva, 2006.

———. *Libro de la generación y regeneración del hombre, o Historia del género humano: Primera parte de la Obra magna, esto es, Alma. Estudio preliminar de Luis Gómez Canseco.* Translated by Fernando Navarro Antolín et al. Huelva: Universidad de Huelva, 1999.

———. *Naturae historia: Prima in magni operis corpore pars.* Antwerp: Jan Moretus, 1601.

———. *Phaleg, sive De gentium sedibus primis orbisque terrae situ liber.* In *Biblia Sacra Hebraice, Chaldaice, Graece, et Latine: Philippi II. Reg. Cathol. pietate, et studio ad Sacrosanctae Ecclesiae usum*, edited by Benito Arias Montano, Guido Fabricius Boderianus, F. Raphelengius, Andreas Masius, Lucas of Bruges, et al. Vol. 8. Antwerp: Christophe Plantin 1572.

———. *Prefacios de Benito Arias Montano a la Biblia Regia de Felipe II.* Estudio introductorio, edición, traducción y notas de María Asunción Sánchez Manzano. Humanistas españoles 32. Salamanca: Junta de Castilla y León, Consejería de educación y cultura, 2006.

———. *Sermones castellanos.* Edición y estudio de Valentín Núñez Rivera. Bibliotheca Montaniana 18. Huelva: Universidad de Huelva, 2008.

———. *Tratado sobre la fe que había que revelarse/Tractatus tertius de fide, quae revelanda erat; Adán, o de la lengua, intérprete del pensamiento humano, y de los rudimentos comunes a todas las lenguas/Adam, sive de humani sensus interprete lengua.* Edited by Fernando Navarro Antolín and Luis Gómez Canseco. Huelva: Universidad de Huelva, 2009.

Arias Montano, Benito, José de Sigüenza, and Ignacio García Aguilar. *Poesía castellana.* Bibliotheca Montaniana 28. Huelva: Universidad de Huelva, 2014.

Asencio, F. "Juan de Mariana y la Políglota de Amberes: Censura oficial y sugerencias de M. Bataillon." *Gregorianum* 36 (1955): 50–80.

Ashworth, E. J. "Traditional Logic." In *The Cambridge History of Renaissance Philosophy*, edited by Charles Schmitt and Quentin Skinner, 143–72. Cambridge: Cambridge University Press, 1988.

Augustine. *On Christian Doctrine*. Translated by D. W. Robertson. New York: Macmillan, 1958.

———. "On the Literal Interpretation of Genesis, an Unfinished Book." Translated by Roland J. Teske. In *On Genesis: Two Books on Genesis against the Manichees and on the Literal Interpretation of Genesis, an Unfinished Book*, edited by Roland J. Teske. Washington, DC: Catholic University of America Press, 1990.

Baldini, Ugo. "The Academy of Mathematics of the Collegio Romano from 1553 to 1612." In *Jesuit Science and the Republic of Letters*, edited by Mordechai Feingold, 47–98. Cambridge, MA: MIT Press, 2003.

———. "L'astronomia del cardinale Bellarmino." In *Novità celesti e crisi del sapere*, edited by P. Galluzzi, 293–305. Florence: Giunti Barbèra, 1984.

Baldini, Ugo, and George V. Coyne. "The Louvaine Lectures (Lectiones Lovanienses) of Bellarmine and the Autograph Copy of His 1616 Declaration to Galileo." *Studi Galileiani: Vatican Publications* 1, no. 2 (1984): 1–48.

Baldini, Ugo, and Leen Spruit, eds. *Catholic Church and Modern Science Documents from the Archives of the Roman Congregations of the Holy Office and the Index*. Vol. 1, *16th-Century Documents*. Rome: Libreria Editrice Vaticana, 2010.

Barker, Peter. "Stoic Contributions to Early Modern Science." In *Atoms, Pneuma, and Tranquillity: Epicurean and Stoic Themes in European Thought*, edited by Margaret J. Osler, 135–54. Cambridge: Cambridge University Press, 1991.

Barker, Peter, and Bernard R. Goldstein. "Realism and Instrumentalism in Sixteenth Century Astronomy: A Reappraisal." *Perspectives on Science* 6, no. 3 (1998): 223–58.

Barona, Josep Lluís. "Clusius' Exchange of Botanical Information with Spanish Scholars." In *Carolus Clusius in a New Context: Towards a Cultural History of a Renaissance Naturalist*, edited by Florike Egmond, Paul Hoftijzer, and Robert P. W. Visser, 99–113. Amsterdam: Edita, 2007.

———. *Sobre medicina y filosofía natural en el Renacimiento*. Valencia: Seminari d'estudis sobre la ciencia, 1993.

Barrera-Osorio, Antonio. *Experiencing Nature: The Spanish American Empire and the Early Scientific Revolution*. Austin: University of Texas Press, 2006.

Basil, Saint. *St. Basil: Letters and Select Works: Hexameron*. A Select Library of Nicene and Post-Nicene Fathers of the Christian Church, series 2. Edited by Philip Schaff and Henry Wace. Vol. 8. Edinburgh: T. and T. Clark, 1997.

Batiffol, Pierre. *La Vaticane de Paul III à Paul V d'après des documents nouveaux*. Paris: Leroux, 1890.

Bauer, Emmanuel J. "Francisco Suárez (1548–1617): Scholasticism after Humanism." In *Philosophers of the Renaissance*, edited by Paul Richard Blum, 236–55. Washington, DC: Catholic University of America Press, 2010.

Bécares Botas, Vicente, ed. *Arias Montano y Plantino: El libro flamenco de la España de Felipe II*. León: Secretariado de publicaciones de la Universidad de León, 1999.

Bell, Aubrey F. G. *Benito Arias Montano*. London: Oxford University Press, 1922.

Bellarmino, Roberto Francesco Romolo. *Autobiografia*. 1613. Edited by Pasquale Giustiniani. Brescia: Morcelliana, 1999.

———. *A Commentary on the Book of Psalms*. Translated by John O'Sullivan. Booneville, NY: Preserving Christian Publications, 1999.

———. *De controversiis christianae fidei adversus huius temporis haereticos*. 4 vols. Lyon: Pillehotte, 1586–93.

———. "The Mind's Ascent to God by the Ladder of Created Things." In *Spiritual Writings*, edited by John Patrick Donnelly and Roland J. Teske, 47–230. New York: Paulist Press, 1989.

Bellarmino, Roberto, and Federico Cesi. "Federico Cesi a Roberto Bellarmino [Acquaparta/ Roma, 1618]." *Rendiconti della R. Accademia nazionale dei Lincei, Classe di scienze morali, storiche e filologiche*, ser. 6, 7, no. 2 (1938): 648–63.

Bellarmino, Roberto, and Xavier-Marie Le Bachelet. *Auctarium Bellarminianum*. Paris: Beauchesne, 1913.

Berns, Andrew D. *The Bible and Natural Philosophy in Renaissance Italy: Jewish and Christian Physicians in Search of Truth*. New York: Cambridge University Press, 2015.

Biblia Sacra Hebraice, Chaldaice, Graece, et Latine: Philippi II. Reg. Cathol. pietate, et studio ad Sacrosanctae Ecclesiae usum. Edited by Benito Arias Montano, Guido Fabricius Boderianus, F. Raphelengius, Andreas Masius, Lucas of Bruges, et al. 8 vols. Antwerp: Christophe Plantin, 1569–73.

Blackwell, Richard J. *Galileo, Bellarmine, and the Bible: Including a Translation of Foscarini's Letter on the Motion of the Earth*. Edited by Paolo Antonio Foscarini. Notre Dame, IN: University of Notre Dame Press, 1991.

Blair, Ann. "Mosaic Physics and the Search for a Pious Natural Philosophy in the Late Renaissance." *Isis* 91, no. 1 (2000): 32–58.

Blüher, Karl Alfred. *Séneca en España: Investigaciones sobre la recepción de Séneca en España desde el siglo XIII hasta el siglo XVII*. Translated by Juan Conde. Madrid: Gredos, 1983.

Boner, Patrick J. "Life in the Liquid Fields: Kepler, Tycho and Gilbert on the Nature of the Heavens and Earth." *History of Science* 46, no. 153 (2008): 275–97.

Bono, James J. *The Word of God and the Languages of Man: Interpreting Nature in Early Modern Science and Medicine*. Madison: University of Wisconsin Press, 1995.

Bouwsma, William J. *Concordia Mundi: The Career and Thought of Guillaume Postel, 1510– 1581*. Cambridge, MA: Harvard University Press, 1957.

Brahe, Tycho. *Astronomiae instauratae progymnasmata: Quorum haec prima pars De restitutione motuum solis et lunae, stellarumque inerrantium tractat; Et praeterea De admiranda noua stella anno 1572. Exorta Luculenter Agit*. Frankfurt: Gottfried Tambach, 1610.

Brendecke, Arndt. *Empirical Empire: Spanish Colonial Rule and the Politics of Knowledge*. Berlin: De Gruyter, 2016.

Brient, Elizabeth. "How Can the Infinite Be the Measure of the Finite? Three Mathematical Metaphors from *De docta ignorantia*." In *Cusanus: The Legacy of Learned Ignorance*, edited by Peter J. Casarella, 210–25. Washington, DC: Catholic University of America Press, 2006.

Burnett, Stephen G. *Christian Hebraism in the Reformation Era (1500–1660)*. Leiden: Brill, 2012.

Bustamante García, Jesús. "Los círculos intelectuales y las empresas culturales de Felipe II: Tiempos, lugares y ritmos del humanismo en la España del siglo XVI." In *Elites intelectuales y modelos colectivos: Mundo ibérico (siglos XVI-XIX)*, edited by Mónica Quijada and Jesús Bustamante García, 33–58. Madrid: CSIC, 2003.

———. "Francisco Hernández, Plinio del Nuevo Mundo: Tradición clásica, teoría nominal y sistema terminológico indígena en una obra renacentista." In *Entre dos mundos: Fronteras culturales y agentes mediadores*, edited by B. Ares Queija and S. Gruzinski, 243–68. Seville: Escuela de Estudios Hispano-Americanos, 1997.

Byrne, Susan. *Ficino in Spain*. Toronto: University of Toronto Press, 2015.

Cabanelas, Darío. "Arias Montano y los libros plúmbeos de Granada." *Miscelánea de Estudios Árabes y Hebraicos* 18–19 (1969–70): 7–41.

Cabrera de Córdoba, Luis. *Historia de Felipe II, rey de España*. Edited by José Martínez Millán and Carlos J. de Carlos Morales. 4 vols. Valladolid: Consejería de Educación y Cultura, 1998.

Calero, Francisco, and José María López Piñero. *Los temas polémicos de la medicina renacentista: Las controversias (1556) de Francisco Vallés*. Introducción por José María López Piñero. Madrid: CSIC, 1988.

Campos y Fernández de Sevilla, F. Javier. "Bibliografía de y sobre el p. José de Sigüenza, OSH." *La ciudad de Dios: Revista Agustiniana* 219, no. 1 (2006): 293–313.

Cantarero de Salazar, Alejandro. "Reexamen crítico de la biografía del humanista Sebastián Fox Morcillo (c. 1526-c. 1560)." *Studia aurea* 9 (2015): 531–64.

Caro, Rodrigo. *Varones insignes en letras naturales de la ilustrísima ciudad de Sevilla*: Estudio y edición crítica de Luis Gómez Canseco. Seville: Diputación Provincial de Sevilla, 1992.

Caro Baroja, Julio. *Vidas mágicas e inquisición*. 2 vols. Madrid: Taurus, 1967.

Caso Amador, Rafael. "El origen judeoconverso del humanista Benito Arias Montano." *Revista de estudios extremeños* 71, no. 3 (2015): 1665–1711.

Castro, León de. *Apologeticus pro lectione apostolica, et evangelica, pro vulgata Divi Hieronymi, pro translatione LXX virorum, proque omni ecclesiastica lectione contra earum obtrectatores*. Salamanca: Haeredes Mathiae Gastii, 1585.

Cesi, Federico. "De caeli unitate, tenuitate fusaque et pervia stellarum motibus natura ex Sacris Litteris epistola (1618)." In *Scienziati del seicento*, edited by Maria Luisa Altieri Biagi and Bruno Basile, 9–35. Milan: Ricciardi, 1980.

Chabrán, Rafael, and Simon Varey. "'An Epistle to Arias Montano': An English Translation of a Poem by Francisco Hernández." *Huntington Library Quarterly* 55, no. 4 (1992): 621–34.

Charlo Brea, Luis, ed. *Levino Torrencio: Correspondencia con Benito Arias Montano*. Madrid: CSIC, 2007.

Clusius, Carolus. *La correspondencia de Carolus Clusius con los científicos españoles*. Edited by Josep Lluís Barona and Xavier Gómez Font. Valencia: Universitat de València, 1998.

Clusius, Carolus, and Peter F. X. de Ram. *Caroli Clusii Atrebatis ad Thomam Redigerum et Joannem Cratonem epistolae: Accedunt Remberti Dodonaei, Abrahami Ortelii, Gerardi Mercatoris et Ariae Montani ad eumdem Cratonem epistolae*. Brussels: M. Hayez, 1847.

Cobos Bueno, J. M., Andrés Oyola Fabián, and José R. Vallejo. "La dimensión botánica del humanista extremeño Benito Arias Montano." *Llull: Boletín de la Sociedad Española de historia de las ciencias* 38, no. 81 (2015): 127–65.

Cobos Bueno, J. M., and J. M. Vaquero Martínez. "Benito Arias Montano y el estudio de los fluidos." *Llull: Boletín de la Sociedad Española de historia de las ciencias* 22, no. 43 (1999): 75–106.

Collins, John J. "Natural Theology and Biblical Tradition: The Case of Hellenistic Judaism." *Catholic Biblical Quarterly* 60, no. 1 (1998): 1–15.

Colmeiro, Miguel. *La botánica y los botánicos de la península hispano-lusitana: Estudios bibliográficos y biográficos.* Madrid: Rivadeneyra, 1858.

Copenhaver, Brian P., and Charles B. Schmitt. *Renaissance Philosophy.* Oxford: Oxford University Press, 1992.

Coudert, Allison P. *Religion, Magic, and Science in Early Modern Europe and America.* Santa Barbara, CA: ABC-CLIO, 2011.

Coudert, Allison, and Jeffrey S. Shoulson, eds. *Hebraica Veritas? Christian Hebraists and the Study of Judaism in Early Modern Europe.* Philadelphia: University of Pennsylvania Press, 2004.

Crowther, Kathleen M. "Sacred Philosophy, Secular Theology: The Mosaic Physics of Levinus Lemnius (1505–1568) and Francisco Valles (1524–1592)." In *Nature and Scripture in the Abrahamic Religions: Up to 1700*, edited by Jitse M. van der Meer and Scott Mandelbrote, 2:397–428. Leiden: Brill, 2008.

Dávila Pérez, Antonio. "La apología de la *Biblia Regia* escrita por Benito Arias Montano, un documento en paradero desconocido." *Euphrosyne: Revista de filología clássica* 44 (2016): 279–90.

———. "Arias Montano y Amberes: Enlaces espirituales, bibliófilos y comerciales entre España y los Países Bajos." *Excerpta philologica* 9 (1999): 199–212.

———. *Correspondencia conservada en el Museo Plantin-Moretus de Amberes/Benito Arias Montano.* Estudio introductorio, edición crítica, traducción anotada e índices a cargo de Antonio Dávila Pérez. 2 vols. Madrid: Instituto de estudios humanísticos, Ediciones del Laberinto, CSIC, 2002.

———. "La correspondencia inédita de Benito Arias Montano: Nuevas prospecciones y estudio." In *Benito Arias Montano y los humanistas de su tiempo*, edited by José María Maestre Maestre, Eustaquio Sánchez Salor, Manuel Antonio Díaz Gito, Luis Charlo Brea, and Pedro Juan Galán Sánchez, 1:65–78. Mérida: Editora Regional de Extremadura, Instituto de estudios humanísticos, 2006.

———. "Correspondencia Latina inédita de Pedro de Valencia con la imprenta plantiniana (1598–1604)." *Humanistica Lovaniensia: Journal of Neo-Latin Studies* 54 (2005): 213–65.

———. "Las dos versiones de la 'De Psalterii Anglicani exemplari animaduersio' de Benito Arias Montano en la Biblia Políglota de Amberes." *Sefarad* 74, no. 1 (2014): 185–254.

———. "El epistolario de Benito Arias Montano: Catálogo provisional." *De gulden passer* 80 (2002): 63–129.

———. "'Pro Hebraicis exemplaribus et lingua': Carta latina inédita de Benito Arias Montano a Gilberto Genebrardo (BNE, Ms. 149)." *Ágora: Estudios clássicos em debate* 17, no. 1 (2015): 337–412.

———. "'Regnavit a ligno Deus, affirmat Arias Montano, negat Lindanus': Revisión de la polémica Arias Montano-Lindanus a la luz de los nuevos documentos." *Humanistica Lovaniensia: Journal of Neo-Latin Studies* 58 (2009): 125–89.

———. "Retractación o pertinacia: Vicisitudes de un tratado parcialmente perdido de Arias Montano al hilo de la polémica en torno a la Biblia Políglota de Amberes." *Sefarad* 71, no. 2 (2011): 369–412.

Debus, Allen G. *The Chemical Philosophy: Paracelsian Science and Medicine in the Sixteenth and Seventeenth Centuries.* New York: Science History Publications, 1977.

———. *The French Paracelsians: The Chemical Challenge to Medical and Scientific Tradition in Early Modern France.* Cambridge: Cambridge University Press, 1991.

Del Rio, Martín Antoine. *Disquisitionum magicarum libri sex: Quibus continetur accurata curiosarum artium, et vanarum superstitionum confutatio, utilis theologis, iurisconsultis, medicis, philologis.* Mannheim: Joannes Albinus, 1612.

———. *Martín Del Rio: Investigations into Magic.* Manchester: Manchester University Press, 2000.

Demonet, Marie-Luce. *Les voix du signe: Nature et origine du langage à la Renaissance (1480–1580).* Paris: H. Champion, 1992.

Denzler, Georg. *Kardinal Guglielmo Sirleto (1514–1585): Leben und Werk. Ein Beitrag zur nachtridentinischen Reform.* Münchener theologische Studien 1. Historische Abt. 17. Munich: M. Hueber, 1964.

Domenichini, Daniele. *Analecta montaniana: Il Tractatus de perfectione Christiana, a cura di Daniele Domenichini.* Collana di testi e studi ispanici 5. Studi e testi di letteratura religiosa del Cinque e Seicento 4. Pisa: Giardini, 1985.

———. "Quattro inediti di Benito Arias Montano sulla questione sacromontana (1596/1598)." *Anales de literatura española* 5 (1986–87): 51–66.

Domínguez Domínguez, Juan Francisco. *Arias Montano y sus maestros.* Madrid: Ediciones Clásicas, 2013.

———. "Carta de Arias Montano a fray Luis de León: Comentario, edición y traducción." *Cuadernos de pensamiento* 12 (1998): 285–312.

———. "Correspondencia de Pedro Chacón (II), Carta no. 3, Pedro Chacón a León de Castro." *La Ciudad de Dios: Revista Agustiniana* 226, no. 1 (2013): 203–44.

———. "Correspondencia de Pedro Chacón (III), Comentario a la Carta no. 3." *La ciudad de Dios: Revista Agustiniana* 226, no. 2 (2013): 379–420.

———. "Pedro Chacón." In *Diccionario biográfico y bibliográfico del humanismo español (siglos XV-XVII)*, 193–219. Madrid: Ediciones clásicas, 2012.

Donahue, William H. "The Solid Planetary Spheres in Post-Copernican Natural Philosophy." In *The Copernican Achievement*, edited by Robert S. Westman, 244–75. Berkeley: University of California Press, 1975.

Doyle, John P. *Collected Studies on Francisco Suárez, S.J. (1548–1617).* Edited by Victor M. Salas. Leuven: Leuven University Press, 2010.

Drayson, Elizabeth. *The Lead Books of Granada.* Basingstoke, UK: Palgrave Macmillan, 2013.

Duhem, Pierre M. *To Save the Phenomena: An Essay on the Idea of Physical Theory from Plato to Galileo.* Translated by Edmund Doland and Chaninah Maschler. Chicago: University of Chicago Press, 1969.

Dunkelgrün, Theodor William. "The Multiplicity of Scripture: The Confluence of Textual Traditions in the Making of the Antwerp Polyglot Bible (1568–1573)." PhD diss., University of Chicago, 2012.

Eamon, William, and Víctor Navarro Brotóns, eds. *Más allá de la Leyenda Negra: España y la Revolución Científica/Beyond the Black Legend: Spain and the Scientific Revolution.* Valencia: Instituto de historia de la ciencia y documentación López Piñero, Universitat de València, CSIC, 2007.

Efron, Noah J., and Menachem Fisch. "Astronomical Exegesis: An Early Modern Jewish Interpretation of the Heavens." *Osiris* 16 (2001): 72–87.

Egmond, Florike. *The World of Carolus Clusius: Natural History in the Making, 1550–1610.* London: Pickering and Chatto, 2010.

Fagius, Paul. *Exegesis sive expositio dictionum hebraicarum literalis et simplex, in quatuor capita Geneseos pro studiosis linguae Hebraicae.* Isny, 1542.

Fantoli, Annibale. *Galileo: For Copernicanism and for the Church.* 3rd ed. Vatican City: Vatican Observatory Publications, 2003.

Feijoo, Benito Jerónimo. "Cartas eruditas y curiosas." In *Edición digital de las Obras de Feijoo,* edited by Biblioteca Filosofía en Español and Fundación Gustavo Bueno—Oviedo: Filosofia.org, 1997.

Fernández de Navarrete, Martín, et al., eds. *Colección de documentos inéditos para la historia de España.* 113 vols. Vaduz: Kraus Reprint, 1964–75.

Fernández Fernández, Laura. "El 'arte mágica' en el scriptorium Alfonsí: Del Picatrix al Libro de astromágia." In *De cuerpos y almas en el Judaísmo hispanomedieval: Entre la ciencia médica y la magia sanadora,* edited by Yolanda Moreno Koch and Ricardo Izquierdo Benito, 73–109. Cuenca: Ediciones de la Universidad de Castilla-La Mancha, 2011.

Fernández López, Sergio, ed. *Alfonso de Zamora y Benito Arias Montano, traductores: Los comentarios de David Qimhi a Isaías, Jeremías y Malaquías.* Huelva: Universidad de Huelva, 2011.

Fernández López, Sergio. "Exégesis, erudición y fuentes en el *Apparatus* de la *Biblia Regia.*" In *Antigüedades Hebráicas: Antiquitatum Iudaicarum libri IX. Tratados exegéticos de la Biblia Regia,* edited by Luis Gómez Canseco and Sergio Fernández López, 45–85. Bibliotheca Montaniana 25. Huelva: Universidad de Huelva, 2013.

Fernández Marcos, Natalio. "De los nombres de Cristo de fray Luis de León y *De arcano sermone* de Arias Montano." *Sefarad* 48, no. 2 (1988): 245–70.

———. "La exégesis bíblica de Cipriano de la Huerga." In *Obras completas de Cipriano de la Huerga,* edited by Gaspar Morocho Gayo, Francisco Javier Fuente Fernández, and J. F Domínguez Domínguez, 9:13–31. León: Secretariado de Publicaciones de la Universidad de León, 1996.

———. "Lenguaje arcano y lenguaje del cuerpo: La hermenéutica bíblica de Arias Montano." *Sefarad* 62, no. 1 (2002): 57–83.

Fernández Marcos, Natalio, and Emilia Fernández Tejero. *Biblia y humanismo: Textos, talantes y controversias del siglo XVI español.* Madrid: Fundación Universitaria Española, 1997.

Fernández Nieva, Julio. "Un extremeño en Trento." *Revista de estudios extremeños* 52, no. 3 (1996): 937–67.

Finocchiaro, Maurice A. *Defending Copernicus and Galileo: Critical Reasoning in the Two Affairs*. Dordrecht: Springer, 2010.

———. *The Trial of Galileo: Essential Documents*. Indianapolis: Hackett, 2014.

Flórez, Ramiro. "Actitud ante la filosofía y el humanismo cultural de Arias Montano." *Cuadernos de pensamiento* 12 (1998): 111–33.

Flórez, Ramiro, and Isabel Balsinde. *El Escorial y Arias Montano: Ejercicios de comprensión*. Madrid: Fundación Universitaria Española, 2000.

Formigari, Lia. *A History of Language Philosophies*. Philadelphia: John Benjamins, 2004.

Forshaw, Peter J. "The Genesis of Christian Kabbalah: Early Modern Speculations on the Work of Creation." In *Hidden Truths from Eden*, edited by Caroline Vander Stichele and Scholz Susanne, 121–44. Atlanta: SBL Press, 2014.

———. "Vitriolic Reactions: Orthodox Response to the Alchemical Exegesis of Genesis." In *The Word and the World: Biblical Exegesis and Early Modern Science*, edited by Kevin Killeen and Peter J. Forshaw, 111–36. New York: Palgrave Macmillan, 2007.

Fox Morcillo, Sebastián. *Sebastiani Foxii Morzilli Hispalensis De naturae philosophia, seu De Platonis et Aristotelis consensione, libri V*. Paris: Jacob Putean, 1560.

Fragnito, Gigliola. "The Central and Peripheral Organization of Censorship." In *Church Censorship and Culture in Early Modern Italy*, edited by Gigliola Fragnito, 13–49. New York: Cambridge University Press, 2001.

Freedberg, David. *The Eye of the Lynx: Galileo, His Friends, and the Beginnings of Modern Natural History*. Chicago: University of Chicago Press, 2002.

Freedman, H., and Maurice Simon, eds. *Midrash rabbah*. 3rd ed. Vol. 1. London: Soncino Press, 1983.

French, R. K., and Andrew Cunningham. *Before Science: The Invention of the Friars' Natural Philosophy*. Aldershot, UK: Scolar Press, 1996.

Freudenthal, Gad. *Aristotle's Theory of Material Substance: Heat and Pneuma, Form and Soul*. Oxford: Clarendon Press, 1995.

———. "The Problem of Cohesion between Alchemy and Natural Philosophy: From Unctuous Moisture to Phlogiston." In *Alchemy Revisited: Proceedings of the International Conference on the History of Alchemy at the University of Groningen, 17–19 April 1989*, edited by Z. R. W. M. von Martels, 107–17. Leiden: Brill, 1990.

Friedman, Jerome. "Sixteenth-Century Christian-Hebraica: Scripture and the Renaissance Myth of the Past." *Sixteenth Century Journal* 11, no. 4 (1980): 67–85.

Fuente Adánez, Alfonso de la. *Una exégesis para el siglo XVI: Antonio de Honcala (1484–1565) y su Comentario al Génesis*. Salamanca: Universidad Pontificia, 1994.

García-Arenal, Mercedes, and Fernando Rodríguez Mediano. *The Orient in Spain: Converted Muslims, the Forged Lead Books of Granada, and the Rise of Orientalism*. Leiden: Brill, 2013.

García de Céspedes, Andrés. *Regimiento de navegación e hydrografía*. Madrid: Juan de la Cuesta, 1606.

García Gutiérrez, Juan. "Arias Montano y Pedro de Valencia: Hitos de una amistad." *Revista de estudios extremeños* 58, no. 1 (2002): 229–58.

Gil, Juan. "Arias Montano en Sevilla." In *Humanismo y pervivencia del mundo clásico: Homenaje al profesor Antonio Fontán*, edited by José María Maestre Maestre, Joaquín Pascual

Barea, and Luis Charlo Brea, 3.1: 263–80. Madrid: Instituto de estudios humanísticos, Ediciones del Laberinto, CSIC, 2002.

———. *Arias Montano y su entorno: Bienes y herederos*. Mérida: Editora Regional de Extremadura, 1998.

———. "Montano y El Escorial." In *Arias Montano y su tiempo*, edited by Catalina Pulido Corrales, 173–87. Badajoz: Editora Regional de Extremadura, 1998.

Gil, L. "Advertimiento del maestro León de Castro sobre la impresión de la Biblia quinquelingue." In *Stephanion: Homenaje a María C. Giner*, edited by C. Codoñer Merino, M. P. Fernández Álvarez, and J. A. Fernández Delgado, 45–53. Salamanca: Ediciones Universidad de Salamanca, 1988.

Godman, Peter. *The Saint as Censor: Robert Bellarmine between Inquisition and Index*. Leiden: Brill, 2000.

Goldstein, Bernard R. "Saving the Phenomena: The Background to Ptolemy's Planetary Theory." *Journal for the History of Astronomy* 28, no. 1 (1997): 1–12.

Gómez Canseco, Luis. "Ciencia, religión y poesía en el humanismo: Benito Arias Montano." *Edad de oro* 27 (2008): 127–45.

———. *El humanismo después de 1600: Pedro de Valencia*. Seville: Universidad de Sevilla, 1993.

———. "Los sentidos del lenguaje arcano: Para una lectura del Liber Ioseph." In *Libro de José o sobre el lenguaje arcano*, edited by Luis Gómez Canseco and Fernando Navarro Antolín, 43–86. Huelva: Universidad de Huelva, 2006.

González Carvajal, Tomás. "Elogio histórico del Doctor Benito Arias Montano." *Memorias de la Real Academia de Historia* 7 (1832): 1–199.

González de la Calle, Pedro Urbano. *Sebastián Fox Morcillo: Estudio histórico-crítico de sus doctrinas*. Madrid: Asilo de Huérfanos del Sagrado Corazón de Jesús, 1903.

Gonzalo Sánchez-Molero, José Luis. "La biblioteca de Arias Montano en el Escorial." In *Benito Arias Montano y los humanistas de su tiempo*, edited by José María Maestre Maestre, Eustaquio Sánchez Salor, Manuel Antonio Díaz Gito, Luis Charlo Brea, and Pedro Juan Galán Sánchez, 1:91–110. Mérida: Editora Regional de Extremadura, 2006.

Gorman, Michael, and Jonathan J. Sanford, eds. *Categories: Historical and Systematic Essays*. Washington, DC: Catholic University of America Press, 2004.

Gracia, Jorge, and Lloyd Newton. *Medieval Theories of the Categories*. In *The Stanford Encyclopedia of Philosophy* (Winter 2016), ed. Edward N. Zalta. https://plato.stanford.edu/archives/win2016/entries/medieval-categories/.

Grafton, Anthony. "Christian Hebraism and the Rediscovery of Hellenistic Judaism." In *Jewish Culture in Early Modern Europe*, edited by Richard I. Cohen, Natalie B. Dohrmann, Adam Shear, and Elchanan Reiner, 169–80. Pittsburgh: University of Pittsburgh Press, 2014.

Granada, Luis de. *Introducción del símbolo de la fe* (1583). Edited by José María Balcells. Madrid: Cátedra, 1989.

Granada, Miguel A. "Agostino Steuco y la perennis philosophia: Sobre algunos aspectos y dificultades de la concordia entre prisca theologia y cristianismo." *Revista de filosofía* 8 (1994): 23–38.

———. "New Visions of the Cosmos." In *The Cambridge Companion to Renaissance Philosophy*, edited by James Hankins, 270–86. Cambridge: Cambridge University Press, 2007.

Grant, Edward. *The Foundations of Modern Science in the Middle Ages: Their Religious, Insti-
tutional, and Intellectual Contexts.* Cambridge: Cambridge University Press, 1996.

———. *Planets, Stars, and Orbs: The Medieval Cosmos, 1200–1687.* Cambridge: Cambridge
University Press, 1996.

Greene, Edward Lee. *Landmarks of Botanical History.* 2 vols. Stanford, CA: Stanford University Press, 1983.

Groh, Dieter. "The Emergence of Creation Theology: The Doctrine of the Book of Nature
in the Early Church Fathers in the East and the West up to Augustine." In *The Book of
Nature in Antiquity and the Middle Ages,* edited by Arie Johan Vanderjagt and Klaas van
Berkel, 21–56. Leuven: Peeters, 2005.

Guanzelli, Giovanni Maria. *Index librorum expurgandorum: In quo quinquaginta auctorum
libri prae ceteris desiderati emendantur; Per Fr. Io. Mariam Brasichellen.* Rome: R. Cam.
Apost., 1607.

Hahm, David E. *The Origins of Stoic Cosmology.* Columbus: Ohio State University Press,
1977.

Hamilton, Alastair. *The Family of Love.* Cambridge: Lutterworth Press, 1987.

———. "From Familism to Pietism: The Fortunes of Pieter van der Brocht's Biblical Illustrations and Hiël Commentaries from 1584—1717." *Quaerendo* 11, no. 4 (1977): 271–301.

———. "Hiël and the Hiëlists: The Doctrines and Followers of Hendrik Jansen van Barrefelt."
Quaerendo 7 (1977): 243–86.

Hänsel, Sylvaine. *Benito Arias Montano, 1527–1598: Humanismo y arte en España.* Huelva:
Editora Regional de Extremadura, Diputación Provincial de Huelva, Universidad de
Huelva, 1999.

Harris, A. Katie. *From Muslim to Christian Granada: Inventing a City's Past in Early Modern
Spain.* Baltimore: Johns Hopkins University Press, 2007.

Harrison, Peter. *The Bible, Protestantism, and the Rise of Natural Science.* Cambridge: Cambridge University Press, 1998.

———. *The Fall of Man and the Foundation of Science.* Cambridge: Cambridge University
Press, 2007.

Hattab, Helen. "Suárez's Last Stand for the Substantial Form." In *The Philosophy of Francisco
Suárez,* edited by Benjamin Hill and Henrik Lagerlund, 101–20. Oxford: Oxford University Press, 2012.

Haydn, Hiram Collins. *The Counter-Renaissance.* New York: Grove Press, 1960.

Hedesan, Georgiana. "The Mystery of the *Mysterium Magnum*: Paracelsus's Alchemical
Interpretation of Creation in *Philosophia ad Atheniesis* and Its Early Modern Commen-
tators." In *Hidden Truths from Eden,* edited by Caroline Vander Stichele and Susanne
Scholz, 145–66. Atlanta: SBL Press, 2014.

Heide, Albert van der. *Hebraica Veritas: Christopher Plantin and the Christian Hebraists.
Catalogue to the Exhibition Hebraica Veritas. Did God speak Hebrew?,* 16 May-17 August
2008. Antwerp Plantin-Moretus Museum, Print Room, UNESCO World Heritage, 2008.

Helmont, Jean Baptiste van. *Oriatrike, or Physick Refined: The Common Errors Therein
Refuted, and the Whole Art Reformed and Rectified. Being a New Rise and Progress of
Phylosophy and Medicine for the Destruction of Diseases and Prolongation of Life.* London:
Lodowick Lloyd, 1662.

Hernández, Francisco. *The Mexican Treasury: The Writings of Dr. Francisco Hernández.* Edited by Simon Varey. Stanford, CA: Stanford University Press, 2000.

Hill, Benjamin, and Henrik Lagerlund, eds. *The Philosophy of Francisco Suárez.* Oxford: Oxford University Press, 2012.

Honcala, Antonio. *Commentaria in Genesim.* Alcalá: Ioannis Brocarii, 1555.

Höpfl, Hildebrand. *Beiträge zur Geschichte der Sixto-Klementinischen Vulgata, nach gedruckten und ungedruckten Quellen.* Freiburg im Breisgau: Herder, 1913.

Hopkins, Jasper. *Nicholas of Cusa on Learned Ignorance: A Translation and an Appraisal of "De docta ignorantia."* 2nd ed. Minneapolis: A. J. Banning Press, 1985.

Horace. *Odes and Epodes.* Translated by Niall Rudd. Loeb Classical Library. Cambridge, MA: Harvard University Press, 2004.

Howell, Kenneth J. "The Hermeneutics of Nature and Scripture in Early Modern Science and Theology." In *Nature and Scripture in the Abrahamic Religions: Up to 1700*, edited by Jitse M. van der Meer and Scott Mandelbrote, 1:275–98. Leiden: Brill, 2008.

Hudry, Françoise. *Le livre des vingt-quatre philosophes: Résurgence d'un texte du IVe siècle.* Introduction, texte latin, traduction et annotations par Françoise Hudry. Histoire des doctrines de l'antiquité classique 39. Paris: J. Vrin, 2009.

Ibn Ezra, Abraham ben Meïr. *Ibn Ezra's Commentary on the Pentateuch: Genesis (Bereshit).* Translated and annotated by H. Norman Strickman and Arthur M. Silver. Vol. 1. New York: Menorah, 1988.

Idel, Moshe. *Kabbalah in Italy, 1280–1510: A Survey.* New Haven, CT: Yale University Press, 2011.

———. "The Magical and Neoplatonic Interpretation of the Kabbalah in the Renaissance." In *Jewish Thought in the Sixteenth Century*, edited by Bernard Dov Cooperman, 186–242. Cambridge, MA: Harvard University Center for Jewish Studies, 1983.

Inwood, Brian. "Why Physics?" In *God and Cosmos in Stoicism*, edited by Ricardo Salles, 201–23. Oxford: Oxford University Press, 2009.

James, Thomas. *A Treatise of the Corruption of Scripture, Councels, and Fathers, by the Prelats, Pastors, and Pillars of the Church of Rome, for Maintenance of Popery and Irreligion by Thomas Iames [. . .].* London: Mathew Lownes, 1611.

Jiménez Guijarro, Pedro. *Juan de Mariana (1535–1624).* Madrid: Ediciones del Orto, 1997.

Jones, John A. "Las advertencias de Pedro de Valencia y Juan Ramírez acerca de la impresión de la Paraphrasis chaldaica de la Biblia Regia." *Bulletin Hispanique* 84, no. 3–4 (1982): 328–46.

———. "Pedro de Valencia en su correspondencia: Carta y relación de unos papeles de Alonso Sánchez." *Boletín de la Real Academia Española* 65, no. 234 (1985): 133–42.

———. "Pedro de Valencia's Defence of Arias Montano: The Expurgatory Indexes of 1607 (Rome) and 1612 (Madrid)." *Bibliothèque d'humanisme et Renaissance* 40 (1978): 121–36.

———. "Pedro de Valencia's Defence of Arias Montano: A Note on the Spanish Indexes of 1632, 1640 and 1667." *Bibliothèque d'humanisme et Renaissance* 57, no. 1 (1995): 83–88.

Jorge López, Juan José. *El pensamiento filosófico de Benito Arias Montano: Una reflexión sobre su Opus magnum.* Mérida: Editora Regional de Extremadura, 2002.

Kagan, Richard L., and Abigail Dyer. *Inquisitorial Inquiries: Brief Lives of Secret Jews and Other Heretics.* Baltimore: Johns Hopkins University Press, 2004.

Kelter, Irving A. "Reading the Book of God as the Book of Nature: The Case of the Louvain Humanist Cornelius Valerius (1512–1578)." In *The Word and the World: Biblical Exegesis and Early Modern Science*, edited by Kevin Killeen and Peter J. Forshaw, 174–87. Basingstoke, UK: Palgrave Macmillan 2007.

Killeen, Kevin. *Biblical Scholarship, Science and Politics in Early Modern England: Thomas Browne and the Thorny Place of Knowledge*. Farnham, UK: Ashgate, 2009.

Killeen, Kevin, and Peter J. Forshaw, eds. *The Word and the World: Biblical Exegesis and Early Modern Science*. New York: Palgrave Macmillan 2007.

Kuntz, Marion L. "Guillaume Postel, Prophet of the Restitution of All Things: His Life and Thought." The Hague: Nijhoff, 1981.

Landtsheer, Jeanine de. "Benito Arias Montano and the Friends from His Antwerp Sojourn." *De gulden passer* 80 (2002): 39–61.

Lanoye, Diederik. "Benito Arias Montano (1527–1598) and the University of Louvain." *Lias* 29 (2002): 23–44.

Lara Ródenas, Manuel José de. "Arias Montano en Portugal: La revisión de un tópico sobre la diplomacia secreta de Felipe II." In *Anatomía del humanismo: Benito Arias Montano, 1598–1998. Homenaje al profesor Melquiades Andrés Martín*. Actas del simposio internacional celebrado en la Universidad de Huelva del 4 al 6 del noviembre de 1998, edited by Luis Gómez Canseco, 343–67. Huelva: Universidad de Huelva, 1998.

Lattis, James M. *Between Copernicus and Galileo: Christoph Clavius and the Collapse of Ptolemaic Cosmology*. Chicago: University of Chicago Press, 1994.

Laursen, John Christian. "Pedro de Valencia's *Academica* and Scepticism in Late Renaissance Spain." In *Renaissance Scepticisms*, 111–23. Archives of the History of Ideas 199. Dordrecht: Springer Netherlands, 2008.

Lazure, Guy. "Building Bridges between Antwerp and Seville: Friends and Followers of Benito Arias Montano, 1579–1598." De *gulden passer* 89, no. 1 (2011): 31–43.

Lefèvre de la Boderie, Guy. "Dictionarium Syro-chaldaicum, Guidone Fabricio Boderiano collectore et auctore." In *Biblia Sacra Hebraice, Chaldaice, Graece, et Latine*, edited by Benito Arias Montano, vol. 6. Antwerp: Christophe Plantin, 1573.

Leo, Annibale de, and Vito Guerriero. *Dell'antichissima città di Brindisi e suo celebre porto: Memoria inedita, seguita da un Articolo storico de'vescovi di quella chiesa per V. Guerriero*. Naples: Stamperia della Società filomatica, 1846.

León, Luis de. *Comentario sobre el Génesis (Expositio in Genesim): Introducción, transcripción, versión y notas de Hipólito Navarro Rodríguez*. Madrid: Ediciones Escurialenses, Real Monasterio de El Escorial, 2009.

Levitin, Dmitri. *Ancient Wisdom in the Age of the New Science: Histories of Philosophy in England, c. 1640–1700*. New York: Cambridge University Press, 2015.

Lewis, Jack P. "The Days of Creation: An Historical Survey of Interpretation." *Journal of the Evangelical Theological Society* 32 (1989): 433–55.

Lindberg, David C. "The Medieval Church Encounters the Classical Tradition: Saint Augustine, Roger Bacon and the Handmaiden Metaphor." In *When Science and Christianity Meet*, edited by David C. Lindberg and Ronald L. Numbers, 7–32. Chicago: University of Chicago Press, 2003.

Lipsius, Justus. *Epistolario de Justo Lipsio y los españoles, 1577–1606*. Edited by Alejandro Ramírez. St. Louis, MO: Washington University Press, 1966.

———. *Iusti Lipsi Physiologiae stoicorum libri tres*. Antwerp: Jan Moretus, 1604.

———. *Juste Lipse et la restauration du stoïcisme: Étude et traduction des traités stoïciens. De la constance, Manuel de philosophie stoïcienne, Physique des stoïciens (extraits)*. Edited by Jacqueline Lagrée. Paris: Vrin, 1994.

López Guillamón, Ignacio. "Benito Arias Montano y la Biblioteca escurialense." *Revista de estudios extremeños* 52, no. 3 (1996): 969–85.

López Pérez, Miguel. "Ciencia y pensamiento hermético en la Edad Moderna Temprana." In *Más allá de la Leyenda Negra: España y la Revolución Científica/Beyond the Black Legend: Spain and the Scientific Revolution*, edited by William Eamon and Víctor Navarro Brotóns, 57–72. Valencia: Instituto de historia de la ciencia y documentación López Piñero, Universitat de València-CSIC, 2007.

López Piñero, José María. *Ciencia y técnica en la sociedad española de los siglos XVI y XVII*. Barcelona: Labor, 1979.

López Piñero, José María, and José Pardo Tomás. *La influencia de Francisco Hernández (1515–1587) en la constitución de la botánica y la materia médica modernas*. Valencia: Instituto de Estudios Documentales e Históricos sobre la Ciencia, Universitat de València-CSIC, 1996.

López Rodríguez, José Ramón. "Sevilla, el nacimiento de los museos, América y la botánica." In *La antigüedad como argumento II*, edited by Fernando Gascó and José Beltrán, 75–97. Seville: Scryptorium; Consejeria de Cultura Junta de Andalucía, 1995.

López Terrada, María José. "Flora and the Hapsburg Crown: Clusius, Spain and American Natural History." In *Silent Messengers: The Circulation of Material Objects of Knowledge in the Early Modern Low Countries*, edited by Sven Dupré and Christoph Herbert Lüthy, 43–68. Berlin: Lit, 2011.

López Toro, J. "Fray Luis de León y Benito Arias Montano." *Revista de archivos, bibliotecas y museos* 61, no. 2 (1955): 531–53.

Lowenthal, David. *The Past Is a Foreign Country*. Cambridge: Cambridge University Press, 1985.

Macías Rosendo, Baldomero. "El *Apparatus sacer* en la *Biblia Regia de Amberes*." In *Antigüedades hebráicas: Antiquitatum iudaicarum libri IX. Tratados exegéticos de la Biblia Regia*, edited by Luis Gómez Canseco and Sergio Fernández López, 13–42. Bibliotheca Montaniana 25. Huelva: Universidad de Huelva, 2013.

———. *La Biblia Políglota de Amberes en la correspondencia de Benito Arias Montano (ms. estoc. A 902)*. Huelva: Universidad de Huelva, 1998.

———. "La correspondencia de Arias Montano con Abraham Ortelio: Nuevos testimonios de una amistad sin fronteras." *La ciudad de Dios: Revista Agustiniana* 217, no. 2 (2004): 551–72.

———. *La correspondencia de Benito Arias Montano con el presidente de Indias Juan de Ovando: Cartas de Benito Arias Montano conservadas en el Instituto de Valencia de don Juan*. Huelva: Universidad de Huelva, 2008.

———. "El *De arcano sermone* en el marco de la *Biblia Políglota* de Amberes." In *Libro de José o sobre el lenguaje arcano*, edited by Luis Gómez Canseco and Fernando Navarro Antolín, 19–42. Huelva: Universidad de Huelva, 2006.

Magnier, Grace. *Pedro de Valencia and the Catholic Apologists of the Expulsion of the Moriscos: Visions of Christianity and Kingship*. Leiden: Brill, 2010.

Magruder, Kerry V. "Jesuit Science after Galileo: The Cosmology of Gabriele Beati." *Centaurus* 51 (2009): 189–212.

Maimonides, Moses. *Guide for the Perplexed*. Translated by M. Friedländer. New York: Pardes, 1904.

Malkiel, David. "The Artifact and Humanism in Medieval Jewish Thought." *Jewish History* 27, no. 1 (2013): 21–40.

Maluenda, Tomás. *Commentaria in Sacram Scripturam una cum nova de verbo ad verbum ex Hebraico translatione variisque lectionibus*. 5 vols. Lyon: Claudius Prost, Petrus et Claudius Rigaud, 1650.

Maravall, José Antonio. *Estudios de historia del pensamiento español*. 4 vols. Madrid: Centro de estudios políticos y constitucionales, 1999.

Mariana, Juan de. *Obras del padre Juan de Mariana*. 2 vols. Madrid: Rivadeneyra, 1854.

———. "Tractatus II—Pro editione Vulgata." In *Ioannis Marianae e Societate Iesu Tractatus VII*, 33–126. Cologne: Agripinae Sumptibus Antonii Hierati, 1609.

Martin, Craig. *Renaissance Meteorology: Pomponazzi to Descartes*. Baltimore: Johns Hopkins University Press, 2011.

Martínez de Bujanda, Jesús. "Indici dei libri prohibiti, Roma." In *Dizionario storico dell'Inquisizione*, edited by Adriano Prosperi, Vincenzo Lavenia, and John A. Tedeschi. 5 vols. 2:780–83. Pisa: Edizioni della Normale, 2010.

———. "L'influence de Sebond en Espagne au XVIe siècle." *Renaissance and Reformation* 10 (1974): 78–84.

Martínez de Bujanda, Jesús, Francis M. Higman, and James K. Farge, eds. *Index des livres interdits*. 9 vols. Sherbrooke, Canada: Centre d'études de la Renaissance, Editions de l'Université de Sherbrooke, 1984.

Martínez Ripoll, Antonio. "La Universidad de Alcalá y la formación humanista, bíblica y arqueológica de Benito Arias Montano." *Cuadernos de pensamiento* 12 (1998): 13–92.

Mayer, Annemarie C. "Llull and the Divine Attributes in 13th Century Context." *Anuario filosófico* 49, no. 1 (2016): 139–54.

Menéndez y Pelayo, Marcelino. "Esplendor y decadencia de la cultura científica española." In *Edición nacional de las obras completas de Menéndez Pelayo*, edited by Enrique Sánchez Reyes, 59:403–38. Santander: CSIC, 1941.

Methuen, Charlotte. "Interpreting the Books of Nature and Scripture in Medieval and Early Modern Thought: An Introductory Essay." In *Nature and Scripture in the Abrahamic Religions: Up to 1700*, edited by Jitse M. van der Meer and Scott Mandelbrote, 1:179–218. Leiden: Brill, 2008.

Momigliano, Arnaldo. *The Classical Foundations of Modern Historiography*. Berkeley: University of California Press, 1990.

Monardes, Nicolás. *Nicolás Monardes, primera y segunda y tercera partes de la Historia medicinal, de las cosas que se traen de nuestras Indias Occidentales [. . .]*. Seville: Alonso Escrivano, 1574.

Morales Lara, Enrique. "Las cartas de Benito Arias Montano a Abraham Ortels: Edición crítica y traducción a español." *Humanistica Lovaniensia: Journal of Neo-Latin Studies* 51 (2002): 153–205.

———. "Españoles en el 'Album amicorum' de Abraham Ortels: Un poema de Arias Montano, Furió Cerol, Alvar Nuñez." *Revista Agustiniana* 39, no. 120 (1998): 1029–56.

———. "Otras tres cartas de Benito Arias Montano a Abraham Ortels: Edición crítica y traducción a español." *Humanistica Lovaniensia: Journal of Neo-Latin Studies* 53 (2004): 219–49.

Morán Turina, José Miguel, and Fernando Checa Cremades. *El coleccionismo en España: De la cámara de maravillas a la galería de pinturas.* Madrid: Cátedra, 1985.

Morocho Gayo, Gaspar. "Estudio introductorio del discurso sobre el pergamino y láminas de Granada." In *Obras completas de Pedro de Valencia: Escritos políticos*, edited by Gaspar Morocho Gayo, 4.2:141-357. Léon: Universidad de Léon, 1999.

———. "Felipe II: Las ediciones litúrgicas y la Biblia Real." *La ciudad de Dios: Revista Agustiniana* 211, no. 3 (1998): 813–81.

———. "'Magnum illum vergensem Cyprinaum monachum. Alium praeterea neminen [. . .]': Cipriano de la Huerga, maestro de Benito Arias Montano." In *Obras completas de Cipriano de la Huerga*, edited by Gaspar Morocho Gayo, Francisco Javier Fuente Fernández, and J. F Domínguez Domínguez, 9:71–112. León: Secretariado de publicaciones de la Universidad de León, 1996.

———. "Transmisión histórica y valoración actual del biblismo de Arias Montano." *Cuadernos de pensamiento* 12 (1998): 135–242.

———. "Trayectoria humanística de Benito Arias Montano, I: Sus cuarenta primeros años (c. 1525/27–1567)." In *El humanismo extremeño: Estudios presentados a las 2as Jornadas organizadas por la Real Academia de Extremadura en Fregenal de la Sierra en 1997*, 157–210. Trujillo: Real Academia de Extremadura de las letras y las artes, 1998.

———. "Trayectoria humanística de Benito Arias Montano, II: Años de plenitud (1568–1598)." In *El humanismo extremeño: Estudios presentados a las 3as jornadas organizadas por la Real Academia de Extremadura en Fregenal de la Sierra, Aracena y Alajar en 1998*, edited by Mariano Fernández-Daza, Carmelo Solís, Francisco Tejanda, Manuel Terrón, and Antonio Viudas, 227–304. Trujillo: Real Academia de Extremadura de las letras y las artes, 1999.

Morocho Gayo, Gaspar, Francisco Javier Fuente Fernández, and J. F. Domínguez Domínguez, eds. *Obras completas de Cipriano de la Huerga.* 10 vols. León: Secretariado de publicaciones de la Universidad de León, 1990–96.

Muccillo, Maria. "La 'prisca theologia' nel 'De perenni philosophia' di Agostino Steuco." *Rinascimento: Rivista dell'Istituto nazionale di studi sul Rinascimento* 28 (1988): 41–111.

Navarro Antolín, Fernando, Luis Gómez Canseco, and Baldomero Macías Rosendo. "Fronteras del humanismo: Arias Montano y el Nuevo Mundo." In *Orbis incognitus: Avisos y legajos del Nuevo Mundo. Homenaje al profesor Luis Navarro García*, edited by Fernando Navarro Antolín and Luis Navarro García, 101–36. Huelva: Universidad de Huelva, 2007.

Navarro Brotóns, Víctor. "La ciencia en la España del siglo XVII: El cultivo de las disciplinas físico-matemáticas." *Arbor: Ciencia, pensamiento y cultura* 153, no. 604-5 (1996): 197–252.

———. "The Reception of Copernicus in Sixteenth-Century Spain: The Case of Diego de Zuniga." *Isis* 86, no. 1 (1995): 52–78.

Navarro Brotóns, Víctor, and Enrique Rodríguez Galdeano. *Matemáticas, cosmología y humanismo en la España del siglo XVI: Los comentarios al segundo libro de la historia natural de Plinio de Jerónimo Muñoz.* Valencia: Instituto de Estudios Documentales e Históricos sobre la ciencia, Universitat de València, 1998.

Newman, William R., and Lawrence Principe. *Alchemy Tried in the Fire: Starkey, Boyle, and the Fate of Helmontian Chymistry.* Chicago: University of Chicago Press, 2002.

Noguera Ramón, Vicente. "Historia de la vida y escritos de p. Juan de Mariana." In *Historia general de España que escribió el p. Juan de Mariana: Ilustrada en esta nueva impresión de tablas cronológicas, notas, y observaciones críticas, con la vida del autor.* 9 vols. Valencia: Benito Monfort, 1783–96.

Ogilvie, Brian W. *The Science of Describing: Natural History in Renaissance Europe.* Chicago: University of Chicago Press, 2006.

Ortega, Eusebio, and Benjamín Marcos. *Francisco de Valles (el Divino): Biografía, datos bibliográficos, sus doctrinas filosóficas y método. Con un prólogo del dr. D. Adolfo Bonilla y San Martín.* Madrid: Clásica española, 1914.

Ortega Sánchez, Delfín. "El enfrentamiento entre Arias Montano y León de Castro en la correspondencia privada del humanismo cristiano: ¿Límites filológicos o divergencias humanísticas?" *Tonos Digital* 11 (2006).

Ozaeta, José M. "Arias Montano, maestro del p. José de Sigüenza." *La ciudad de Dios: Revista Agustiniana* 203, no. 3 (1990): 535–82.

———. "Tres sermones inéditos del p. fr. José de Sigüenza en honor a S. Lorenzo." *La ciudad de Dios: Revista Agustiniana* 219, no. 1 (2006): 153–83.

P., K. M. "A Grand Inquisitor and His Library." *Bodleian Quarterly Record* 3, no. 24 (1922): 239–44.

Pacheco, Francisco. *Libro de descripción de verdaderos retratos de ilustres y memorables varones.* 1599.

Pagano, Sergio. *I documenti vaticani del processo di Galileo Galilei (1611–1741).* New edition enlarged, revised, and annotated by Sergio Pagano. Collectanea Archivi Vaticani 69. Vatican City: Archivio Segreto Vaticano, 2009.

Pagnino, Santes. *Thesaurus linguae sanctae.* Lyon: Sebastian Gryphius, 1529.

Palumbo, M. "Guanzelli, G. M." In *Dizionario storico dell'Inquisizione,* edited by Adriano Prosperi, Vincenzo Lavenia, and John A. Tedeschi, 2:740. Pisa: Edizioni della Normale, 2010.

Paniagua Pérez, Jesús. "Burócratas e intelectuales en la corte de Felipe II: La amistad de Juan de Ovando y Benito Arias Montano." *La ciudad de Dios: Revista Agustiniana* 211, no. 3 (September–December 1998): 919–53.

Paradinas Fuentes, Jesús Luis. "Cipriano de la Huerga y la filosofía del Renacimiento." In *Obras completas de Cipriano de la Huerga,* edited by Gaspar Morocho Gayo, Francisco Javier Fuente Fernández, and J. F Domínguez Domínguez, 9:33–69. León: Secretariado de publicaciones de la Universidad de León, 1996.

Paradinas Fuentes, Jesús Miguel. "Arias Montano, ¿Filósofo?" *La ciudad de Dios: Revista Agustiniana* 219, no. 2 (2006): 449–93.

Pardo Tomás, José. "Censura inquisitorial y lectura de libros científicos: Una propuesta de replanteamiento." *Tiempos modernos* 4, no. 9 (2003): 1–18.

———. *Ciencia y censura: La Inquisición española y los libros científicos en los siglos XVI y XVII*. Madrid: CSIC, 1991.

———. "Francisco Hernández: Medicina e historia natural en el Nuevo Mundo." In *Los orígenes de la ciencia moderna: Actas del XI y XII seminario Orotava de historia de la ciencia*. 215–44. Canary Islands: Fundación Canaria Orotava de Historia de la Ciencia, 2004.

Parente, Fausto. "The Index, the Holy Office, the Condemnation of the Talmud and Publication of Clement VIII's Index." Translated by Adrian Belton. In *Church Censorship and Culture in Early Modern Italy*, edited by Gigliola Fragnito, 163–93. New York: Cambridge University Press, 2001.

Pascual Barea, Joaquín. "El epitafio latino inédito de Arias Montano a un joven médico y astrólogo y el tratado de cirugía de Francisco Arceo." *Excerpta philologica* 10–12 (2000–2002): 357–72.

Pastore, S. "Arias Montano, Benito." In *Dizionario storico dell'Inquisizione*, edited by Adriano Prosperi, Vincenzo Lavenia, and John A. Tedeschi, 1:95–97. Pisa: Edizioni della Normale, 2010.

Peña Diaz, M. "Pineda, Juan de." In *Dizionario storico dell'Inquisizione*, edited by Adriano Prosperi, Vincenzo Lavenia, and John A. Tedeschi, 3:1210. Pisa: Edizioni della Normale, 2010.

Perea Siller, Francisco Javier. "Capacidad referencial e historia de la lengua hebrea en Benito Arias Montano." *Helmantica: Revista de filología clásica y hebrea* 55, no. 166 (2004): 31–46.

Perea Siller, Francisco Javier, and Bartolomé Pozuelo Calero. "El *Phaleg* en su entorno: La concepción montaniana de la geografía e historia primitivas." In *Benito Arias Montano y los humanistas de su tiempo*, edited by José María Maestre Maestre, Eustaquio Sánchez Salor, Manuel Antonio Díaz Gito, Luis Charlo Brea, and Pedro Juan Galán Sánchez, 1:335–48. Mérida: Editora Regional de Extremadura, Instituto de estudios humanísticos, 2006.

Pereira, Benedicto. *Prior tomus Commentariorum et disputationum in Genesim: Continens historiam Mosis ab exordio mundi usque ad Noëticum diluvium septem libris explanatam*. Lyon: Horatio Cardone, 1599.

Pereira, Gómez. *Antoniana Margarita*. Edited by J. L. Barreiro Barreiro, Concepción Souto García, and Juan Luis Camacho Lliteras. Santiago de Compostela: Universidad de Santiago de Compostela, 2000.

Pérez Custodio, Violeta. "Plinio el Viejo y los progymnasmata: La edición complutense de la *Naturalis historia* de 1569." In *Humanismo y pervivencia del mundo clásico: Homenaje al profesor Antonio Prieto (Alcañiz, 2005)*, edited by José María Maestre Maestre, Joaquín Pascual Barea, and Luis Charlo Brea, 4.2: 973–96. Madrid: CSIC, 2008.

Pérez Goyena, A. "Arias Montano y los Jesuitas." *Estudios eclesiasticos* 7 (1928): 273–317.

Peterfreund, Stuart. *Turning Points in Natural Theology from Bacon to Darwin: The Way of the Argument from Design*. New York: Palgrave Macmillan, 2012.

Philo. *On the Creation of the Cosmos according to Moses*. Edited by David T. Runia. Leiden: Brill, 2001.

Pico della Mirandola, Giovanni. "The Heptaplus: On the Sevenfold Narration of the Six Days of Genesis." Translated by Douglas Carmichael. In *On the Dignity of Man, On Being and the One, Heptaplus*. Indianapolis: Bobbs-Merrill, 1965.

Pike, Ruth. "The Converso Origin of Sebastián Fox Morcillo." *Hispania* 51, no. 4 (1968): 877–82.

Pineda, Juan de. *Commentariorum in Job libri tredecim*. 2 vols. Cologne: Wilhelm Metternich and Sons, 1733.

Pinta Llorente, Miguel de la. *Proceso criminal contra el hebraísta salmantino, Martin Martínez de Cantalapiedra*. Edición y estudio por Miguel de la Pinta Llorente. Madrid: CSIC, Instituto Arias Montano de Estudios Hebraicos y Oriente Próximo, 1946.

———. *Procesos inquisitoriales contra los catedráticos hebraístas de Salamanca: Gaspar de Grajal, Martínez de Cantalapiedra y fray Luis de León*. Documentos inéditos para la historia de la cultura española en el siglo XVI. Madrid: Monasterio de El Escorial, 1935.

Pinto Crespo, Virgilio. "El aparato de control censorial y las corrientes doctrinales." *Hispania sacra* 36, no. 73 (1984): 1–33.

———. "Los índices de libros prohibidos." *Hispania sacra* 35, no. 71 (1983): 161–91.

Pizarro Llorente, Henar. *Un gran patrón en la corte de Felipe II: Don Gaspar de Quiroga*. Madrid: Universidad Pontificia Comillas, 2004.

Pomata, Gianna. "Observation Rising: Birth of an Epistemic Genre, 1500–1650." In *Histories of Scientific Observation*, edited by Lorraine Daston and Elizabeth Lunbeck, 45–80. Chicago: University of Chicago Press, 2011.

Pomata, Gianna, and Nancy G. Siraisi, eds. *Historia: Empiricism and Erudition in Early Modern Europe*. Cambridge, MA: MIT Press, 2005.

Poole, William. "Francis Lodwick's Creation: Theology and Natural Philosophy in the Early Royal Society." *Journal of the History of Ideas* 66, no. 2 (2005): 245–63.

Popkin, Richard H. *The History of Scepticism: From Savonarola to Bayle*. New York: Oxford University Press, 2003.

Portuondo, María M. "America and the Hermeneutics of Nature in Renaissance Europe." In *Global Goods and the Spanish Empire, 1492–1824: Circulation, Resistance and Diversity*, edited by Bethany Aram and Bartolomé Yun, 78–98. Basingstoke, UK: Palgrave Macmillan, 2014.

———. "On Early Modern Science in Spain." In *The Early Modern Hispanic World: Transnational and Interdisciplinary Approaches*, edited by Kimberly Lynn and Erin Kathleen Rowe, 193–219. Cambridge: Cambridge University Press, 2017.

———. *Secret Science: Spanish Cosmography and the New World*. Chicago: University of Chicago Press, 2009.

———. "The Study of Nature, Philosophy and the Royal Library of San Lorenzo of the Escorial." *Renaissance Quarterly* 63, no. 4 (2010): 1106–50.

Principe, Lawrence. *The Secrets of Alchemy*. Chicago: University of Chicago Press, 2013.

Prosperi, Adriano. *L'eresia del Libro grande: Storia di Giorgio Siculo e della sua setta*. Milan: Feltrinelli, 2000.

Puig, Jaume de. *Les sources de la pensée philosophique de Raimond Sebond (Ramón Sibiuda)*. Paris: Honoré Champion, 1994.

Quiroga, G. *Index et catalogus librorum prohibitorum*. Madrid: Alphonsus Gomezius, 1583.

———. *Index librorum expurgatorum*. Madrid: Alphonsus Gomezius, 1584.

Randles, W. G. L. "Le ciel chez les Jésuites espagnols et portugais." *Mare liberum* 4 (1992): 307–13.

———. *The Unmaking of the Medieval Christian Cosmos, 1500–1760: From Solid Heavens to Boundless æther*. Aldershot, UK: Ashgate, 1999.

Raphelengius, Franciscus. "Thesauri hebraicae linguae olim a Sante Pagnino lucensi conscripti epitome: Cui accessit Grammatices libellus ex optimis quibusque grammaticis collectus." In *Biblia Sacra Hebraice, Chaldaice, Graece, et Latine*, edited by Benito Arias Montano, vol. 6. Antwerp: Christophe Plantin, 1572.

Raz-Krakotzkin, Amnon. *The Censor, the Editor, and the Text: The Catholic Church and the Shaping of the Jewish Canon in the Sixteenth Century*. Philadelphia: University of Pennsylvania Press, 2007.

Reinharts, Dennis. "The Americas Revealed in the *Theatrum*." In *Abraham Ortelius and the First Atlas*, edited by Marcel van den Broecke, Peter van den Krogt, and Peter Meurer, 209–20. Amsterdam: HES, 1998.

Rekers, Ben. *Arias Montano*. Madrid: Taurus, 1973.

———. *Benito Arias Montano (1527–1598)*. London: Warburg Institute, University of London, 1972.

Remes, Pauliina. *Neoplatonism*. Berkeley: University of California Press, 2008.

Remmert, Volker R. "'Whether the Stars Are Innumerable for Us?': Astronomy and Biblical Exegesis in the Society of Jesus around 1600." In *The Word and the World: Biblical Exegesis and Early Modern Science*, edited by Kevin Killeen and Peter J. Forshaw, 157–73. New York: Palgrave Macmillan 2007.

Repici, Luciana. "Andrea Cesalpino e la botanica antica." *Rinascimento: Rivista dell'Istituto Nazionale di Studi sul Rinascimento* 45 (2005): 47–87.

Rey, Eusebio. "Censura inédita del P. J. de Mariana de la *Políglota Regia de Amberes*." *Razón y fe* 155 (1957): 523–48.

Rey Bueno, Mar. *Los señores del fuego: Destiladores y espagíricos en la corte de los Austrias*. Madrid: Corona Borealis, 2002.

Rey Bueno, Mar, and María Esther Alegre Pérez. "Renovación de la terapéutica real: Los destiladores de su majestad, maestros simplicistas y médicos herbolarios de Felipe II." *Asclepio* 53, no. 1 (2001): 27–55.

Rey Bueno, Mar, and Miguel López Pérez. "Simón de Tovar (1528–1596): Redes familiares, naturaleza americana y comercio de maravillas en la Sevilla del XVI." *Dynamis* 26 (2006): 69–91.

Ricci, Saverio. "Federico Cesi e la 'nova' del 1604: La teoria della fluidità del cielo e un opuscolo dimenticato di Joannes van Heeck." *Atti della Accademia Nazionale dei Lincei* 43, fasc. 5–6 (1988): 111–33.

Riccioli, Giovanni Battista. *Almagestum novum astronomiam veterem novamque complectens observationibus aliorum, et propriis novisque theorematibus, problematibus, ac tabulis promotam*. Bologna: Victorii Benatii, 1651.

Robbins, Frank Egleston. *The Hexaemeral Literature: A Study of the Greek and Latin Commentaries on Genesis.* Chicago: University of Chicago Press, 1912.

Robbins, Jeremy. "The Arts of Perception." *Bulletin of Spanish Studies* 82, no. 8 (2005): 1–289.

Robles, Luis. "Documentación para un estudio sobre Tomás Maluenda, OP (1565–1628)." *Revista española de teología Madrid* 38, no. 1–2 (1978): 113–40.

Rooses, Max. *Christophe Plantin.* 2nd ed. Antwerp: Maes, 1896.

Rooses, Max, and Jean Denucé, eds. *Correspondance de Christophe Plantin.* 9 vols. Antwerp: J. E. Buschmann, 1883.

Rossi, Paolo. "Logic and the Art of Memory: The Quest for a Universal Language." London: Athlone, 2000.

Roth, Norman, ed. *Medieval Jewish Civilization: An Encyclopedia.* New York: Routledge, 2014.

Sabuco de Nantes y Barrera, Oliva, and Miguel Sabuco. *Nueva filosofía de la naturaleza del hombre.* Madrid: Pedro Madrigal, 1588.

———. *The True Medicine.* Edited and translated by Gianna Pomata. Toronto: Iter, 2010.

Sachs, Julius. *History of Botany (1530–1860).* Edited by Henry E. F. Garnsey and Isaac Bayley Balfour. Oxford: Clarendon Press, 1890.

Sáenz-Badillos, Angel. "Benito Arias Montano, Hebraísta." *Thélème: Revista complutense de estudios franceses* 12 (1997): 345–59.

Salazar, Antonio. "Arias Montano y Pedro de Valencia." *Revista de estudios extremeños* 15, no. 3 (1959): 475–93.

Salles, Ricardo. *God and Cosmos in Stoicism.* Oxford: Oxford University Press, 2009.

Sánchez, Francisco. *Quod nihil scitur.* Lyon: Antoine Gryphius, 1581.

Sánchez, Juan Luis. "Arias Montano y la espiritualidad en el siglo XVI." *La ciudad de Dios: Revista Agustiniana* 211 (1998): 33–49.

Sánchez de las Brozas, Francisco. *Doctrina del estóico filósofo Epicteto, que se llama comúnmente Enchiridión.* Edited by Luis M. Gómez Canseco. Badajoz: Diputación de Badajoz, Departamento de publicaciones, 1993.

Sánchez Martínez, Antonio. *La espada, la cruz y el padrón: Soberanía, fe y representación cartográfica en el mundo ibérico bajo la Monarquía hispánica, 1503–1598.* Madrid: CSIC, 2013.

Sánchez Nogales, José Luis. *Camino del hombre a Dios: La teología natural de R. Sibiuda.* Biblioteca teológica granadina 29. Granada: Facultad de Teología, 1995.

Sánchez Salor, Eustaquio. "Contenido de la Biblia Políglota." In *Anatomía del humanismo: Benito Arias Montano, 1598–1998. Homenaje al profesor Melquiades Andrés Martín: Actas del simposio internacional celebrado en la Universidad de Huelva del 4 al 6 del noviembre de 1998,* edited by Luis Gómez Canseco, 279–301. Huelva: Universidad de Huelva, 1998.

Sandoval y Rojas, Bernardo de. *Index librorum prohibitorum et expurgatorum [. . .].* Madrid: Luis Sánchez, 1612.

San Esteban y Falces, Juan de. *Ars ad solvenda omnia argumenta haereticorum cum evidentia falsitatis eorum: Ex doctrina S. Thomae Aquinatis, et sanctorum Patrum [. . .].* 2 vols. Rome: Alfonso Ciaccone, 1623.

Saunders, Jason Lewis. *Justus Lipsius: The Philosophy of Renaissance Stoicism*. New York: Liberal Arts Press, 1955.

Schaffer, Simon, Lissa Roberts, Kapil Raj, and James Delbourgo, eds. *The Brokered World: Go-Betweens and Global Intelligence, 1770–1820*. Sagamore Beach, MA: Science History Publications, 2009.

Schmidt-Biggemann, Wilhelm. *Philosophia Perennis: Historical Outlines of Western Spirituality in Ancient, Medieval and Early Modern Thought*. International Archives of the History of Ideas 189. Dordrecht: Springer, 2004.

Schmitt, Charles B. *Aristotle and the Renaissance*. Cambridge, MA: Harvard University Press, 1983.

———. "Perennial Philosophy: From Agostino Steuco to Leibniz." *Journal of the History of Ideas* 27, no. 4 (1966): 505–32.

Schwartz, Daniel, ed. *Interpreting Suárez: Critical Essays*. Cambridge: Cambridge University Press, 2012.

Sebastián Castellanos de Losada, Basílio, ed. *Biografía eclesiástica completa: Vida de los personajes del Antiguo y Nuevo Testamento, de todos los santos que venera la Iglesia, papas y eclesiásticos célebres por su virtudes y talentos en órden alfabético*. 30 vols. Madrid: Alejandro Gómez Fuentenebro, 1865.

Secret, François. "L'emithologie de Guillaume Postel." In *Umanesimo e esoterismo, atti del V l' Convegno internazionale di studi umanistici, Oberhofen, 16–17 settembre 1960*, edited by Enrico Castelli, 381–437. Padua: CEDAM, 1960.

———. "Guy Le Fèvre de la Boderie, représentant de G. Postel à la Polyglotte d'Anvers." *De gulden passer* 44 (1966): 245–57.

———. *Les kabbalistes chrétiens de la Renaissance*. Paris: Dunod, 1964.

Sellars, John. *Stoicism*. Berkeley: University of California Press, 2006.

Seneca, Lucius Annaeus. *Ad Lucilium epistulae morales*. Edited by Richard M. Gummere. 3 vols. Cambridge, MA: Harvard University Press, 1917.

———. *Natural Questions*. Edited by Harry M. Hine. Chicago: University of Chicago Press, 2010.

Seville, Isidore of. *The Etymologies of Isidore of Seville*. Translated by Stephen A. Barney. Cambridge: Cambridge University Press, 2006.

Shalev, Zur. "Benjamin of Tudela, Spanish Explorer." *Mediterranean Historical Review* 25, no. 1 (2010): 17–33.

———. "Sacred Geography, Antiquarianism and Visual Erudition: Benito Arias Montano and the Maps in the Antwerp Polyglot Bible." *Imago Mundi* 55, no. 1 (October 2003): 56–80.

———. *Sacred Words and Worlds*. Leiden: Brill, 2011.

Sigüenza, José de. "La fundación del Monasterio de El Escorial." In *Historia de la Orden de San Jerónimo*. Nueva biblioteca de autores españoles. 2 vols. Madrid: Bailly Bailliére, 1907–9.

———. *Historia de la orden de San Jerónimo: Estudio preliminar de Francisco J. Campos y Fernández de Sevilla*. Edición actualizada y corregida por Ángel Weruaga Prieto. 2 vols. Valladolid: Junta de Castilla y León, Consejería de Educación y Cultura, 2000.

——. *Historia del Rey de los reyes y Señor de los señores: Prólogo y estudio crítico por P. Luis Villalba Muñoz, OSA*. 3 vols. Madrid: La ciudad de Dios, 1916.

——. *Historia del Rey de los reyes y Señor de los señores*. Ignacio García Aguilar, estudio; Sergio Fernández López, Ignacio García Aguila, Natalia Palomino Tizado, edición. Bibliotheca Montaniana 29. Huelva: Universidad de Huelva, 2016.

——. *Tercera parte de la Historia de la orden de San Gerónimo, doctor de la iglesia*. Madrid: Imprenta Real por Iuan Flamenco 1605.

——. *La vida S. Gerónimo Dotor de la Santa Iglesia*. Madrid: Tomás Iunti, 1595.

Snobelen, Stephen D. "'In the Language of Men': The Hermeneutics of Accommodation in the Scientific Revolution." In *Nature and Scripture in the Abrahamic Religions: Up to 1700*, edited by Jitse M. van der Meer and Scott Mandelbrote, 2:691–732. Brill's Series in Church History 36. Leiden: Brill, 2008.

Somolinos d'Ardois, Germán, and José Mirando. *Vida y obra de Francisco Hernández: Precedida de España y Nueva España en la época de Felipe II*. Mexico City: Universidad Nacional de México, 1960.

Sprat, Thomas. *The History of the Royal-Society of London for the Improving of Natural Knowledge*. London: J. Martyn and J. Allestry, 1667.

Steuco, Agostino. *Cosmopoeia*. Lyon: Sebastian Gryphius, 1535.

Strack, H. L., and G. Stemberger. *Introduction to the Talmud and Midrash*. Translated by Markus Bockmuehl. Edinburgh: T. and T. Clark, 1991.

Suárez, Francisco. *Commentariorum ac disputationum in primam partem divi Thomae partis II de Deo effectore [. . .] tractatus II de Opere sex dierum ac tertius de Anima [. . .]*. Mainz: Hermann Mylius Birckmann, 1622.

Suárez Dobarrio, Fernando. "Filosofía y humanismo crítico en Pedro de Valencia." *Revista de estudios extremeños* 45, no. 2 (1989): 245–68.

Suárez Sánchez de León, Juan Luis. *El pensamiento de Pedro de Valencia: Escepticismo y modernidad en el humanismo español*. Badajoz: Diputación de Badajoz, 1997.

Talmage, Frank. "David Kimhi, the Man and the Commentaries." Cambridge, MA: Harvard University Press, 1975.

Tellechea Idígoras, José Ignacio. "Benito Arias Montano y San Carlos Borromeo." In *Anatomía del humanismo: Benito Arias Montano, 1598–1998. Homenaje al profesor Melquiades Andrés Martín: Actas del simposio internacional celebrado en la Universidad de Huelva del 4 al 6 del noviembre de 1998*, edited by Luis Gómez Canseco, 63–85. Huelva: Universidad de Huelva, 1998.

Thompson, Colin P. "The Lost Works of Luis de León: (2) 'Expositio in Genesim.'" *Bulletin of Hispanic Studies* 57, no. 3 (1980): 199–212.

——. "Reading the Escorial Library: Fray José de Sigüenza and the Culture of Golden-Age Spain." In *Culture and Society in Habsburg Spain*, edited by Clive Griffin Nigel Griffin, Eric Southworth, and Colin P. Thompson, 79–94. London: Tamesis, 2001.

Thorsrud, Harald. *Ancient Scepticism*. Berkeley: University of California Press, 2009.

Tirosh-Samuelson, Hava. "Kabbalah: A Medieval Tradition and Its Contemporary Appeal." *History Compass* 6, no. 2 (2008): 552–87.

——. "Kabbalah and Science in the Middle Ages." In *Science in Medieval Jewish Cultures*, edited by Gad Freudenthal, 476–510. Cambridge: Cambridge University Press, 2012.

Tudela, Benjamín de. *Itinerarium Benjamini Tudelensis . . . ex hebraico latinum factum Bened: Aria Montano interprete.* Antwerp: Christophe Plantin, 1575.

———. *The Itinerary of Benjamin of Tudela: Travels in the Middle Ages.* Introductions by Michael A. Signer, 1983, Marcus Nathan Adler, 1907, and A. Asher, 1840. Malibu, CA: J. Simon, 1983.

Urquiza Herrera, Antonio. *Coleccionismo y nobleza.* Madrid: Marcial Pons, 2007.

Valencia, Pedro de. *Academica, sive De iudicio erga verum: Ex ipsis primis fontibus. Opera Petri Valentiae Zafrensis.* Antwerp: Jan Moretus, 1596.

———. *Academica, sive De iudicio erga verum: Ex ipsis primis fontibus o Cuestiones académicas.* Introduction, translation, and notes by José Oroz Reta. Badajoz: Diputación de Badajoz, 1987.

———. "Cartas de Pedro de Valencia al p. Sigüenza." In *Epistolario español,* edited by Eugenio de Ochoa, 43–45. Biblioteca de autores españoles 62. Madrid: Rivadeneyra, 1856.

———. *Obras completas de Pedro de Valencia.* Edited by Gaspar Morocho Gayo, et al. Humanistas Españoles. 10 vols. Léon: Universidad de Léon, 2006.

Valerius, Cornelius. *Physicae, seu De naturae philosophia institutio.* Antwerp: Christophe Plantin, 1568.

Valles de Covarrubias, Francisco. *De iis, quae scripta sunt physicē in libris sacris sive de sacra philosophia liber singularis.* Turin: Nicolas Beuilaqua, 1587.

———. *Libro singular de Francisco Valles, sobre las cosas que fueron escritas físicamente en los libros sagrados, o de la Sagrada Filosofía.* Translated by Eustasio Sánchez F. Villarán. Biblioteca clásica de la medicina española 17. Madrid: Instituto de España, Real Academia Nacional de Medicina, 1971.

Vanderjagt, Arjo. *"Ad Fontes!* The Early Humanist Concern for the *Hebraica Veritas."* In *Hebrew Bible: Old Testament, the History of Its Interpretation.* Vol. 2, *From the Renaissance to the Enlightenment,* edited by Magne Sæbø, 154–89. Göttingen: Vandenhoeck und Ruprecht, 2008.

van der Meer, Jitse M., and Scott Mandelbrote, eds. *Nature and Scripture in the Abrahamic Religions: Up to 1700.* 2 vols. Brill's Series in Church History 36. Leiden: Brill, 2008.

Vande Walle, W. F. "Dodonaeus: A Bio-bibliographical Summary." In *Dodonaeus in Japan: Translation and the Scientific Mind in the Tokugawa Period,* edited by Willy vande Walle and Kazuhiko Kasaya, 33–41. Leuven: Leuven University Press, 2001.

Varey, Simon, Rafael Chabrán, and Dora B. Weiner, eds. *Searching for the Secrets of Nature: The Life and Works of Dr. Francisco Hernández.* Stanford, CA: Stanford University Press, 2000.

Vasoli, Cesare. "Note su tre teologie platoniche; Ficino, Steuco e Patrizi." *Rinascimento: Rivista dell'Istituto Nazionale di Studi sul Rinascimento,* ser. 2, 48 (2008): 81–100.

Venegas, Alejo. *Primera parte de las diferencias de libros q[ue] ay en el universo.* Toledo: Juan de Ayala, 1540.

———. *Primera parte de las differencias de libros que ay en el uniuerso declaradas por [. . .] Alexio Vanegas [. . .]; agora nueuamente emendada y corregida; los libros que esta primera parte contiene son quatro [. . .].* Madrid: Alonso Gómez, 1569.

Visser, Robert. "Dodonaeus and the Herbal Tradition." In *Dodonaeus in Japan: Transla-*

tion and the Scientific Mind in the Tokugawa Period, edited by Willy Vande Walle and Kazuhiko Kasaya, 45–57. Leuven: Leuven University Press, 2001.

Vives, Juan Luis. *Vives: On Education. A Translation of the De tradendis disciplinis of Juan Luis Vives Together with an Introduction.* Edited by Foster Watson. Cambridge: Cambridge University Press, 1913.

Voet, Léon. *The Golden Compasses: A History and Evaluation of the Printing and Publishing Activities of the Officina Plantiniana at Antwerp.* 2 vols. Amsterdam: Van Gendt, 1969.

Walton, Michael Thomson. *Genesis and the Chemical Philosophy: True Christian Science in the Sixteenth and Seventeenth Centuries.* Brooklyn, NY: AMS Press, 2011.

Waterworth, J., ed. *The Council of Trent: The Canons and Decrees of the Sacred and Oecumenical Council of Trent.* London: Dolman, 1848.

Webb, Clement C. J. *Studies in the History of Natural Theology.* Oxford: Clarendon Press, 1915.

Westman, Robert S. *The Copernican Question: Prognostication, Skepticism, and Celestial Order.* Berkeley: University of California Press, 2011.

Wicks, Jared. "Catholic Old Testament Interpretation in the Reformation and Early Confessional Eras." In *Hebrew Bible: Old Testament, the History of Its Interpretation.* Vol. 2, *From the Renaissance to the Enlightenment*, edited by Magne Sæbø, 617–48. Göttingen: Vandenhoeck and Ruprecht, 2008.

Wilkinson, Robert J. *The Kabbalistic Scholars of the Antwerp Polyglot Bible.* Leiden: Brill, 2007.

Williams, Arnold. *The Common Expositor: An Account of the Commentaries of Genesis, 1527–1633.* Chapel Hill: University of North Carolina Press, 1948.

Woolford, Thomas. "Natural Theology and Natural Philosophy in the Late Renaissance." PhD diss., University of Cambridge, 2011.

Ximeno, Vicente. *Los escritores que han florecido hasta el año M.DC.L. [1650] y una noticia preliminar de los mas antiguos.* Valencia: Joseph Estevan Dolz, 1747.

Ystella, Ludovico. *Commentaria in Genesim et Exodum in parengrapham expositionem et scholia distributa, [. . .]* Rome: S. Paulinum, 1609.

Zanier, Giancarlo. "Il 'De sacra philosophia' (1587) di Francisco Valles." In *Medicina e filosofia tra '500 e '600*, 20–38. Milan: Angeli, 1983.

Zapata, Antonio. *Novus index librorum prohibitorum et expurgatorum/editus autoritate [et] iussu [. . .] Antonii Zapata.* Seville: Francisco de Lyra, 1632.

Zlotowitz, Meir, and Nosson Scherman, eds. *Bereishis/Genesis: A New Translation with a Commentary Anthologized from Talmudic, Midrashic, and Rabbinic sources.* 2nd ed. 2 vols. Brooklyn, NY: Mesorah Publications, 1986.

Zuili, Marc. "Algunas observaciones acerca de un moralista toledano del siglo XVI." *Criticón* 65 (1995): 17–29.

Zuñiga, Diego de. *Física.* Edición de Gerardo Bolado. Barañáin: Ediciones Universidad de Navarra, 2009.

INDEX